FUNDAMENTAL AND ADVANCED TOPICS IN WIND POWER

Edited by **Rupp Carriveau**

INTECHOPEN.COM

Fundamental and Advanced Topics in Wind Power

Edited by Rupp Carriveau

CBS, Edition 2016

Published by InTech

Janeza Trdine 9, 51000 Rijeka, Croatia

Publishing Process Manager Davor Vidic

Technical Editor Teodora Smiljanic

Cover Designer InTech Design Team

Image Copyright Floris Slooff, 2010. Used under license from Shutterstock.com

Additional hard copies can be obtained from orders@intechopen.com

Fundamental and Advanced Topics in Wind Power, Edited by Rupp Carriveau
p. cm.
ISBN 978-953-307-508-2

Contents

Part 1

Aerodynamics and Environmental Loading of Wind Turbines

1

Aerodynamics of Wind Turbines

Emrah Kulunk
New Mexico Institute of Mining and Technology
USA

1. Introduction

A wind turbine is a device that extracts kinetic energy from the wind and converts it into mechanical energy. Therefore wind turbine power production depends on the interaction between the rotor and the wind. So the major aspects of wind turbine performance like power output and loads are determined by the aerodynamic forces generated by the wind. These can only be understood with a deep comprehension of the aerodynamics of steady state operation. Accordingly, this chapter focuses primarily on steady state aerodynamics.

2. Aerodynamics of HAWTs

The majority of the chapter details the classical analytical approach for the analysis of horizontal axis wind turbines and the performance prediction of these machines. The analysis of the aerodynamic behaviour of wind turbines can be started without any specific turbine design just by considering the energy extraction process. A simple model, known as actuator disc model, can be used to calculate the power output of an ideal turbine rotor and the wind thrust on the rotor. Additionally more advanced methods including momentum theory, blade element theory and finally blade element momentum (BEM) theory are introduced. BEM theory is used to determine the optimum blade shape and also to predict the performance parameters of the rotor for ideal, steady operating conditions. Blade element momentum theory combines two methods to analyze the aerodynamic performance of a wind turbine. These are momentum theory and blade-element theory which are used to outline the governing equations for the aerodynamic design and power prediction of a HAWT rotor. Momentum theory analyses the momentum balance on a rotating annular stream tube passing through a turbine and blade-element theory examines the forces generated by the aerofoil lift and drag coefficients at various sections along the blade. Combining these theories gives a series of equations that can be solved iteratively.

2.1 Actuator Disc Model

The analysis of the aerodynamic behaviour of wind turbines can be started without any specific turbine design just by considering the energy extraction process. The simplest model of a wind turbine is the so-called actuator disc model where the turbine is replaced by a circular disc through which the airstream flows with a velocity U_∞ and across which there is a pressure drop from p_u to p_d as shown in Fig. 1. At the outset, it is important to stress that the actuator disc theory is useful in discussing overall efficiencies of turbines but it cannot be utilized to design the turbine blades to achieve a desired performance. Actuator disc model is based on the assumptions like no frictional drag, homogenous, incompressible, steady state fluid flow, constant pressure increment or thrust per unit area over the disk, continuity of velocity through the disk and an infinite number of blades.

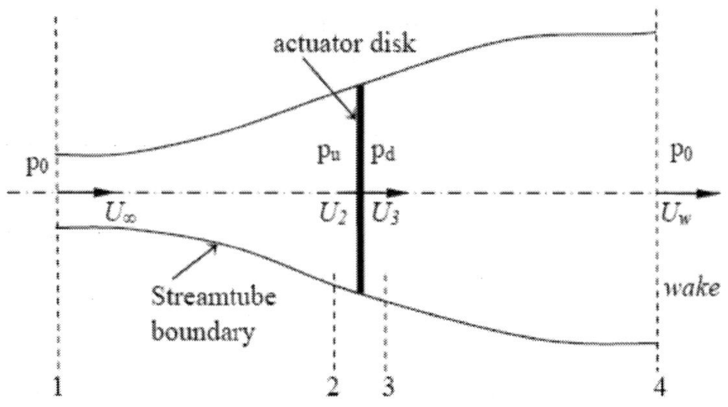

Fig. 1. Actuator Disk Model

The analysis of the actuator disk theory assumes a control volume in which the boundaries are the surface walls of a stream tube and two cross-sections. In order to analyze this control volume, four stations (1: free-stream region, 2: just before the blades, 3: just after the blades, 4: far wake region) need to be considered (Fig. 1). The mass flow rate remains the same throughout the flow. So the continuity equation along the stream tube can be written as

$$\rho A_\infty U_\infty = \rho A_d U_d = \rho A_w U_w \tag{1}$$

Assuming the continuity of velocity through the disk gives Eqn 2.

$$U_2 = U_3 = U_R \tag{2}$$

For steady state flow the mass flow rate can be obtained using Eqn 3.

$$\dot{m} = \rho A U_R \tag{3}$$

Applying the conservation of linear momentum equation on both sides of the actuator disk gives Eqn 4.

$$T = \dot{m}(U_\infty - U_w) \tag{4}$$

Since the flow is frictionless and there is no work or energy transfer is done, Bernoulli equation can be applied on both sides of the rotor. If we apply energy conservation using Bernoulli equation between station 1 and 2, then 3 and 4, Eqn 5 and Eqn 6 can be obtained respectively.

$$p_d + \frac{1}{2}\rho U_R^2 = p_o + \frac{1}{2}\rho U_w^2 \tag{5}$$

$$p_o + \frac{1}{2}\rho U_\infty^2 = p_u + \frac{1}{2}\rho U_R^2 \tag{6}$$

Combining Eqn 5 and 6 gives the pressure decrease p' as

$$p' = \frac{1}{2}\rho(U_\infty^2 - U_w^2) \tag{7}$$

Also the thrust on the actuator disk rotor can be expressed as the sum of the forces on each side

$$T = Ap' \tag{8}$$

where

$$p' = (p_u - p_d) \tag{9}$$

Substituting equation 7 into equation 8 gives the thrust on the disk in more explicit form.

$$T = \frac{1}{2}A\rho(U_\infty^2 - U_w^2) \tag{10}$$

Combining Eqn 3, 4 and 10 the velocity through the disk can be obtained as

$$U_R = \frac{U_\infty + U_w}{2} \tag{11}$$

Defining the axial induction factor a as in Eqn 12

$$a = \frac{U_\infty - U_R}{U_\infty} \tag{12}$$

gives Eqn 13 and 14.

$$U_R = U_\infty(1 - a) \tag{13}$$

$$U_w = U_\infty(1 - 2a) \tag{14}$$

To find the power output of the rotor Eqn 15 can be used.

$$P = TU_R \tag{15}$$

By substituting equation 10 into 15 gives the power output based on the momentum balance on both sides of the actuator disk rotor in more explicit form.

$$P = \frac{1}{2}\rho(U_\infty^2 - U_w^2)U_R \tag{16}$$

Also substituting equations 13 and 14 into equation 15 gives

$$P = 2\rho AaU_\infty^3(1 - a)^2 \tag{17}$$

Finally the performance parameters of a HAWT rotor (power coefficient C_P, thrust coefficient C_T, and the tip-speed ratio λ) can be expressed in dimensionless form which is given in Eqn 18, 19 and 20 respectively.

$$C_P = \frac{2P}{\rho U_\infty^3 \pi R^2} \tag{18}$$

$$C_T = \frac{2T}{\rho U_\infty^2 \pi R^2} \tag{19}$$

$$\lambda = \frac{R\Omega}{U_\infty} \tag{20}$$

Substituting Eqn 17 into Eqn 18, the power coefficient of the rotor can be rewritten as

$$C_P = 4a(1-a)^2 \tag{21}$$

Also using the equations 10, 13 and 17 the axial thrust on the disk can be rewritten as

$$T = 2Aa\rho(1-a)U_\infty^2 \tag{22}$$

Finally substituting equation 22 into equation 19 gives the thrust coefficient of the rotor as

$$C_T = 4a(1-a) \tag{23}$$

2.2 Rotating annular stream tube analysis

Thus far the method is developed on the assumption that there was no rotational motion. To extend the method developed, the effects of this rotational motion needs to be included so it is necessary to modify the qualities of the actuator disk by assuming that it can also impart a

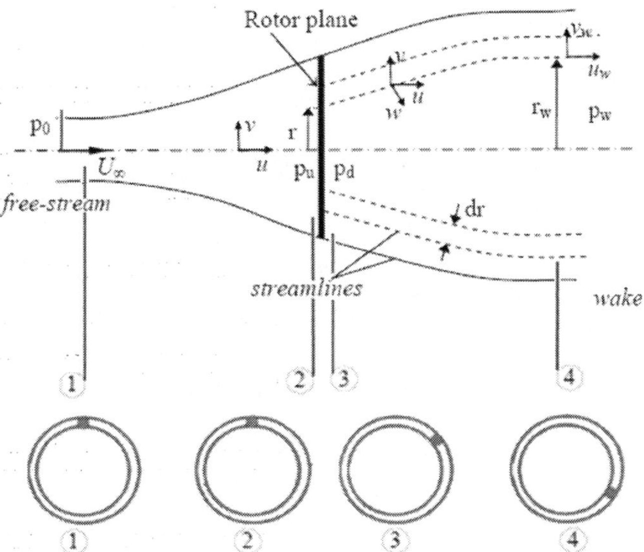

Fig. 2. Title Rotating Annular Stream Tube Analysis

rotational component to the fluid velocity while the axial and radial components remain unchanged. Using a rotating annular stream tube analysis, equations can be written that express the relation between the wake velocities (both axial and rotational) and the corresponding wind velocities at the rotor disk.

This analysis considers the conservation of angular momentum in the annular stream tube (Fig. 2). If the condition of continuity of flow is applied for the annular element taken on the rotor plane Eqn 24 can be written.

$$u_w r_w dr_w = urdr \tag{24}$$

Applying the conservation of the angular momentum on upstream and the wake region of the flow domain gives

$$w_w r_w^2 = wr^2 \tag{25}$$

Also the torque caused by the angular momentum balance on the differential annular element can be obtained using Eqn 26.

$$dQ = \rho uwr^2 dA \tag{26}$$

where $dA = 2r\pi dr$. Also applying the Bernoulli equation between station 1 and 2 then between 3 and 4 gives Bernoulli's constants as

$$H_o = p_o + \frac{1}{2}\rho U_\infty^2 = p_u + \frac{1}{2}\rho(u^2 + v^2)$$

$$H_1 = p_d + \frac{1}{2}\rho(u^2 + v^2 + w^2 r^2) = p_w + \frac{1}{2}\rho(u_w^2 + r_w^2 w_w^2)$$

And taking the difference between these constants gives

$$H_o - H_1 = p' - \frac{1}{2}\rho(w^2 r^2)$$

Which means the kinetic energy of the rotational motion given to the fluid by the torque of the blade is equal to $-(1/2)\rho(w^2 r^2)$. So the total pressure head between both sides of the rotor becomes

$$\begin{aligned} p_o - p_w &= \frac{1}{2}\rho(u_w^2 - U_\infty^2) + \frac{1}{2}w_w^2 \rho r_w^2 + (H_o - H_1) \\ &= \frac{1}{2}\rho(u_w^2 - U_\infty^2) + \frac{1}{2}\rho(w_w^2 r_w^2 - w^2 r^2) + p' \end{aligned} \tag{27}$$

Applying the Bernoulli's equation between station 2 and 3 gives the pressure drop as

$$p' = \frac{1}{2}\rho[-\Omega^2 + (\Omega + w)^2]r^2 = \rho\left(\Omega + \frac{w}{2}\right)wr^2 \tag{28}$$

Substituting this result into the equation 27 gives

$$p_o - p_w = \frac{1}{2}\rho(u_w^2 - U_\infty^2) + \rho\left(\Omega + \frac{w}{2}\right)r_w^2 w_w \tag{29}$$

In station 4, the pressure gradient can be written as

$$\frac{dp_w}{dr_w} = \rho r_w w_w^2 \tag{30}$$

Differentiating Eqn 29 relative to r_w and equating to equation 30 gives

$$\frac{1}{2}\frac{d}{dr_w}(U_\infty^2 - u_w^2) = (\Omega + w_w)\frac{d}{dr_w}(r_w^2 w_w) \tag{31}$$

The equation of axial momentum for the given annular blade element in differential form can be written as

$$dT = \rho u_w (U_\infty - u_w)dA_w + (p_o - p_w)dA_w \tag{32}$$

Since $dT = p'dA$, Eqn 32 can be written as Eqn 33.

$$dT = \rho\left(\Omega + \frac{w}{2}\right)wr^2 dA \tag{33}$$

Finally, combining Eqn 24, 27, 32 and 33 gives

$$\frac{1}{2}(U_\infty - u_w)^2 = \left(\frac{\Omega + w_w/2}{u_w} - \frac{\Omega + w/2}{U_\infty}\right)u_w r_w^2 w_w \tag{34}$$

An exact solution of the stream-tube equations can be obtained when the flow in the slipstream is not rotational except along the axis which implies that the rotational momentum wr^2 has the same value for all radial elements. Defining the axial velocities as $u = U_\infty(1 - a)$ and $u_w = U_\infty(1 - b)$ gives

$$a = \frac{b}{2}\left[1 - \frac{(1-a)b^2}{4\lambda^2(b-a)}\right] \tag{35}$$

Also the thrust on the differential element is equal to

$$\begin{aligned}
dT &= 2\rho u(u - U_\infty)dA \\
&= 4\pi\rho U_\infty^2 a(1-a)rdr
\end{aligned} \tag{36}$$

Using Eqn 28 Eqn 36 can be rewritten as

$$\begin{aligned}
dT &= p'dA \\
&= 2\pi\rho(\Omega + w/2)wr^3 dr
\end{aligned} \tag{37}$$

If the angular induction factor is defined as $a' = \frac{w}{2\Omega}$, then dT becomes

$$dT = 4\pi\rho\Omega^2 a'(1 + a')r^3 dr \tag{38}$$

In order to obtain a relationship between axial induction factor and angular induction factor, Eqn 37 and 38 can be equated which gives

$$\frac{a'}{(1-a)} = \frac{(\sigma C_L)}{4\lambda_r \sin\varphi}\left[1-(C_D/C_L)\cot\varphi\right] \tag{63}$$

By rearranging Eqn 63 and combining it with Eqn 50

$$\frac{a'}{(1+a')} = \frac{(\sigma C_L)}{4\cos\varphi}\left[1-(C_D/C_L)\cot\varphi\right] \tag{64}$$

can be written. In order to calculate the induction factors a and a', C_D can be set to zero. Thus the induction factors can be determined independently from airfoil characteristics. Subsequently, Eqn 62, 63 and 64 can be rewritten as Eqn 65, 66, and 67 respectively.

$$\frac{a}{(1-a)} = (\sigma C_L)\frac{\cos\varphi}{4\sin^2\varphi} \tag{65}$$

$$\frac{a'}{(1-a)} = \frac{(\sigma C_L)}{4\lambda_r \sin\varphi} \tag{66}$$

$$\frac{a'}{(1+a')} = \frac{(\sigma C_L)}{4\cos\varphi} \tag{67}$$

Finally, by rearranging Eqn 65, 66 and 67 and solving it for a and a', the following useful analytical relationships can be obtained.

$$a = \frac{1}{\left[1+[4\sin^2\varphi/(\sigma C_L)\cos\varphi]\right]} \tag{68}$$

$$a' = \frac{1}{\left[[4\cos\varphi/(\sigma C_L)]-1\right]} \tag{69}$$

$$a/a' = \lambda_r/\tan\varphi \tag{70}$$

$$C_L = \frac{4\sin\varphi}{\sigma}\frac{(\cos\varphi - \lambda_r \sin\varphi)}{(\sin\varphi + \lambda_r \cos\varphi)} \tag{71}$$

The total power of the rotor can be calculated by integrating the power of each differential annular element from the radius of the hub to the radius of the rotor.

$$P = \int_{r_h}^{R} dP = \int_{r_h}^{R} \Omega dQ \tag{72}$$

And rewriting the power coefficient given in Eqn 18 using Eqn 72 gives

$$C_p = \frac{P}{1/2\,\rho U_\infty^3 A} = \frac{\int_{\lambda_h}^{R} \Omega dQ}{1/2\,\rho U_\infty^3 \pi R^2}$$

Using Eqn 60, 65 and 70 the power coefficient relation can be rearranged as

$$C_p = \frac{8}{\lambda^2} \int_{\lambda_h}^{\lambda} \lambda_r^3 a'(1-a)\left[1-(C_D/C_L)\cot\varphi\right]d\lambda_r \tag{73}$$

2.4.1 Tip losses

At the tip of the turbine blade losses are introduced. The ratio of the average value of tip loss factor to that at a blade position is given in Fig. 6. As it is shown in the figure only near the tip the ratio begins to fall to zero so it is called 'the tip-loss factor'.

With uniform circulation the azimuthal average value of a is also radially uniform but that implies a discontinuity of axial velocity at the wake boundary with a corresponding discontinuity in pressure. Whereas such discontinuities are acceptable in the idealized actuator disc situation they will not occur in practice with a finite number of blades.

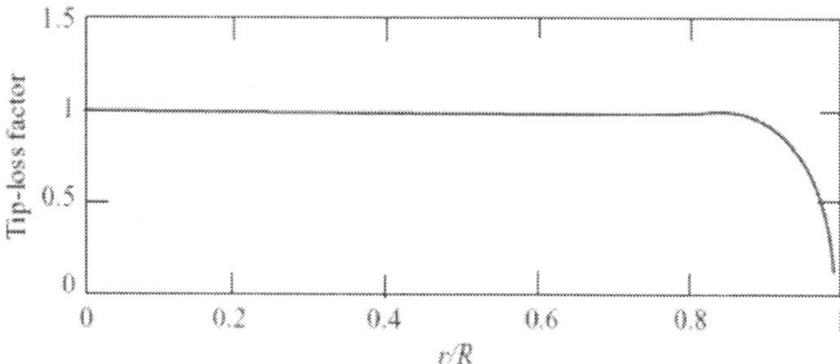

Fig. 6. Span-wise Variation of the Tip-loss Factor for a Blade with Uniform Circulation

The losses at the blade tips can be accounted for in BEM theory by means of a correction factor, f which varies from 0 to 1 and characterizes the reduction in forces along the blade. An approximate method of estimating the effect of tip losses has been given by L. Prandtl and the expression obtained by Prandtl for tip-loss factor is given by Eqn 74.

$$f = \frac{2}{\pi}\cos^{-1}\left\{\exp\left[\frac{-(B/2)\left[1-(\frac{r}{R})\right]}{(\frac{r}{R})\sin\varphi}\right]\right\} \tag{74}$$

The application of this equation for the losses at the blade tips is to provide an approximate correction to the system of equations for predicting rotor performance and blade design. Carrying the tip-loss factor through the calculations, the changes will be as following:

$$dQ = 4f\pi\rho U_\infty \Omega a'(1-a)r^3 dr \tag{75}$$

$$dT = 4f\pi\rho U_\infty^2 a(1-a)r\,dr \tag{76}$$

$$\frac{a'}{1-a} = \frac{\sigma C_L}{4f\lambda_r \sin\varphi} \tag{78}$$

$$\frac{a'}{1-a'} = \frac{\sigma C_L}{4f\cos\varphi} \tag{79}$$

$$C_L = \frac{4f\sin\varphi}{\sigma}\frac{(\cos\varphi - \lambda_{r\sin\varphi})}{(\sin\varphi + \lambda_{r\cos\varphi})} \tag{80}$$

$$a = \frac{1}{1 + \left[\dfrac{4f\sin^2\varphi}{(\sigma C_L)\cos\varphi}\right]} \tag{81}$$

$$a' = \frac{1}{\left[\dfrac{4f\cos\varphi}{(\sigma C_L)} - 1\right]} \tag{82}$$

$$C_P = \frac{8}{\lambda^2}\int_{\lambda_h}^{\lambda} f\lambda_r^{\,3}a'(1-a)\left[1 - \frac{C_D}{C_L}\tan\beta\right]d\lambda_r \tag{83}$$

The results for the span-wise variation of power extraction in the presence of tip-loss for a blade with uniform circulation on a three-bladed HAWT operating at a tip speed ratio of 6 is shown in Fig. 7 and it clearly demonstrates the effects of tip-loss.

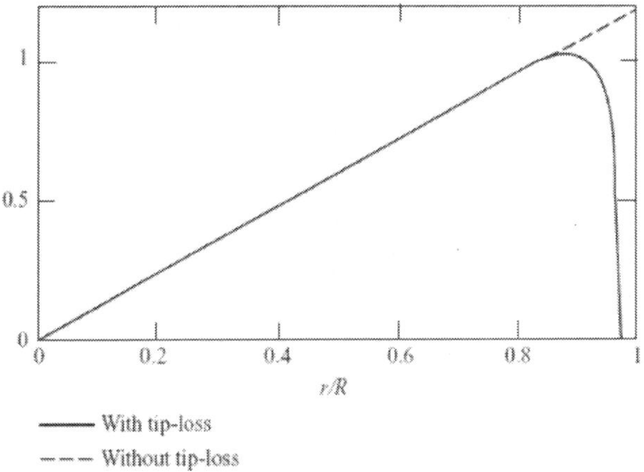

Fig. 7. Span-wise Variation of Power Extraction

Accordingly, for a selected airfoil type and a specified tip-speed ratio with blade length, the blade geometry can be designed for optimum rotor. And using these geometric parameters determined the aerodynamic performance of the rotor can be analyzed.

2.5 Blade design procedure

The aerodynamic design of optimum rotor blades from a known airfoil type means determining the geometric parameters (such as chord length and twist angle distribution along the blade span) for a certain tip-speed ratio at which the power coefficient of the rotor is maximum. For this reason firstly the change of the power coefficient of the rotor with respect to tip-speed ratio should be figured out in order to determine the design tip-speed ratio λ_d where the rotor has a maximum power coefficient. The blade design parameters will then be according to this design tip-speed ratio. Examining the plots between relative wind angle and local tip-speed ratio for a wide range of glide ratios gives us a unique relationship when the maximum elemental power coefficient is considered. And this relationship can be found to be nearly independent of glide ratio and tip-loss factor. Therefore a general relationship can be obtained between optimum relative wind angle and local tip-speed ratio which will be applicable for any airfoil type.

$$\frac{\partial}{\partial\varphi}\left\{\sin^2\varphi\left(\cos\varphi - \lambda_r\sin\varphi\right)\left(\sin\varphi + \lambda_r\cos\varphi\right)\right\} = 0 \tag{85}$$

Eqn 85 reveals after some algebra;

$$\varphi_{opt} = \left(2/3\right)\tan^{-1}\left(1/\lambda_r\right) \tag{86}$$

Having found the solution of determining the optimum relative wind angle for a certain local tip-speed ratio, the rest is nothing but to apply the equations from 80 to 83 derived from the blade-element momentum theory and modified including the tip loss factor. Dividing the blade length into N elements, the local tip-speed ratio for each blade element can then be calculated as

$$\lambda_{r,i} = \lambda\left(r_i/R\right) \tag{87}$$

Then rewriting Eqn 86 for each blade element gives

$$\varphi_{opt,i} = \left(2/3\right)\tan^{-1}\left(1/\lambda_{r,i}\right) \tag{88}$$

In addition the tip loss correction factor for each element can be calculated as

$$f_i = \frac{2}{\pi}cos^{-1}\left\{exp\left[\frac{-(B/2)\left[1 - \left(\frac{r_i}{R}\right)\right]}{\left(\frac{r_i}{R}\right)\sin\varphi_{opt,i}}\right]\right\} \tag{89}$$

The local chord-length for each blade element can then be determined using the following expression

$$c_i = \frac{8\pi r_i F_i \sin\varphi_{opt,i}}{BC_{L,design}} \frac{\left(\cos\varphi_{opt,i} - \lambda_{r,i}\sin\varphi_{opt,i}\right)}{\left(\sin\varphi_{opt,i} + \lambda_{r,i}\cos\varphi_{opt,i}\right)} \tag{90}$$

where $C_{L_{design}}$ is chosen such that the glide ratio is minimum at each blade element. Also the twist distribution can easily be calculated by Eqn 91

$$\theta_i = \varphi_{opt,i} - \alpha_{design} \tag{91}$$

where α_{design} is again the design angle of attack at which $C_{L_{design}}$ is obtained. Now chord-length and twist distribution along the blade span are known and in this case lift coefficient and angle of attack have to be determined from the known blade geometry parameters. This requires an iterative solution in which for each blade element the axial and angular induction factors are firstly taken as the values equal to the corresponding designed blade elements. Then the actual induction factors are determined within an acceptable tolerance of the previous guesses during iteration.

3. Conclusion

The kinetic energy extracted from the wind is influenced by the geometry of the rotor blades. Determining the aerodynamically optimum blade shape, or the best possible approximation to it, is one of the main tasks of the wind turbine designer. Accordingly this chapter sets out the basis of the aerodynamics of HAWTs and the design methods based on these theories to find the best possible design compromise for the geometric shape of the rotor which can only be achieved in an iterative process.

Performance analysis of HAWT rotors has been performed using several methods. In between these methods BEM model is mainly employed as a tool of performance analysis due to its simplicity and readily implementation. Most wind turbine design codes are based on this method. Accordingly the chapter explains the aerodynamics of HAWTs based on a step-by-step approach starting from the simple actuator disk model to more complicated and accurate BEM method.

The basic of BEM method assumes that the blade can be analyzed as a number of independent elements in span-wise direction. The induced velocity at each element is determined by performing the momentum balance for an annular control volume containing the blade element. Then the aerodynamic forces on each element are calculated using the lift and drag coefficient from the empirical two-dimensional wind tunnel test data at the geometric angle of attack of the blade element relative to the local flow velocity. BEM theory-based methods have aspects by reasonable tool for designer, but they are not suitable for accurate estimation of the wake effects, the complex flow such as three-dimensional flow or dynamic stall because of the assumptions being made.

4. Acknowledgment

I would like to express my deepest gratitude and appreciation to Dr. Sidney Xue and Dr. Carsten H. Westengaard from VESTAS Americas R&D Center in Houston, TX and Matthew F. Barone from Sandia National Labs Albuquerque, NM; also to Dr. Nilay Sezer Uzol from TOBB University of Economics and Technology, Ankara, Turkey and my advisors at New Mexico Tech Mechanical Engineering Department Dr. Warren J. Ostergren and Dr. Sayavur I. Bakhtiyarov for their unwavering support, invaluable guidance and encouragement throughout my formation in wind energy field.

Finally I would like to dedicate this chapter to Rachel A. Hawthorne. Thank you for all the love and happiness you have brought into my life!

5. Nomenclature

A: Area of wind turbine rotor
a: Axial induction factor at rotor plane

r: Radial coordinate at rotor plane
r_i : Blade radius for the ith blade element

a': Angular induction factor

B: Number of blades of a rotor

C_D: Drag coefficient of an airfoil

C_L: Lift coefficient of an airfoil

C_P: Power coefficient of wind turbine rotor

C_T: Thrust coefficient of wind turbine rotor

C_{T_r} :Local thrust coefficient of each annular rotor section

c_i: Blade chord length for the ith blade element

F_D: Drag force on an annular blade element

F_L: Lift force on an annular blade element

f : Tip-loss factor

f_i: Tip-loss factor for the ith blade element

H: Bernoulli's constant

$\dot{m}$: Air mass flow rate through rotor plane

N: Number of blade elements

P: Power output from wind turbine rotor

p': Pressure drop across rotor plane

Q: rotor torque

R: Radius of wind turbine rotor

T: rotor thrust

U_∞: Free stream velocity of wind

U_{rel}: Relative wind velocity

U_R: Uniform wind velocity at rotor plane

α : Angle of attack

λ: Tip-speed ratio of rotor

λ_d: Design tip-speed ratio

λ_r: Local tip-speed ratio

$\lambda_{r,i}$: Local tip-speed ratio for the ith blade element

θ_i: Pitch angle for the ith blade element

$\varphi_{opt,i}$: Optimum relative wind angle for the ith blade element

ρ : Air density

Ω : Angular velocity of wind turbine rotor

σ: Solidity ratio

v: Kinematic viscosity of air

γ: Glide ratio

Re: Reynolds number

HAWT: Horizontal-axis wind turbine

BEM: Blade element momentum

6. References

Bertagnolio, F., Niels Sorensen, N., Johansen, J., Fuglsang, P., *Wind Turbine Airfoil Catalogue*. Riso National Laboratory, Roskilde, 2001.

Burton, T. and Sharpe, D., *Wind Energy Handbook*. John Wiley & Sons Ltd, ISBN 0-471-48997-2, Chichester, 2006.

Glauert, H., (1935b). *Windmills and Fans*. Aerodynamic theory (ed. Durand, W. F.). Julius Springer, Berlin, Germany.

Hau, E., *Wind Turbines: Fundamentals, Technologies, Application, Economics*. Krailling, Springer, 2006.

Hoerner, S. F., (1965). *'Pressure drag on rotating bodies'*. Fluid dynamic drag (ed. Hoerner, Hunt, B. R. and Lipsman, R. L., A Guide to MATLAB: For Beginners and Experienced Users. Cambridge University Press, New York, 2001.

Kulunk, E., Yilmaz, N., *HAWT Rotor Design and Performance Analysis*. ASME 3rd International Conference on Energy Sustainability, ES2009-90441. San Francisco, CA, USA. July 19-23, 2009.

Mathews, J. H. and Fink, K. D., *Numerical Methods Using MATLAB*. Prentice Hall, Upper Saddle River, 1999.

Meijer Drees, J., (1949). *A theory of airflow through rotors and its application to some helicopter problems'*. J. Heli. Ass. G.B., 3, 2, 79–104. S. F.), pp. 3–14.

Uzol, N., Long, L., *3-D Time-Accurate CFD Simulations of Wind Turbine Rotor Flow Fields*. AIAA 2006-0394. The Pennsylvania State University, 2006.

Wilson, R. E., Lissaman, P. B. S., Walker, S. N., *"Aerodynamic Performance of Wind Turbines"*. Energy Research and Development Administration, 224, 1976.

2

Wind Turbines Theory - The Betz Equation and Optimal Rotor Tip Speed Ratio

Magdi Ragheb[1] and Adam M. Ragheb[2]
[1]*Department of Nuclear, Plasma and Radiological Engineering*
[2]*Department of Aerospace Engineering*
University of Illinois at Urbana-Champaign, 216 Talbot Laboratory,
USA

1. Introduction

The fundamental theory of design and operation of wind turbines is derived based on a first principles approach using conservation of mass and conservation of energy in a wind stream. A detailed derivation of the "Betz Equation" and the "Betz Criterion" or "Betz Limit" is presented, and its subtleties, insights as well as the pitfalls in its derivation and application are discussed. This fundamental equation was first introduced by the German engineer Albert Betz in 1919 and published in his book "Wind Energie und ihre Ausnutzung durch Windmühlen," or "Wind Energy and its Extraction through Wind Mills" in 1926. The theory that is developed applies to both horizontal and vertical axis wind turbines.

The power coefficient of a wind turbine is defined and is related to the Betz Limit. A description of the optimal rotor tip speed ratio of a wind turbine is also presented. This is compared with a description based on Schmitz whirlpool ratios accounting for the different losses and efficiencies encountered in the operation of wind energy conversion systems. The theoretical and a corrected graph of the different wind turbine operational regimes and configurations, relating the power coefficient to the rotor tip speed ratio are shown. The general common principles underlying wind, hydroelectric and thermal energy conversion are discussed.

2. Betz equation and criterion, performance coefficient C_p

The Betz Equation is analogous to the Carnot cycle efficiency in thermodynamics suggesting that a heat engine cannot extract all the energy from a given source of energy and must reject part of its heat input back to the environment. Whereas the Carnot cycle efficiency can be expressed in terms of the Kelvin isothermal heat input temperature T_1 and the Kelvin isothermal heat rejection temperature T_2:

$$\eta_{Carnot} = \frac{T_1 - T_2}{T_1} = 1 - \frac{T_2}{T_1},$$

(1)

the Betz Equation deals with the wind speed upstream of the turbine V_1 and the downstream wind speed V_2.

The limited efficiency of a heat engine is caused by heat rejection to the environment. The limited efficiency of a wind turbine is caused by braking of the wind from its upstream speed V_1 to its downstream speed V_2, while allowing a continuation of the flow regime. The additional losses in efficiency for a practical wind turbine are caused by the viscous and pressure drag on the rotor blades, the swirl imparted to the air flow by the rotor, and the power losses in the transmission and electrical system.

Betz developed the global theory of wind machines at the Göttingen Institute in Germany (Le Gouriérès Désiré, 1982). The wind rotor is assumed to be an ideal energy converter, meaning that:

1.   It does not possess a hub,
2.   It possesses an infinite number of rotor blades which do not result in any drag resistance to the wind flowing through them.

In addition, uniformity is assumed over the whole area swept by the rotor, and the speed of the air beyond the rotor is considered to be axial. The ideal wind rotor is taken at rest and is placed in a moving fluid atmosphere. Considering the ideal model shown in Fig. 1, the cross sectional area swept by the turbine blade is designated as S, with the air cross-section upwind from the rotor designated as S_1, and downwind as S_2.

The wind speed passing through the turbine rotor is considered uniform as V, with its value as V_1 upwind, and as V_2 downwind at a distance from the rotor. Extraction of mechanical energy by the rotor occurs by reducing the kinetic energy of the air stream from upwind to downwind, or simply applying a braking action on the wind. This implies that:

$$V_2 < V_1.$$

Consequently the air stream cross sectional area increases from upstream of the turbine to the downstream location, and:

$$S_2 > S_1.$$

If the air stream is considered as a case of incompressible flow, the conservation of mass or continuity equation can be written as:

$$\dot{m} = \rho S_1 V_1 = \rho S V = \rho S_2 V_2 = \text{constant} \tag{2}$$

This expresses the fact that the mass flow rate is a constant along the wind stream. Continuing with the derivation, Euler's Theorem gives the force exerted by the wind on the rotor as:

$$\begin{aligned} F &= ma \\ &= m\frac{dV}{dt} \\ &= \dot{m}\Delta V \\ &= \rho S V.(V_1 - V_2) \end{aligned} \tag{3}$$

The incremental energy or the incremental work done in the wind stream is given by:

$$dE = F dx \tag{4}$$

From which the power content of the wind stream is:

$$P = \frac{dE}{dt} = F\frac{dx}{dt} = FV \tag{5}$$

Substituting for the force F from Eqn. 3, we get for the extractable power from the wind:

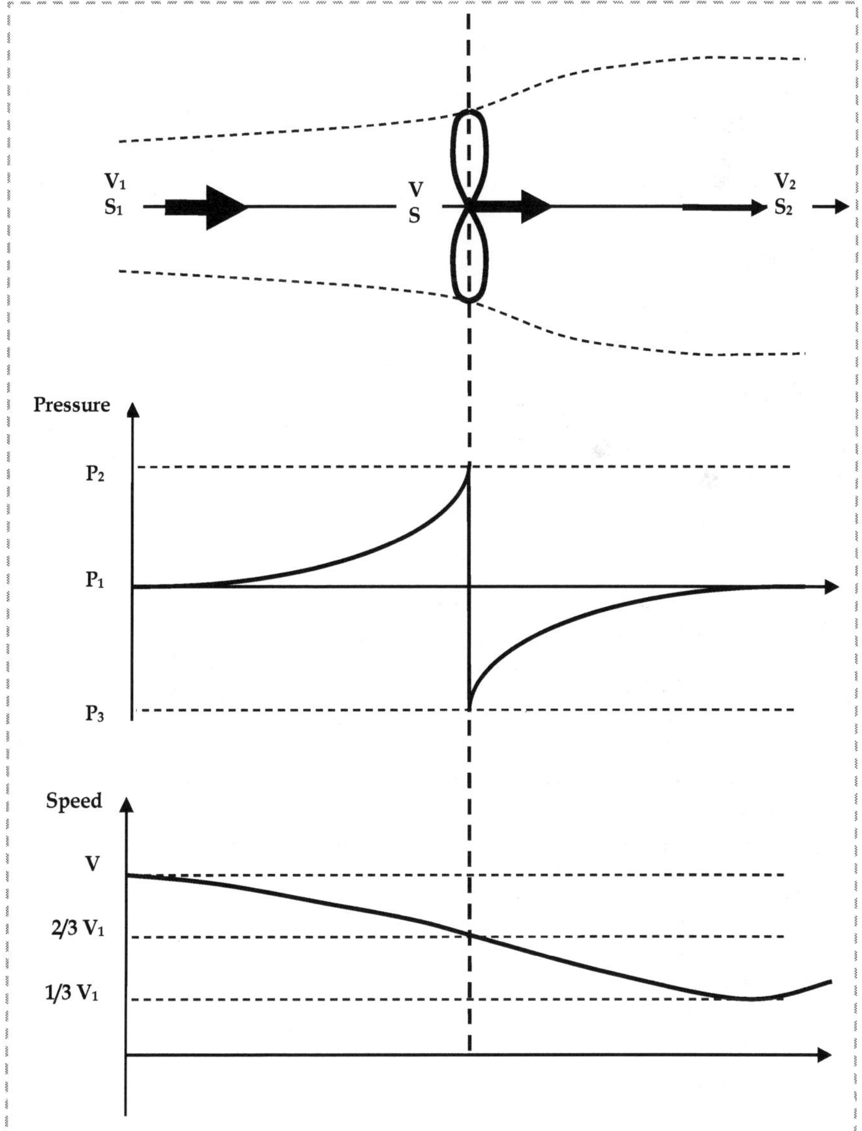

Fig. 1. Pressure and speed variation in an ideal model of a wind turbine.

$$P = \rho S V^2 . (V_1 - V_2) \tag{6}$$

The power as the rate of change in kinetic energy from upstream to downstream is given by:

$$
\begin{aligned}
P &\approx \frac{\Delta E}{\Delta t} \\
&\approx \frac{\frac{1}{2} m V_1^2 - \frac{1}{2} m V_2^2}{\Delta t} \\
&= \frac{1}{2} \dot{m} \left(V_1^2 - V_2^2 \right)
\end{aligned}
\tag{7}
$$

Using the continuity equation (Eqn. 2), we can write:

$$P = \frac{1}{2} \rho S V \left(V_1^2 - V_2^2 \right) \tag{8}$$

Equating the two expressions for the power P in Eqns. 6 and 8, we get:

$$P = \frac{1}{2} \rho S V \left(V_1^2 - V_2^2 \right) = \rho S V^2 (V_1 - V_2)$$

The last expression implies that:

$$
\begin{aligned}
\frac{1}{2}(V_1^2 - V_2^2) &= \frac{1}{2}(V_1 - V_2)(V_1 + V_2) \\
&= V(V_1 - V_2), \qquad \forall V, S, \rho \neq 0
\end{aligned}
$$

or:

$$V = \frac{1}{2}(V_1 + V_2),\ \forall (V_1 - V_2) \neq 0 \ or \ V_1 \neq V_2 \tag{9}$$

This in turn suggests that the wind velocity at the rotor may be taken as the average of the upstream and downstream wind velocities. It also implies that the turbine must act as a brake, reducing the wind speed from V_1 to V_2, but not totally reducing it to V = 0, at which point the equation is no longer valid. To extract energy from the wind stream, its flow must be maintained and not totally stopped.

The last result allows us to write new expressions for the force F and power P in terms of the upstream and downstream velocities by substituting for the value of V as:

$$
\begin{aligned}
F &= \rho S V (V_1 - V_2) \\
&= \frac{1}{2} \rho S (V_1^2 - V_2^2)
\end{aligned}
\tag{10}
$$

$$
\begin{aligned}
P &= \rho S V^2 (V_1 - V_2) \\
&= \frac{1}{4} \rho S (V_1 + V_2)^2 (V_1 - V_2) \\
&= \frac{1}{4} \rho S (V_1^2 - V_2^2)(V_1 + V_2)
\end{aligned}
\tag{11}
$$

We can introduce the "downstream velocity factor," or "interference factor," b as the ratio of the downstream speed V_2 to the upstream speed V_1 as:

$$b = \frac{V_2}{V_1}$$ (12)

From Eqn. 10 the force F can be expressed as:

$$F = \frac{1}{2}\rho S.V_1^2 \left(1 - b^2\right)$$ (13)

The extractable power P in terms of the interference factor b can be expressed as:

$$P = \frac{1}{4}\rho S \left(V_1^2 - V_2^2\right)\left(V_1 + V_2\right)$$
$$= \frac{1}{4}\rho S V_1^3 \left(1 - b^2\right)\left(1 + b\right)$$ (14)

The most important observation pertaining to wind power production is that the extractable power from the wind is proportional to the cube of the upstream wind speed V_1^3 and is a function of the interference factor b.

The "power flux" or rate of energy flow per unit area, sometimes referred to as "power density" is defined using Eqn. 6 as:

$$P^{'} = \frac{P}{S}$$
$$= \frac{\frac{1}{2}\rho S V^3}{S}$$
$$= \frac{1}{2}\rho V^3 , [\frac{Joules}{m^2.s}], [\frac{Watts}{m^2}]$$ (15)

The kinetic power content of the undisturbed upstream wind stream with $V = V_1$ and over a cross sectional area S becomes:

$$W = \frac{1}{2}\rho S V_1^3 , [\frac{Joules}{m^2.s} m^2], [Watts]$$ (16)

The performance coefficient or efficiency is the dimensionless ratio of the extractable power P to the kinetic power W available in the undisturbed stream:

$$C_p = \frac{P}{W}$$ (17)

The performance coefficient is a dimensionless measure of the efficiency of a wind turbine in extracting the energy content of a wind stream. Substituting the expressions for P from Eqn. 14 and for W from Eqn. 16 we have:

$$C_p = \frac{P}{W}$$

$$= \frac{\frac{1}{4}\rho S V_1^3 \left(1-b^2\right)\left(1+b\right)}{\frac{1}{2}\rho S V_1^3}$$

$$= \frac{1}{2}\left(1-b^2\right)\left(1+b\right) \tag{18}$$

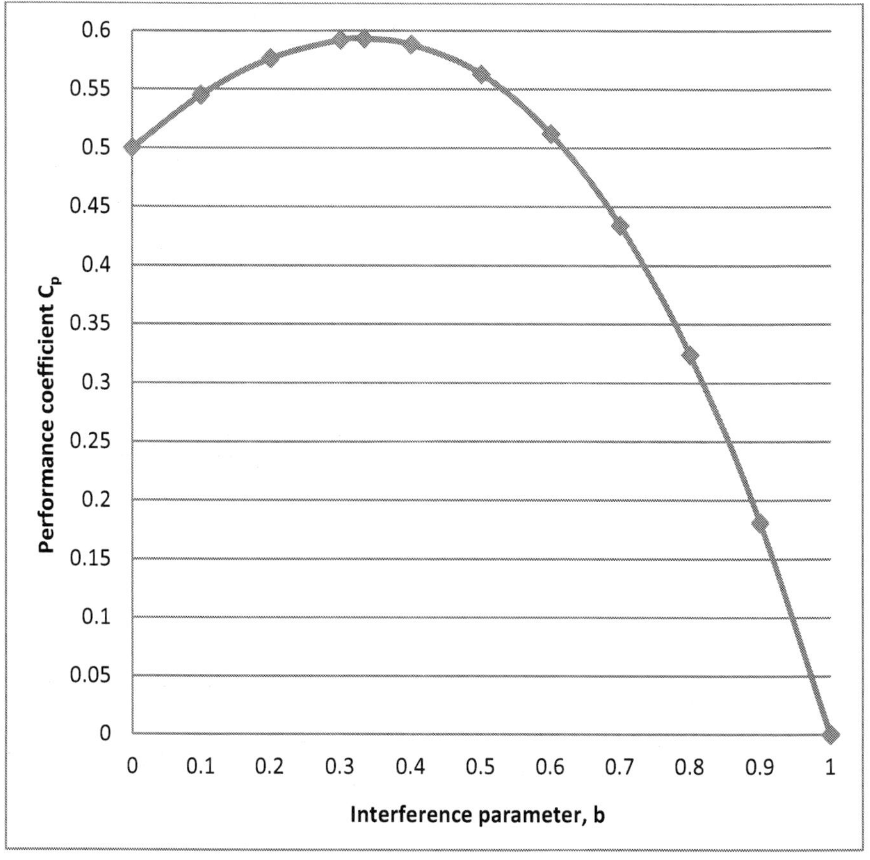

b	0.0	0.1	0.2	0.3	1/3	0.4	0.5	0.6	0.7	0.8	0.9	1.0
C_p	0.500	0.545	0.576	0.592	0.593	0.588	0.563	0.512	0.434	0.324	0.181	0.00

Fig. 2. The performance coefficient C_p as a function of the interference factor b.

When $b = 1$, $V_1 = V_2$ and the wind stream is undisturbed, leading to a performance coefficient of zero. When $b = 0$, $V_1 = 0$, the turbine stops all the air flow and the performance coefficient is equal to 0.5. It can be noticed from the graph that the performance coefficient reaches a maximum around $b = 1/3$.

A condition for maximum performance can be obtained by differentiation of Eq. 18 with respect to the interference factor b. Applying the chain rule of differentiation (shown below) and setting the derivative equal to zero yields Eq. 19:

$$\frac{d}{dx}(uv) = u\frac{dv}{dx} + v\frac{du}{dx}$$

$$\frac{dC_p}{db} = \frac{1}{2}\frac{d}{db}[(1-b^2)(1+b)]$$

$$= \frac{1}{2}[(1-b^2) - 2b(1+b)]$$

$$= \frac{1}{2}(1 - b^2 - 2b - 2b^2)$$

$$= \frac{1}{2}(1 - 3b^2 - 2b)$$

$$= \frac{1}{2}(1 - 3b)(1 + b)$$

$$= 0$$

$$(19)$$

Equation 19 has two solutions. The first is the trivial solution:

$$(1 + b) = 0$$

$$b = \frac{V_2}{V_1} = -1, \Rightarrow V_2 = -V_1$$

The second solution is the practical physical solution:

$$(1 - 3b) = 0$$

$$b = \frac{V_2}{V_1} = \frac{1}{3}, \Rightarrow V_2 = \frac{1}{3}V_1$$

$$(20)$$

Equation 20 shows that for optimal operation, the downstream velocity V_2 should be equal to one third of the upstream velocity V_1. Using Eqn. 18, the maximum or optimal value of the performance coefficient C_p becomes:

$$C_{p,opt} = \frac{1}{2}(1 - b^2)(1 + b)$$

$$= \frac{1}{2}\left(1 - (\tfrac{1}{3})^2\right)\left(1 + \frac{1}{3}\right)$$

$$= \frac{16}{27}$$

$$= 0.59259$$

$$= 59.26\ percent$$

$$(21)$$

This is referred to as the Betz Criterion or the Betz Limit. It was first formulated in 1919, and applies to all wind turbine designs. It is the theoretical power fraction that can be extracted from an ideal wind stream. Modern wind machines operate at a slightly lower practical non-ideal performance coefficient. It is generally reported to be in the range of:

$$C_{p,prac.} \approx \frac{2}{5} = 40 \; percent \tag{22}$$

Result I

From Eqns. 9 and 20, there results that:

$$\begin{aligned} V &= \frac{1}{2}(V_1 + V_2) \\ &= \frac{1}{2}(V_1 + \frac{V_1}{3}) \\ &= \frac{2}{3}V_1 \end{aligned} \tag{23}$$

Result II

From the continuity Eqn. 2:

$$\dot{m} = \rho S_1 V_1 = \rho S V = \rho S_2 V_2 = constant$$
$$S = S_1 \frac{V_1}{V} = \frac{3}{2}S_1 \tag{24}$$
$$S_2 = S_1 \frac{V_1}{V_2} = 3S_1$$

This implies that the cross sectional area of the airstream downwind of the turbine expands to 3 times the area upwind of it.

Some pitfalls in the derivation of the previous equations could inadvertently occur and are worth pointing out. One can for instance try to define the power extraction from the wind in two different ways. In the first approach, one can define the power extraction by an ideal turbine from Eqns. 23, 24 as:

$$\begin{aligned} P^1_{ideal} &= P_{upwind} - P_{downwind} \\ &= \frac{1}{2}\rho S_1 V_1^3 - \frac{1}{2}\rho S_2 V_2^3 \\ &= \frac{1}{2}\rho S_1 V_1^3 - \frac{1}{2}\rho 3 S_1 (\frac{1}{3}V_1)^3 \\ &= \frac{1}{2}\rho(\frac{8}{9}S_1 V_1^3) \\ &= \frac{8}{9}\frac{1}{2}\rho S_1 V_1^3 \end{aligned}$$

This suggests that fully 8/9 of the energy available in the upwind stream can be extracted by the turbine. That is a confusing result since the upwind wind stream has a cross sectional area that is smaller than the turbine intercepted area.

The second approach yields the correct result by redefining the power extraction at the wind turbine using the area of the turbine as $S = 3/2\,S_1$:

$$
\begin{aligned}
P_{ideal} &= \frac{1}{2}\rho\left(\frac{8}{9}S_1 V_1^3\right)\\[4pt]
&= \frac{1}{2}\rho\left(\frac{8}{9}\frac{2}{3}S V_1^3\right)\\[4pt]
&= \frac{1}{2}\rho\left(\frac{16}{27}S V_1^3\right)\\[4pt]
&= \frac{16}{27}\frac{1}{2}\rho S V_1^3
\end{aligned}
\tag{25}
$$

The value of the Betz coefficient suggests that a wind turbine can extract **at most** 59.3 percent of the energy in an undisturbed wind stream.

$$
Betz\,coefficient = \frac{16}{27} = 0.592593 = 59.26\,percent
\tag{26}
$$

Considering the frictional losses, blade surface roughness, and mechanical imperfections, between 35 to 40 percent of the power available in the wind is extractable under practical conditions.

Another important perspective can be obtained by estimating the maximum power content in a wind stream. For a constant upstream velocity, we can deduce an expression for the maximum power content for a constant upstream velocity V_1 of the wind stream by differentiating the expression for the power P with respect to the downstream wind speed V_2, applying the chain rule of differentiation and equating the result to zero as:

$$
\begin{aligned}
\left.\frac{dP}{dV_2}\right|_{V_1} &= \frac{1}{4}\rho S\frac{d}{dV_2}[(V_1+V_2)^2(V_1-V_2)]\\[4pt]
&= \frac{1}{4}\rho S\frac{d}{dV_2}[(V_1^2-V_2^2)(V_1+V_2)]\\[4pt]
&= \frac{1}{4}\rho S[(V_1^2-V_2^2)-2V_2(V_1+V_2)]\\[4pt]
&= \frac{1}{4}\rho S(V_1^2-V_2^2-2V_1V_2-2V_2^2)\\[4pt]
&= \frac{1}{4}\rho S(V_1^2-3V_2^2-2V_1V_2)\\[4pt]
&= 0
\end{aligned}
\tag{27}
$$

Solving the resulting equation by factoring it yield Eqn. 28.

$$
\begin{aligned}
(V_1^2-3V_2^2-2V_1V_2) &= 0\\
(V_1+V_2)(V_1-3V_2) &= 0
\end{aligned}
\tag{28}
$$

Equation 28 once again has two solutions. The trivial solution is shown in Eqn. 29.

$$(V_1 + V_2) = 0$$
$$V_2 = -V_1 \tag{29}$$

The second physically practical solution is shown in Eqn 30.

$$(V_1 - 3V_2) = 0$$
$$V_2 = \frac{1}{3}V_1 \tag{30}$$

This implies the simple result that that the most efficient operation of a wind turbine occurs when the downstream speed V_2 is one third of the upstream speed V_1. Adopting the second solution and substituting it in the expression for the power in Eqn. 16 we get the expression for the maximum power that could be extracted from a wind stream as:

$$
\begin{aligned}
P_{max} &= \frac{1}{4}\rho S\left(V_1^2 - V_2^2\right)\left(V_1 + V_2\right) \\
&= \frac{1}{4}\rho S\left(V_1^2 - \frac{V_1^2}{9}\right)\left(V_1 + \frac{V_1}{3}\right) \\
&= \frac{1}{4}\rho S V_1^3\left(1 - \frac{1}{9}\right)\left(1 + \frac{1}{3}\right) \\
&= \frac{16}{27}\frac{\rho}{2}V_1^3 S \; [Watt]
\end{aligned}
\tag{31}
$$

This expression constitutes the formula originally derived by Betz where the swept rotor area S is:

$$S = \frac{\pi D^2}{4} \tag{32}$$

and the Betz Equation results as:

$$P_{max} = \frac{16}{27}\frac{\rho}{2}V_1^3 \frac{\pi D^2}{4} \; [Watt] \tag{33}$$

The most important implication from the Betz Equation is that there must be a wind speed change from the upstream to the downstream in order to extract energy from the wind; in fact by braking it using a wind turbine.

If no change in the wind speed occurs, energy cannot be efficiently extracted from the wind. Realistically, no wind machine can totally bring the air to a total rest, and for a rotating machine, there will always be some air flowing around it. Thus a wind machine can only extract a fraction of the kinetic energy of the wind. The wind speed on the rotors at which energy extraction is maximal has a magnitude lying between the upstream and downstream wind velocities.

The Betz Criterion reminds us of the Carnot cycle efficiency in Thermodynamics suggesting that a heat engine cannot extract all the energy from a given heat reservoir and must reject part of its heat input back to the environment.

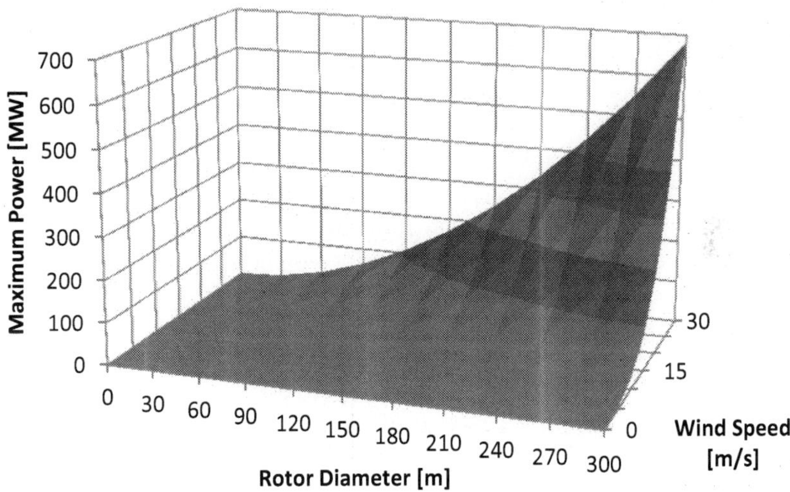

Fig. 3. Maximum power as a function of the rotor diameter and the wind speed. The power increases as the square of the rotor diameter and more significantly as the cube of the wind speed (Ragheb, M., 2011).

3. Rotor optimal Tip Speed Ratio, TSR

Another important concept relating to the power of wind turbines is the optimal tip speed ratio, which is defined as the ratio of the speed of the rotor tip to the free stream wind speed. If a rotor rotates too slowly, it allows too much wind to pass through undisturbed, and thus does not extract as much as energy as it could, within the limits of the Betz Criterion, of course.

On the other hand, if the rotor rotates too quickly, it appears to the wind as a large flat disc, which creates a large amount of drag. The rotor Tip Speed Ratio, TSR depends on the blade airfoil profile used, the number of blades, and the type of wind turbine. In general, three-bladed wind turbines operate at a TSR of between 6 and 8, with 7 being the most widely-reported value.

In addition to the factors mentioned above, other concerns dictate the TSR to which a wind turbine is designed. In general, a high TSR is desirable, since it results in a high shaft rotational speed that allows for efficient operation of an electrical generator. Disadvantages however of a high TSR include:

a. Blade tips operating at 80 m/s of greater are subject to leading edge erosion from dust and sand particles, and would require special leading edge treatments like helicopter blades to mitigate such damage,
b. Noise, both audible and inaudible, is generated,
c. Vibration, especially in 2 or 1 blade rotors,

d. Reduced rotor efficiency due to drag and tip losses,
e. Higher speed rotors require much larger braking systems to prevent the rotor from reaching a runaway condition that can cause disintegration of the turbine rotor blades.

The Tip Speed Ratio, TSR, is dimensionless factor defined in Eqn. 34

$$\text{TSR} = \lambda = \frac{\text{speed of rotor tip}}{\text{wind speed}} = \frac{v}{V} = \frac{\omega r}{V} \tag{34}$$

where:

$$
\begin{aligned}
V & = \text{wind speed [m/sec]} \\
v = \omega r & = \text{rotor tip speed [m/sec]} \\
r & = \text{rotor radius [m]} \\
\omega = 2\pi f & = \text{angular velocity [rad/sec]} \\
f & = \text{rotational frequency [Hz], [sec}^{-1}\text{]}
\end{aligned}
$$

Example 1

At a wind speed of 15 m/sec, for a rotor blade radius of 10 m, rotating at 1 rotation per second:

$$f = 1[\frac{\text{rotation}}{\text{sec}}],$$

$$\omega = 2\pi f = 2\pi[\frac{\text{radian}}{\text{sec}}]$$

$$v = \omega r = 2\pi.10 = 20\pi[\frac{\text{m}}{\text{sec}}]$$

$$\lambda = \frac{\omega r}{V} = \frac{20\pi}{15} = \frac{62.83}{15} = 4$$

Example 2

The Suzlon S.66/1250, 1.25 MW rated power at 12 m/s rated wind speed wind turbine design has a rotor diameter of 66 meters and a rotational speed of 13.9-20.8 rpm.
Its angular speed range is:

$$\omega = 2\pi f$$

$$= 2\pi\frac{13.9 - 20.8}{60}[\text{radian.}\frac{\text{revolutions}}{\text{minute}}.\frac{\text{minute}}{\text{second}}]$$

$$= 1.46 - 2.18[\frac{\text{radian}}{\text{sec}}]$$

The range of its rotor's tip speed can be estimated as:

$$v = \omega r$$

$$= (1.46 - 2.18)\frac{66}{2}$$

$$= 48.18 - 71.94[\frac{\text{m}}{\text{sec}}]$$

The range of its tip speed ratio is thus:

$$\lambda = \frac{\omega r}{V}$$
$$= \frac{48.18 - 71.94}{12}$$
$$\approx 4 - 6$$

The optimal TSR for maximum power extraction is inferred by relating the time taken for the disturbed wind to reestablish itself to the time required for the next blade to move into the location of the preceding blade. These times are t_w and t_b, respectively, and are shown below in Eqns. 35 and 36. In Eqns. 35 and 36, n is the number of blades, ω is the rotational frequency of the rotor, s is the length of the disturbed wind stream, and V is the wind speed.

$$t_s = \frac{2\pi}{n\omega}[\text{sec}] \tag{35}$$

$$t_w = \frac{s}{V}[\text{sec}] \tag{36}$$

If $t_s > t_w$, some wind is unaffected. If $t_w > t_s$, some wind is not allowed to flow through the rotor. The maximum power extraction occurs when the two times are approximately equal. Setting t_w equal to t_s yields Eqn. 37 below, which is rearranged as:

$$t_s \approx t_w$$
$$\frac{2\pi}{n\omega} \approx \frac{s}{V} \Rightarrow \frac{n\omega}{V} \approx \frac{2\pi}{s} \tag{37}$$

Equation 37 may then be used to define the optimal rotational frequency as shown in Eqn. 38:

$$\omega_{optimal} \approx \frac{2\pi V}{ns} \tag{38}$$

Consequently, for optimal power extraction, the rotor blade must rotate at a rotational frequency that is related to the speed of the oncoming wind. This rotor rotational frequency decreases as the radius of the rotor increases and can be characterized by calculating the optimal TSR, $\lambda_{optimal}$ as shown in Eqn. 39.

$$\lambda_{optimal} \approx \frac{\omega_{optimal} r}{V} \approx \frac{2\pi}{n}\left(\frac{r}{s}\right) \tag{39}$$

4. Effect of the number of rotor blades on the Tip Speed Ratio, TSR

The optimal TSR depends on the number of rotor blades, n, of the wind turbine. The smaller the number of rotor blades, the faster the wind turbine must rotate to extract the maximum power from the wind. For an n-bladed rotor, it has empirically been observed that s is approximately equal to 50 percent of the rotor radius. Thus by setting:

$$\frac{s}{r} \approx \frac{1}{2},$$

Eqn. 39 is modified into Eqn. 40:

$$\lambda_{optimal} \approx \frac{2\pi}{n}\left(\frac{r}{s}\right) \approx \frac{4\pi}{n} \tag{40}$$

For n = 2, the optimal TSR is calculated to be 6.28, while it is 4.19 for three-bladed rotor, and it reduces to 3.14 for a four-bladed rotor. With proper airfoil design, the optimal TSR values may be approximately 25 – 30 percent above these values. These highly-efficient rotor blade airfoils increase the rotational speed of the blade, and thus generate more power. Using this assumption, the optimal TSR for a three-bladed rotor would be in the range of 5.24 – 5.45. Poorly designed rotor blades that yield too low of a TSR would cause the wind turbine to exhibit a tendency to slow and stall. On the other hand, if the TSR is too high, the turbine will rotate very rapidly, and will experience larger stresses, which may lead to catastrophic failure in highly-turbulent wind conditions.

5. Power coefficient, C_p

The power generated by the kinetic energy of a free flowing wind stream is shown in Eqn. 41.

$$P = \frac{1}{2}\rho S V^3 \ [Watt] \tag{41}$$

Defining the cross sectional area, S, of the wind turbine, in terms of the blade radius, r, Eqn. 41 becomes Eqn. 42.

$$P = \frac{1}{2}\rho\pi R^2 V^3 \tag{42}$$

The power coefficient (Jones, B., 1950), Eqn. 43, is defined as the ratio of the power extracted by the wind turbine relative to the energy available in the wind stream.

$$C_p = \frac{P_t}{P} = \frac{P_t}{\frac{1}{2}\rho\pi R^2 V^3} \tag{43}$$

As derived earlier in this chapter, the maximum achievable power coefficient is 59.26 percent, the Betz Limit. In practice however, obtainable values of the power coefficient center around 45 percent. This value below the theoretical limit is caused by the inefficiencies and losses attributed to different configurations, rotor blades profiles, finite wings, friction, and turbine designs. Figure 4 depicts the Betz, ideal constant, and actual wind turbine power coefficient as a function of the TSR.

As shown in Fig. 4, maximum power extraction occurs at the optimal TSR, where the difference between the actual TSR (blue curve) and the line defined by a constant TSR is the lowest. This difference represents the power in the wind that is not captured by the wind turbine. Frictional losses, finite wing size, and turbine design losses account for part of the

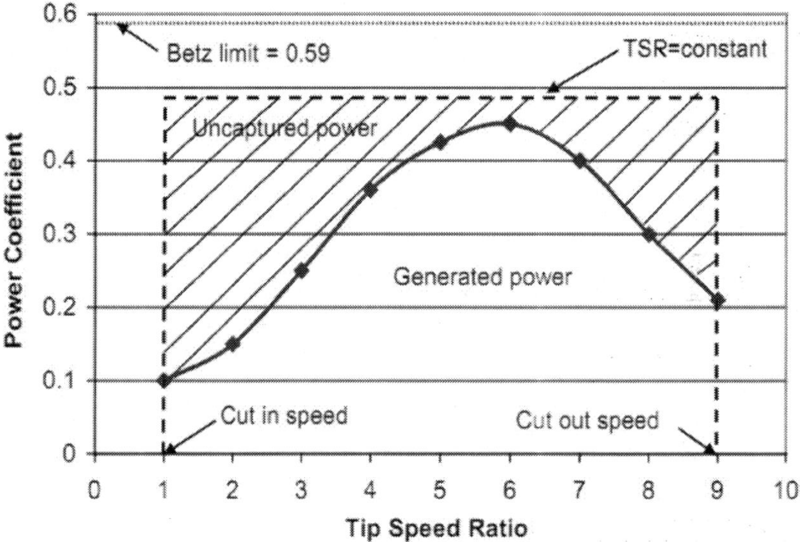

Fig. 4. Power coefficient as a function of TSR for a two-bladed rotor.

uncaptured wind power, and are supplemented by the fact that a wind turbine does not operate at the optimal TSR across its operating range of wind speeds.

6. Inefficiencies and losses, Schmitz power coefficient

The inefficiencies and losses encountered in the operation of wind turbines include the blade number losses, whirlpool losses, end losses and the airfoil profile losses (Çetin, N. S. et. al. 2005).

Airfoil profile losses

The slip or slide number s is the ratio of the uplift force coefficient of the airfoil profile used C_L to the drag force coefficient C_D is:

$$s = \frac{C_L}{C_D} \quad s = \frac{C_L}{C_D} \tag{44}$$

Accounting for the drag force can be achieved by using the profile efficiency that is a function of the slip number s and the tip speed ratio λ as:

$$\eta_{profile} = \frac{s - \lambda}{s} = 1 - \frac{\lambda}{s} \quad \eta_{profile} = \frac{s - \lambda}{s} = 1 - \frac{\lambda}{s} \tag{45}$$

Rotor tip end losses

At the tip of the rotor blade an air flow occurs from the lower side of the airfoil profile to the upper side. This air flow couples with the incoming air flow to the blade. The combined air flow results in a rotor tip end efficiency, $\eta_{tip\ end}$.

Whirlpool Losses

In the idealized derivation of the Betz Equation, the wind does not change its direction after it encounters the turbine rotor blades. In fact, it does change its direction after the encounter.

This is accounted-for by a modified form of the power coefficient known as the Schmitz power coefficient $C_{pSchmitz}$ if the same airfoil design is used throughout the rotor blade.

Tip speed ratio TSR λ	Whirlpool Schmitz Power Coefficient $C_{pSchmitz}$
0.0	0.000
0.5	0.238
1.0	0.400
1.5	0.475
2.0	0.515
2.5	0.531
3.0	0.537
3.5	0.538
4.0	0.541
4.5	0.544
5.0	0.547
5.5	0.550
6.0	0.553
6.5	0.556
7.0	0.559
7.5	0.562
8.0	0.565
8.5	0.568
9.0	0.570
9.5	0.572
10.0	0.574

Table 1. Whirlpool losses Schmitz power coefficient as a function of the tip speed ratio.

Rotor blade number losses

A theory developed by Schmitz and Glauert applies to wind turbines with four or less rotor blades. In a turbine with more than four blades, the air movement becomes too complex for a strict theoretical treatment and an empirical approach is adopted. This can be accounted for by a rotor blades number efficiency η_{blades}.

In view of the associated losses and inefficiencies, the power coefficient can be expressed as:

$$C_p \approx C_{pSchmitz}\,\eta_{profile}\,\eta_{tip\,end}\,\eta_{blades} \tag{46}$$

There are still even more efficiencies involved:

1. Frictional losses in the bearings and gears: $\eta_{friction}$

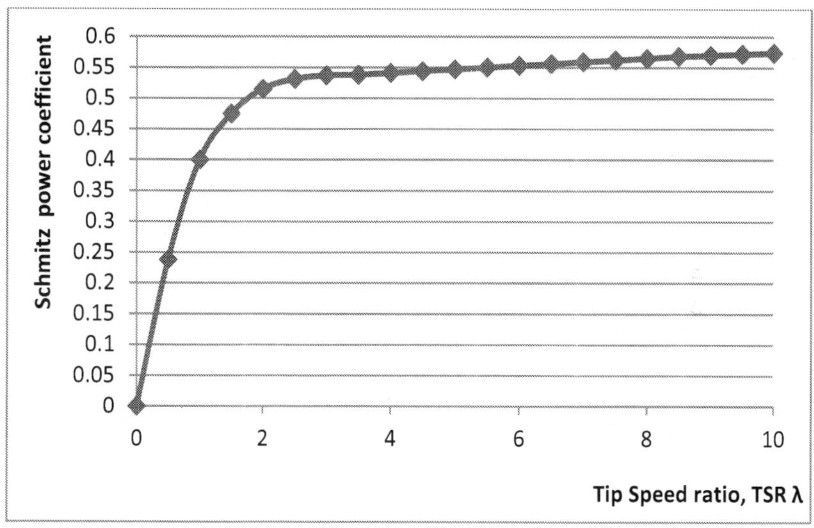

Fig. 5. Schmitz power coefficient as a function of the tip speed ratio, TSR.

2. Magnetic drag and electrical resistance losses in the generator or alternator: $\eta_{electrical}$.

$$C'_p \approx C_{pSchmitz}\eta_{profile}\eta_{tip\,end}\eta_{blades}\eta_{friction}\eta_{electrical} \tag{47}$$

In the end, the Betz Limit is an idealization and a design goal that designers try to reach in a real world turbine. A C_p value of between 0.35 – 0.40 is a realistic design goal for a workable wind turbine. This is still reduced by a capacity factor accounting for the periods of wind flow as the intermittency factor.

7. Power coefficient and tip speed ratio of different wind converters designs

The theoretical maximum efficiency of a wind turbine is given by the Betz Limit, and is around 59 percent. Practically, wind turbines operate below the Betz Limit. In Fig. 4 for a two-bladed turbine, if it is operated at the optimal tip speed ratio of 6, its power coefficient would be around 0.45. At the cut-in wind speed, the power coefficient is just 0.10, and at the cut-out wind speed it is 0.22. This suggests that for maximum power extraction a wind turbine should be operated around its optimal wind tip ratio.

Modern horizontal axis wind turbine rotors consist of two or three thin blades and are designated as low solidity rotors. This implies a low fraction of the area swept by the rotors being solid. Its configuration results in an optimum match to the frequency requirements of modern electricity generators and also minimizes the size and weight of the gearbox or transmission required, as well as increases efficiency.

Such an arrangement results in a relatively high tip speed ratio in comparison with rotors with a high number of blades such as the highly successful American wind mill used for water pumping in the American West and all over the world. The latter required a high starting torque.

The relationship between the rotor power coefficient C_p and the tip speed ratio is shown for different types of wind machines. It can be noticed that it reaches a maximum at different positions for different machine designs.

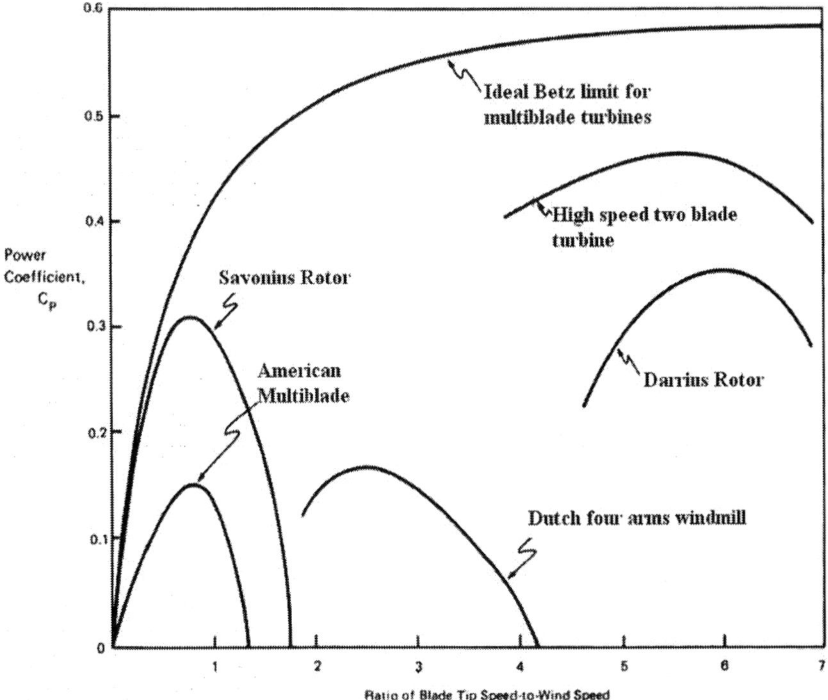

Fig. 6. The power coefficient C_p as a function of the tip speed ratio for different wind machines designs. Note that the efficiency curves of the Savonius and the American multiblade designs were inadvertently switched (Eldridge, F. R., 1980) in some previous publications, discouraging the study of the Savonius design.

The maximum efficiencies of the two bladed design, the Darrieus concept and the Savonius reach levels above 30 percent but below the Betz Limit of 59 percent.
The American multiblade design and the historical Dutch four bladed designs peak at 15 percent. These are not suited for electrical generation but are ideal for water pumping.

8. Discussion and conclusions

Wind turbines must be designed to operate at their optimal wind tip speed ratio in order to extract as much power as possible from the wind stream. When a rotor blade passes through the air stream it leaves a turbulent wake in its path. If the next blade in the rotating rotor arrives at the wake when the air is still turbulent, it will not be able to extract power from the wind efficiently, and will be subjected to high vibration stresses. If the rotor rotated slower, the air hitting each rotor blade would no longer be turbulent. This is another reason for the tip speed ratio to be selected so that the rotor blades do not pass through turbulent air.

Wind power conversion is analogous to other methods of energy conversion such as hydroelectric generators and heat engines. Some common underlying basic principles can guide the design and operation of wind energy conversion systems in particular, and of other forms of energy conversion in general. A basic principle can be enunciated as:

"Energy can be extracted or converted only from a flow system."

In hydraulics, the potential energy of water blocked behind a dam cannot be extracted unless it is allowed to flow. In this case only a part of it can be extracted by a water turbine.

In a heat engine, the heat energy cannot be extracted from a totally insulated reservoir. Only when it is allowed to flow from the high temperature reservoir, to a low temperature one where it is rejected to the environment; can a fraction of this energy be extracted by a heat engine.

Totally blocking a wind stream does not allow any energy extraction. Only by allowing the wind stream to flow from a high speed region to a low speed region can energy be extracted by a wind turbine.

A second principle of energy conversion can be elucidated as:

"Natural or artificial asymmetries in an aerodynamic, hydraulic, or thermodynamic system allow the extraction of only a fraction of the available energy at a specified efficiency."

Ingenious minds conceptualized devices that take advantage of existing natural asymmetries, or created configurations or situations favoring the creation of these asymmetries, to extract energy from the environment.

A corollary ensues that the existence of a flow system necessitates that only a fraction of the available energy can be extracted at an efficiency characteristic of the energy extraction process with the rest returned back to the environment to maintain the flow process.

In thermodynamics, the ideal heat cycle efficiency is expressed by the Carnot cycle efficiency. In a wind stream, the ideal aerodynamic cycle efficiency is expressed by the Betz Equation.

9. References

Ragheb, M., "Wind Power Systems. Harvesting the Wind."
        https://netfiles.uiuc.edu/mragheb/www, 2011.

Thomas Ackerman, Ed. "Wind Power in Power Systems," John Wiley and Sons, Ltd., 2005.

American Institute of Aeronautics and Astronautics (AIAA) and American Society of Mechanical Engineers (ASME), "A Collection of the 2004 ASME Wind Energy Symposium Technical Papers," 42nd AIAA Aerospace Sciences Meeting and Exhibit, Reno Nevada, 5-8 January, 2004.

Le Gouriérès Désiré, "Wind Power Plants, Theory and Design," Pergamon Press, 1982.

Brown, J. E., Brown, A. E., "Harness the Wind, The Story of Windmills," Dodd, Mead and Company, New York, 1977.

Eldridge, F. R., "Wind Machines," 2nd Ed., The MITRE Energy Resources and Environmental Series, Van Nostrand Reinhold Company, 1980.

Calvert, N. G., "Windpower Principles: Their Application on the Small Scale," John Wiley and Sons, 1979.

Torrey, V., "Wind-Catchers, American Windmills of Yesterday and Tomorrow," The Stephen Greene Press, Brattleboro, Vermont, 1976.

Walker, J. F., Jenkins, N., "Wind Energy Technology," John Wiley and Sons, 1997.

Schmidt, J., Palz, W., "European Wind Energy Technology, State of the Art Wind Energy Converters in the European Community," D. Reidel Publishing Company, 1986.

Energy Research and Development Administration (ERDA), Division of Solar Energy, "Solar Program Assessment: Environmental Factors, Wind Energy Conversion," ERDA 77-47/6, UC-11, 59, 62, 63A, March 1977.

Çetin, N. S., M. A Yurdusev, R. Ata and A. Özdemir, "Assessment of Optimum Tip Speed Ratio of Wind Turbines," Mathematical and Computational Applications, Vol. 10, No.1, pp.147-154, 2005.

Hau, E.,"Windkraftanlangen," Springer Verlag, Berlin, Germany, pp. 110-113, 1996.

Jones, B., "Elements of Aerodynamics," John Wiley and Sons, New York, USA, pp. 73-158, 1950.

3

Inboard Stall Delay Due to Rotation

Horia Dumitrescu and Vladimir Cardoş
Institute of Statistical Mathematics and Applied Mathematics of the Romanian Academy
Romania

1. Introduction

In the design process of improved rotor blades the need for accurate aerodynamic predictions is very important. During the last years a large effort has gone into developing CFD tools for prediction of wind turbine flows (Duque et al., 2003; Fletcher et al., 2009; Sørensen et al.,2002). However, there are still some unclear aspects for engineers regarding the practical application of CFD, such as computational domain size, reference system for different computational blocks, mesh quality and mesh number, turbulence, etc. Thus, in the design process and in the power curve prediction of wind turbines, the aerodynamic forces are calculated with some form of the blade element method (BEM) and its extensions to the three-dimensional wing aerodynamics. The results obtained by the standard methods are reasonably accurate in the proximity of the design point, but in stalled condition the BEM is known to underpredict the forces acting on the blades (Himmelskamp, 1947). The major disadvantage of these methods is that the airflow is reduced to axial and circumferential flow components (Glauert, 1963). Disregarding radial flow components present in the bottom of separated boundary layers of rotating wings leads to alteration of lift and drag characteristics of the individual blade sections with respect to the 2-D airfoils (Bjorck, 1995).

Airfoil characteristics of lift (C_L) and drag (C_D) coefficients are normally derived from two-dimensional (2-D) wind tunnel tests. However, after stall the flow over the inboard half of the rotor is strongly influenced by poorly understood 3-D effects (Banks & Gadd, 1963; Tangler, 2002). The 3-D effects yield delayed stall with C_L higher than 2.0 near the blade root location and with correspondingly high C_D. Now the design of constant speed, stall-regulated wind turbines lacks adequate theory for predicting their peak and post-peak power and loads.

During the development of stall-regulated wind turbines, there were several attempts to predict 3-D post-stall airfoil characteristics (Corrigan & Schlichting, 1994; Du & Schling, 1998; Snel et al., 1993), but these methods predicted insufficient delayed stall in the root region and tended to extend the delayed stall region too far out on the blade.

The present work aims at giving a conceptualization of the complex 3-D flow field on a rotor blade, where stall begins and how it progresses, driven by the needs to formulate a reasonably simple model that complements the 2-D airfoil characteristics used to predict rotor performance.

Understanding wind turbine aerodynamics (Hansen & Butterfield, 1993) in all working states is one of the key factors in making improved predictions of their performance. The flow field associated with wind turbines is highly three-dimensional and the transition to two-dimensional outboard separated flow is yet not well understood. A continued effort is necessary to improve the delayed stall modeling and bring it to a point where the prediction becomes acceptable. In the sequel, based on previous computed and measured results, a comprehensible model is devised to explain in physical terms the different phenomena that play a role and to clarify what can be modeled quantitatively in a scientific way and what is possible in an engineering environment.

In 1945 Himmelskamp (Himmelskamp, 1947) first described through measurements the 3-D and rotational effects on the boundary layer of a rotating propeller, finding lift coefficients much higher moving towards the rotation axis. Further experimental studies confirmed these early results, indicating in stall-delay and post-stalled higher lift coefficient values the main effects of rotation on wings. Measurements on wind turbine blades were performed by Ronstend (Ronsten, 1992), showing the differences between rotating and non-rotating pressure coefficients and aerodynamic loads, and by Tangler and Kocurek (Tangler & Kocurek, 1993), who combined results from measurements with the classical BEM method to properly compute lift and drag coefficients and the rotor power in stalled conditions.

The theoretical foundations for the analysis of the rotational effects on rotating blades come at the late 40's with Sears (Sears, 1950), who derived a set of equations for the potential flow field around a cylindrical blade of infinite span in pure rotation. He stated that the spanwise component of velocity is dependent only upon the potential flow and it is independent of the span (the so-called independence principle). Then, Fogarty and Sears (Fogarty & Sears, 1950) extended the former study to the potential flow around a rotating and advancing blade. They confirmed that, for a cylindrical blade advancing like a propeller, the tangential and axial velocity components are the same as in the 2-D motion at the local relative speed and incidence. A more comprehensive work was made once more by Fogarty (Fogarty, 1951), consisting of numerical computations on the laminar boundary layer of a rotating plate and blade with thickness. Here he showed that the separation line is unaffected by rotation and that the spanwise velocities in the boundary layer appeared small compared to the chordwise, and no large effects of rotation were observed in contrast to (Himmelskamp, 1947). A theoretical analysis done by Banks and Gadd (Banks &Gadd, 1963), focused on demonstrating how rotation delays laminar separation. They found that the separation point is postponed due to rotation, and for extreme inboard stations the boundary layer is completely stabilized against separation.

In the NASA report done by McCroskey and Dwyer (McCroskey & Dwyer, 1969), the so-called secondary effects in the laminar incompressible boundary layer of propeller and helicopter rotor blades are widely studied, by means of a combined, numerical and analytical approach. They showed that approaching the rotational axis, the Coriolis force in the cross flow direction becomes more important. On the other hand the centrifugal pumping effect is much weaker than was generally thought before, but its contribution increases particularly in the region of the separated flow. The last two decades have known the rising of computational fluid dynamics and the study of the boundary layer on rotating blade has often been carried on through a numerical approach. Sørensen (Sørensen, 1986) numerically solved the 3-D equations of the boundary layer on a rotating surface, using a viscous-inviscid interaction model. In his results the position of the separation line still appears the same as for 2-D predictions, but where separations are more pronounced a larger difference between the lift coefficient calculated for the 2-D and 3-D case is noticed. A quasi 3-D approach, based on the viscous-inviscid interaction method, was introduced by Snel et al (Snel et al., 1993) and results were compared with measurements. They proposed a semi-empirical law for the correction of the 2-D lift curve, identifying the local chord to radii

(c/r) ratio of the blade section as the main parameter of influence. This result has been confirmed by Shen and Sørensen (Shen & Sørensen, 1999), and by Chaviaropoulos and Hansen (Chaviaropoulos & Hansen, 2000), who performed airfoil computations applying a quasi 3-D Navier-Stokes model, based on the streamfunction-vorticity formulation and respectively a primitive variables form. Du and Seling (Du & Selling, 2000), and Dumitrescu and Cardoş (Dumitrescu & Cardoş, 2010), investigated the effects of rotation on blade boundary layers by solving the 3-D integral boundary-layer equations with the assumed velocity profiles and a closure model. Dumitrescu and Cardoş studying the boundary layer behavior very close to the rotation center ($r/c < 1$) stated that the stall delay depends strongly on the leading edge separation bubbles formed on inboard blade segment due to a suction effect at the root area of the blade (Dumitrescu & Cardoş, 2009).

Recently, with the advent of the supercomputer, computational fluid dynamics (CFD) tools have been employed to investigate the stall-delay for wind turbines (Duque et al., 2003; Fletcher et al., 2009; Sørensen et al., 2002). These calculations in addition to Narramore and Vermeland's results (Narramore & Vermeland, 1992) showed that the 3-D rotational effect was particularly pronounced for the inboard sections, but the genesis of this phenomenon is a problem still open. The present concern aims at giving explanations in physics terms of the different features widely-observed experimentally and computationally in wind turbine flow.

2. Flow at low wind speeds

At low wind speed conditions, i.e. at high speed ratios (TSR>3), the visualization of the computed flow indicates that the flow is well-behaved and attached over much of the rotor, Fig. 1. Figure 1 shows the separated area and radial flow on the suction side of a commercial blade with 40 m length at the design tip speed ratio (TSR=5); the secondary flow is strongest at approximately $0.17R$ and reaches up to $0.31R$, where R is the rotor radius. The local air velocity relative to a rotor blade consists of free-wind velocity V_w defined as the wind speed if there were no rotor present, that due to the blade motion $\Omega_b r$ and the wake induced velocities; at high TSR, a weak wake (Glauert, 1963) occurs and its rotational induction velocity can be neglected. Wind turbine blade sections can operate in two main flow regimes depending on the size of the rotation parameter defined as the ratio of wind velocity to the local tangential velocity $V_w / \Omega_b r$. If the rotation parameter is less than unity along the entire span, and for properly twisted blades, the flow is mostly two-dimensional and attached, while for rotation parameter greater than unity the flow is neither two-dimensional nor steady, and is strongly affected by rotation. At high tip speed ratios the subunitary $V_w / \Omega_b r$ condition is accomplished and the blade sections usually operate at prestall incidences. Then, the boundary layer is attached all the way to the trailing edge on the outside blade and is separated at trailing edge only on the inside blade, Fig. 1. At the root area of the blade the flow close to the hub behaves like a rotating disk in a fluid at rest (Fig. 2a), where the centrifugal forces induce a spinning motion in the separated flow; and a radial velocity field is more or less uniform (Fig. 2b). Beginning at the hub, this secondary flow generates the so-called centrifugal pumping mechanism, acting in separated trailing edge flow. On the other hand, the Coriolis force acts in the chordwise direction as a favorable pressure that mitigates separation at the trailing edge along the whole span.

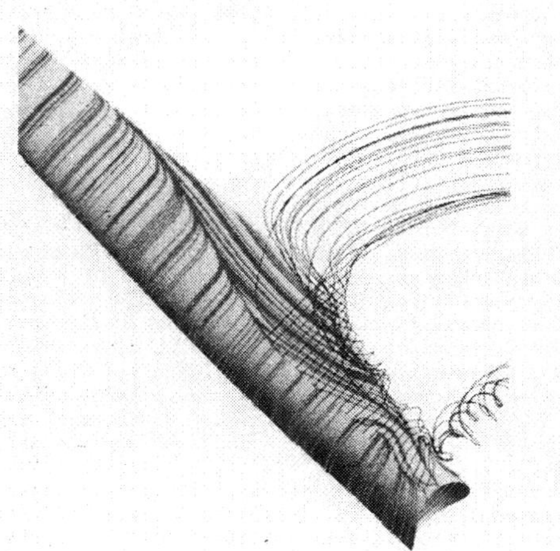

Fig. 1. Typical surface and 3-D streamlines on the blade suction side at low wind speeds (TSR=5)

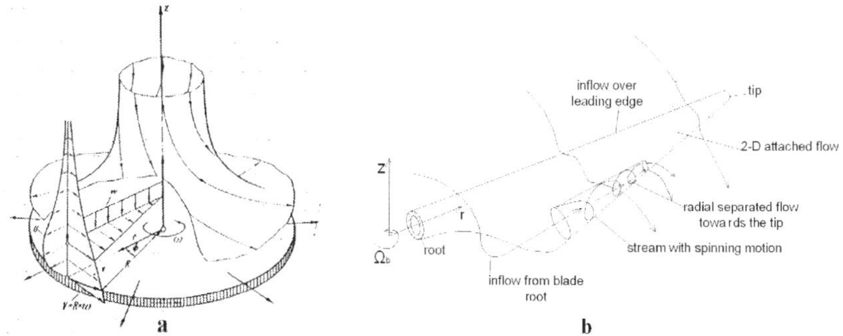

Fig. 2. Pumping-work mode of a wind turbine at low wind speeds (TSR>3.0): a) conceptof flow close to a rotating disk in a fluid at rest; b) model of the separation flow (Corten, 2001).

Therefore, at low wind speed, the main rotational effect is due to the Coriolis force which delays the occurrence of separation to a point further downstream towards the trailing edge, and by this the suction pressures move towards higher levels as r/c decreases. The pumping effect is much weaker than was generally thought before.

The pressure field created by the presence of the turbine is related to the incoming flow field around the blade, taken as being composed of the free wind velocity and the so-called induction velocity due to the rotor and its wake. Thus, the incoming field results from a weak interaction between two different flows: one axial and the other rotational $\left(V_w / \Omega_b r < 1\right)$. In such a weak interaction flow, the basic assumptions made are:

- the radial independence principle is applied to flow effects, i.e. induction velocities used at a certain radial station depend only on the local aerodynamic forces at that same station;
- the mathematical description of the air flow over the blades is based on the 2-D flow potential independent of the span, and on corrections for viscosity and 3-D rotational effects.

These assumptions suitable to BEM methods reduce the complexity of the problem by an order of magnitude yielding reliable results for the local forces and the overall torque in the proximity of the design point, at high tip speed ratios. In order to estimate the 3-D rotational effects, usually neglected in the traditional BEM model, the flow around a hypothetic blade with prestall/stall incidence and chord constant along the whole span is considered in the sequel.

2.1 Representation of flow elements

The set of equations including a simplified form of the inviscid flow and the full three-dimensional boundary-layer equations are used to identify the influence of the3-D rotational effects at low wind speeds.

A. Inviscid flow. In order to find the velocity at the airfoil surface in absence of viscous effects, the reference velocity at a point on a rotating wind turbine blade is

$$U_r = V_w \sqrt{\left(\lambda \frac{r}{R}\right)^2 + (1-a)^2} \tag{1}$$

where V_w is the wind speed, $\lambda = \dfrac{\Omega_b R}{V_w}$ is the tip-speed ratio (TSR), R is the radius of the turbine and a is the axial induced velocity interference, a function of the speed ratio λ (Burton et al., 2001). Starting from the idea of Fogarty and Sears (Fogarty & Sears, 1950), an inviscid edge velocity can be calculated as

$$U = \Omega_b r \frac{\partial \phi}{\partial \theta}, \ V = \Omega_b (\phi - 2\theta), \ W = \Omega_b r \frac{\partial \phi}{\partial z} \tag{2}$$

where U, V and W represent the velocity components in the cylindrical coordinate system (θ, r, z), which rotates with the blade with a constant-rotational speed Ω_b (Fig. 3). $\phi = \phi(\theta, z)$ denotes the 2-D potential solution, that is constant at all radial positions. The interesting point regarding this set of equations (2) is that the spanwise component V can be derived from the local 2-D velocity potential. However, this spanwise component is very small and thus neglected in the present work. The potential edge velocity components can be approached as

$$U_e = U_r U_{2D}, \ V = 0 \tag{3}$$

where the non-dimensional velocity U_{2D} could be obtained by a viscous-inviscid interaction procedure for flow past a 2-D airfoil (Drela, 1989). Since the primary concern of the present work is to investigate the rotational 3-D effects on prestalled and stalled blades by means of the boundary layer method, the inviscid pressure distribution is simply considered as

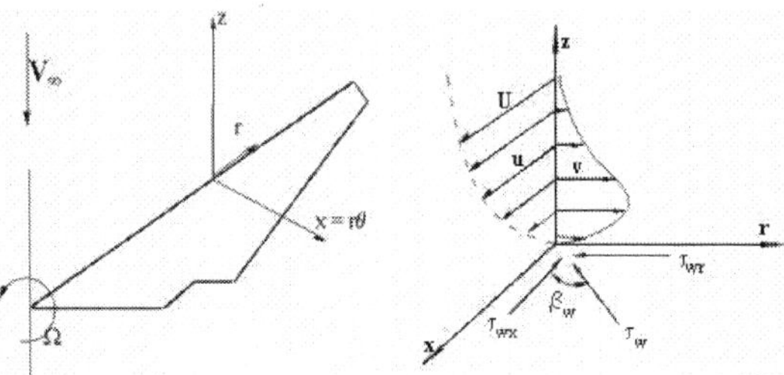

Fig. 3. Cylindrical coordinate system and notation used.

$$C_{pU} = \left|C_{pm}\right|\left(\sqrt{\frac{x}{c}\left(2-\frac{x}{c}\right)}-1\right), C_{pL} = 1-\sqrt{\frac{x}{c}\left(2-\frac{x}{c}\right)} \tag{4}$$

where $C_p = \dfrac{p-p_\infty}{\rho/2\left(V_w^2+\Omega_b r^2\right)}$ is the local pressure coefficient, x/c is the airfoil abscissa by

airfoil chord and subscripts U and L indicate properties on the upper surface and lower surface respectively. The scale parameter $\left|C_{pm}\right|$ is equal to maximum value of C_p and is only dependent on the local aerodynamic field: the airfoil shape, incidence and Reynolds number. This one-parameter pressure distribution family ranges a large gamut from the prestall distributions, values of $\left|C_{pm}\right| = 2-6$, until the stall and post-stall distributions with $\left|C_{pm}\right| = 7-12$. Examples of distributions and experimental comparisons are shown in Fig. 4.

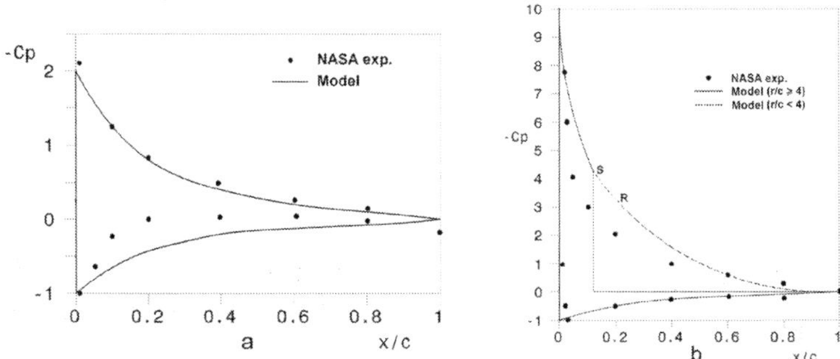

Fig. 4. Pressure distributions compared to experimental data for a NACA 0012 airfoil:

a) $\alpha = 6.75\,\mathrm{deg}(\left|C_{pm}\right| = 2)$; b) $\alpha = 19.35\,\mathrm{deg}(\left|C_{pm}\right| = 9)$

Because a wind turbine can operate a long time at low tip speed ratios, the inboard regions of the blade are stalled, and there leading-edge separation bubbles can occur (Dumitrescu & Cardoş, 2010). Behind the mild separation it is assumed that the flow relaxes with the vanishing skin friction. Therefore, the following flow is applied to solve

$$U_e = U_r \sqrt{1 - C_p}, \text{ for } \frac{x}{c} < \left(\frac{x}{c}\right)_{sep} \tag{5}$$

$$U_e \delta_{2x}^{1/(H+2)} = const., \text{ for } \frac{x}{c} \geq \left(\frac{x}{c}\right)_{sep} \tag{6}$$

where δ_{2x} and H are, respectively, the momentum thickness and the boundary-layer shape parameter in the streamwise direction.

Figures 5 and 6 illustrate such chordwise inviscid velocity distributions and the corresponding variations of the peripheral skin-friction coefficient C_{fx} and the boundary layer shape parameter H, calculated for various values of $|C_{pm}|$ (a velocity gradient like parameter) and $r/c = \infty$ (2-D). Laminar separation takes place at $C_{fx} = 0$, and a given value of the shape factor, $H = 3.4$, is used as the criterion for turbulent separation/reattachment (Cebeci & Cousteix, 1999). The "transition point" x_t is the point that corresponds to the minimum skin-friction, which sometimes can be the point of laminar separation.

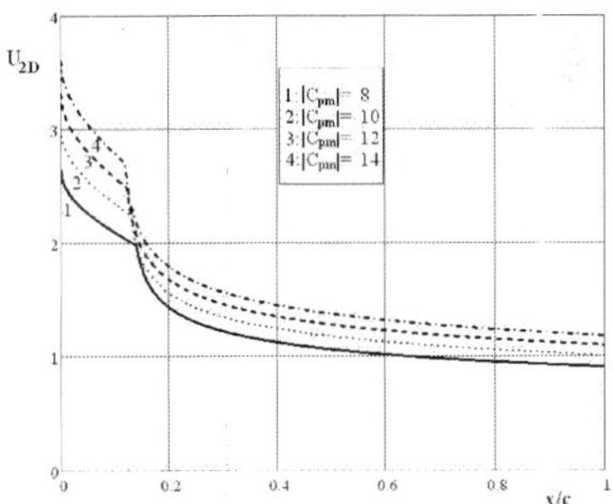

Fig. 5. Chordwise inviscid velocity distribution for various velocity gradient parameters $|C_{pm}|$

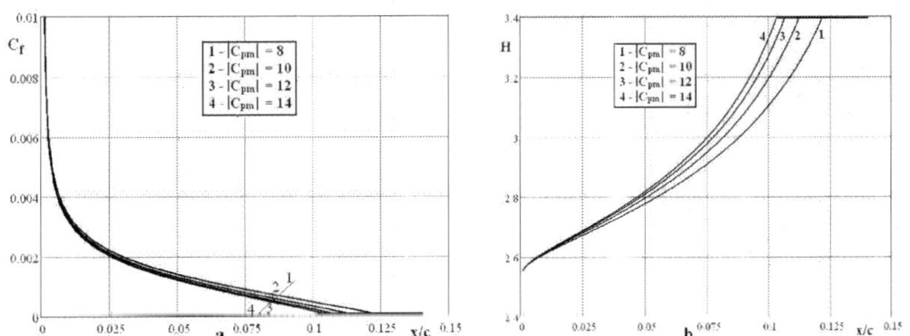

Fig. 6. Variation of a) the peripheral skin-friction coefficient; b) the boundary-layer shape parameter.

B. Viscous flow

Momentum integral equations. The flow in the boundary layer on a rotating blade in attached as well as in stalled conditions is represented using the integral formulation developed in (Dumitrescu & Cardoş, 2010) for analyzing separated and reattaching turbulent flows involving leading-edge separation bubbles. The 3-D incompressible steady boundary layer equations are written in the cylindrical coordinate system (θ, r, z) which rotate with the blade at a constant rotational speed $\Omega_z = \Omega_b$ (Fig. 3); θ denotes the peripheral, z the axial and r the radial (spanwise) direction. The equations used early (Cebeci & Cousteix, 1999) are shown in their conservative laminar form:

Continuity

$$\frac{\partial u}{\partial \theta} + \frac{\partial (rv)}{\partial r} + \frac{\partial (rw)}{\partial z} = 0, \tag{7}$$

Momentum, θ component:

$$\frac{\partial (u^2)}{r \partial \theta} + \frac{\partial (uv)}{\partial r} + \frac{\partial (uw)}{\partial z} + \frac{2v}{r}(u - \Omega_z r) = -\frac{1}{\rho}\frac{\partial P}{r \partial \theta} + \frac{1}{\rho}\frac{\partial \tau_x}{\partial z}, \quad \frac{1}{\rho}\frac{\partial \tau_x}{\partial z} = v\frac{\partial^2 u}{\partial z^2}, \tag{8}$$

Momentum, r component:

$$\frac{\partial (uv)}{r \partial \theta} + \frac{\partial (v^2)}{\partial r} + \frac{\partial (vw)}{\partial z} + \frac{v^2}{r} - \frac{u}{r}(u - 2\Omega_z r) = -\frac{1}{\rho}\frac{\partial P}{\partial r} + \frac{1}{\rho}\frac{\partial \tau_r}{\partial z}, \quad \frac{1}{\rho}\frac{\partial \tau_r}{\partial z} = v\frac{\partial^2 v}{\partial z^2}, \tag{9}$$

where u, v and w stand for velocity components in θ, r and z directions, respectively, r for the local radius measured from the centre of rotation, ρ for the fluid density, v for the kinematic viscosity, and P is a pressure like term including the centrifugal effect

$$P = \frac{p}{\rho} - \frac{1}{2}(\Omega_z r)^2 \tag{10}$$

with p denoting the static pressure.

The equations (7)-(9) are integrated with respect to z (normal to blade) from 0 to δ (boundary layer thickness) and the integral forms of equations are obtained as

$$\frac{\partial \delta_{2x}}{\partial x} + \frac{\partial \delta_{2xr}}{\partial r} + \frac{1}{U_e}\frac{\partial U_e}{\partial x}(2\delta_{2x} + \delta_{1x}) - \frac{\zeta}{Ue}(2\delta_{2xr} + \delta_{1r}) - 2\frac{\Omega_z}{U_e}\delta_{1r} = \frac{\tau_{wx}}{\rho U_e^2}, \tag{11}$$

$$\frac{\partial (\delta_{2xr} + \delta_{1r})}{\partial x} + \frac{\partial \delta_{2r}}{\partial r} + \frac{2}{U_e}\frac{\partial U_e}{\partial x}(\delta_{2xr} + \delta_{1r}) + \frac{1}{U_e}\frac{\partial U_e}{\partial r}(\delta_{2r} + \delta_{1x} + \delta_{2x}) + 2\frac{\Omega_z}{U_e}\delta_{1x} -$$
$$-\frac{\zeta}{U_e}(\delta_{2r} - \delta_{1x} - \delta_{2x}) = \frac{\tau_{wr}}{\rho U_e^2} \tag{12}$$

where $(U_e, V_e=0)$ are the inviscid freestream velocity components and $\left(C_{fx}, C_{fr}\right)$ are the skin-friction coefficient components. The various boundary layer thicknesses are defined as

$$\delta_{1x} = \int_0^\delta \left(1 - \frac{u}{U_e}\right) dz, \quad \delta_{1r} = -\int_0^\delta \frac{v}{U_e} dz,$$

$$\delta_{2x} = \int_0^\delta \left(1 - \frac{u}{U_e}\right)\frac{u}{U_e} dz, \quad \delta_{2r} = -\int_0^\delta \frac{v^2}{U_e^2} dz, \quad \delta_{2xr} = \int_0^\delta \left(1 - \frac{u}{U_e}\right)\frac{v}{U_e} dz,$$

$$(13)$$

An order of magnitude analysis shows that $\dfrac{\partial \delta_{2xr}}{\partial r} = O(\varepsilon \delta_{2x})$ and $\dfrac{\partial \delta_{2r}}{\partial r} = O(\varepsilon^2 \delta_{2x})$, and

thereby these terms can be neglected as a first approximation. The widely used Pohlhausen (Schlichting, 1979) and Mager (Mager, 1951) velocity profiles and the associated closure relations are introduced to solve the laminar integral boundary layer equations (Dumitrescu et al., 2007). For the turbulent boundary layer a power law-type of velocity profile is assumed for the mean stream velocity profile

$$\frac{u}{U_e} = \left(\frac{z}{\delta}\right)^{(H-1)/2} \tag{14}$$

where H is the local boundary layer shape factor.
The Mager cross flow profile is assumed (Mager, 1951)

$$\frac{v}{u} = \varepsilon_w \left(1 - \frac{z}{\delta}\right)^2,$$

where ε_w is the limiting streamline parameter (tan β_w).
Equations (11) and (12) can now be written in terms of the parameters δ_{2x}, ε_w, H and C_{fx},

$$\frac{\partial \delta_{2x}}{\partial x} + (2+H)\delta_{2x}\frac{1}{U_e}\frac{\partial U_e}{\partial x} + \frac{\partial}{\partial r}(L\varepsilon_w \delta_{2x}) - \frac{\zeta}{U_e}(2L+M)\varepsilon_w \delta_{2x} - 2\frac{\Omega_z}{U_e}M\varepsilon_w \delta_{2x} = \frac{1}{2}C_{fx} \tag{15}$$

$$\frac{\partial}{\partial x}\left[(L+M)\varepsilon_w \delta_{2x}\right] + \frac{2}{U_e}\frac{\partial U_e}{\partial x}(L+M)\varepsilon_w \delta_{2x} + \frac{\partial}{\partial r}\left(N\varepsilon_w^2 \delta_{2x}\right) + \frac{1}{U_e}\frac{\partial U_e}{\partial r}\left(N\varepsilon_w^2 + H + 1\right)\delta_{2x} -$$

$$-\frac{\zeta}{U_e}\left(N\varepsilon_w^2 - H - 1\right)\delta_{2x} + 2\frac{\Omega_z}{U_e}H\delta_{2x} = \frac{1}{2}C_{fx}\varepsilon_w, \tag{16}$$

$$L = \frac{\delta_{2xr}}{\varepsilon_w \delta_{2x}} = \frac{2(7H+15)}{(H+2)(H+3)(H+5)},$$

$$M = \frac{\delta_{1r}}{\varepsilon_w \delta_{2x}} = -\frac{16H}{(H-1)(H+3)(H+5)}, \tag{17}$$

$$N = \frac{\delta_{2r}}{\varepsilon_w^2 \delta_{2x}} = -\frac{24}{(H-1)(H+2)(H+3)(H+4)}.$$

The skin-friction relation for flows with pressure gradients and rotation effects is based on the experimental data for a turbulent boundary layer in a rotating channel (Lakshminarayana & Govindan, 1981),

$$C_{fx} = 0.172\,\mathrm{Re}_{\delta_{2x}}^{-0.268}\,10^{-0.678H}\left(1 + B_1\sqrt{\varepsilon_w\left(x - x_t\right)/c}\right) \tag{18}$$

This correlation is a modified version of the correlation developed by Ludwieg and Tillmann (Ludwieg & Tillman, 1949), which includes the effect of rotation. In this relation B_1 is an empiric constant (a value of 0.52 is used), $\mathrm{Re}_{\delta_{2x}}$ is the Reynolds number, based on the streamwise velocity at the edge of the boundary layer and the streamwise momentum thickness δ_{2x}, and x_t is the distance between the leading edge and the transition point along the streamwise direction (the laminar separation point is used).

Closure model-entrainment

The pressure gradients cause large changes in velocity profiles and, consequently, in the shape parameter H. The variation of H cannot be neglected and an additional equation is required. Out of the available auxiliary equations, only the energy integral equation and the entrainment equation have been suitable for the turbulent boundary layers.

The entrainment equation is chosen to model the rotor boundary layer, which in the coordinate system used here can be written as

$$\frac{\partial\left(\delta - \delta_{1x}\right)}{\partial x} + \left(\delta - \delta_1\right)\frac{1}{U_e}\frac{\partial U_e}{\partial x} - \frac{\partial \delta_{1r}}{\partial r} + \frac{\zeta}{U_e}\delta_r = C_E \tag{19}$$

where the entrainment coefficient $C_E \equiv \left(\dfrac{\partial \delta}{\partial x} - \dfrac{W_e}{U_e}\right)$ is a function of the factor

$H_1 = \left(\delta - \delta_{1x}\right)/\delta_{2x}$. C_E represents the volume flow rate per unit area through the surface $\delta(x,r)$ and is the rate of entrainment of inviscid external flow into the boundary layer.

The entrainment function C_E for 3-D flow is not yet available and hence, the correlation for 2-D flow is used (Head, 1958):

$$C_E = 0.0306\left(H_1 - 3.0\right)^{-0.653}. \tag{20}$$

Also, the similarity solutions have shown that H_1 is a function of the streamwise boundary-layer shape parameter H. This relationship, which results from a best fit to experimental data (Lock & Williams, 1987) is

$$H_1 = 2 + 1.5\left(\frac{1.12}{H-1}\right)^{1.093} + 0.5\left(\frac{H-1}{1.12}\right)^{1.093}, \text{ for } H < 4. \tag{21}$$

Then, it is assumed that the variation of the entrainment rate with H_1 follows the same relationship for three-dimensional flows.

Equation (19) is written in a form similar to Eqs. (15) and (16)

$$\frac{\partial}{\partial x}\left(\delta_{2x}H_1\right) + \left(\delta_{2x}H_1\right)\frac{1}{U_e}\frac{\partial U_e}{\partial x} - \frac{\partial}{\partial r}\left(M\varepsilon_w\delta_{2x}\right) + \left(M\varepsilon_w\delta_{2x}\right)\frac{\zeta}{U_e} = C_E\left(H_1\right) \tag{22}$$

Equations (15), (16) and (22) are to be solved for δ_{2x}, ε_w, and H (H_1 is related to H, Eq. (21)) with the prescribed boundary conditions. At the leading edge, δ_{2x} and ε_w are assumed to be zero, and an initial value of 2.55 (laminar flow) is assumed for H.

2.2 Application

A hypothetic blade, with constant incidence and chord along the span is chosen to analyze the rotational effects on the boundary layer of a wind turbine blade. In this example, an external flow with the distribution of velocity on the surface of blade given by Eqs. (5) and (6) is imposed.

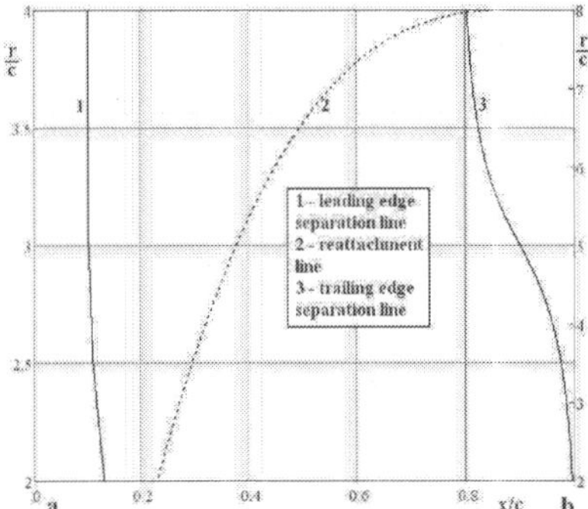

Fig. 7. Relationship between the separation points and the blade radius: a) at stall incidence; b) at prestall incidence.

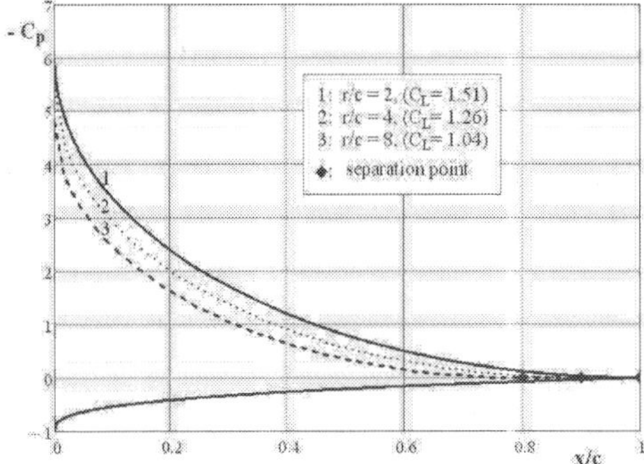

Fig. 8. Influence of the r/c ratio on the suction pressure coefficient and lift coefficient at prestall incidence, $\left|C_{pm}\right| = 6 \,(\alpha \sim 14 \deg)$.

The curves plotted in Fig. 7 are calculated on the prestall ($\left|C_{pm}\right| = 6$) and stall ($\left|C_{pm}\right| = 10$) conditions, with a Reynolds number of 10^6. Relationships between the separation points and the blade radius are shown in Fig. 7, in which two apparent trends can be seen. Firstly, as compared to the 2-D prestall condition ($r/c = \infty$), the separation point is postponed because of the effect of Coriolis force, which acts as a favorable pressure gradient tending to delay the separation to a point further downstream towards the trailing edge. It is shown that, as the radial location r/c decreases, the separation point moves towards the trailing edge, and consequently the suction side distribution of the pressure coefficient moves towards lower levels (Chaviaropoulos & Hansen, 2000), Fig. 8. The drop of the pressure coefficient along the suction side can be related to the separated area on the blades, which reduces in comparison with the 2-D ($r/c = \infty$).

Secondly, in the stall incidence condition, leading edge separation bubbles are formed on the upper surface of the blade, at the root area, which delays the occurrence of massive separation. Beginning at the root of the blade, the bubble continues to stretch towards the trailing edge and at approximately $r/c = 4$ where the bubble stretches all the way to the trailing edge, the flow separates over the whole airfoil (Dumitrescu & Cardoş, 2010) In the next section it is shown that there is a suction effect at the hub up to the mid-span which reduces the leading-edge bubble volume and produces a significant pressure drop along the suction side of the airfoils increasing, thus, the loading of the blade.

3. Flow at high wind speeds

At high wind speeds conditions (TSR ≤ 3.0), though the stall incidences are exceeded on the whole blade, the flow is partially separated on the inboard half-span of the blade and massively separated on the outboard half-span, Fig. 9.

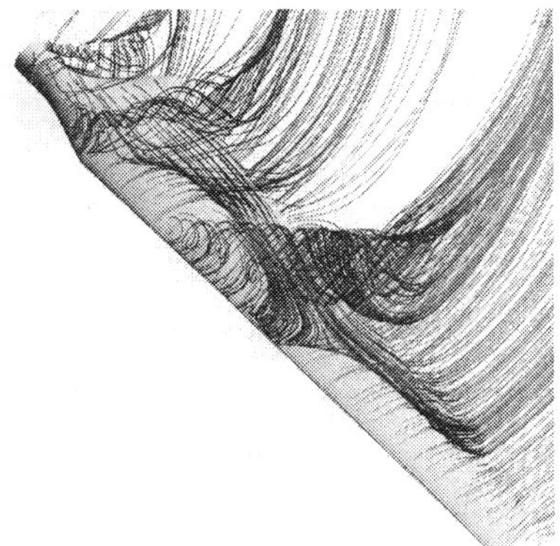

Fig. 9. Typical surface and 3-D streamlines on the blade suction side at high wind speeds (TSR=3.0).

This behavior, which is characterized by significantly increased lift coefficients as compared to the corresponding 2-D case, and by a delay of the occurrence of flow separation to higher angles of attack, is caused by the rotational augmentation at the root area of the blade. The knowledge extracted from experimental and computational visualizations of these rotational effects on the flow field along the blade can be used to develop a model for their prediction or the so-called stall-delay phenomenon.

The wind turbine blade sections, often operate with stall at low tip speed ratios and can undergo a delayed stall phenomenon on the inner part of span. As the wind velocity increases, the inboard regions of the blades are stalled, having $V_w / \Omega_b r > 1$ for much of their operational time. The flow is neither two-dimensional nor steady, and is affected by rotation. As the root is approached the flow causes high values of tangential velocity and it behaves rather like the rotating flow over a stationary disk, Fig. 10 a), b). This is the result of a strong nonlinear interaction between the high wind speed and the rotational flow induced by a constant-speed rotor, which can be written as

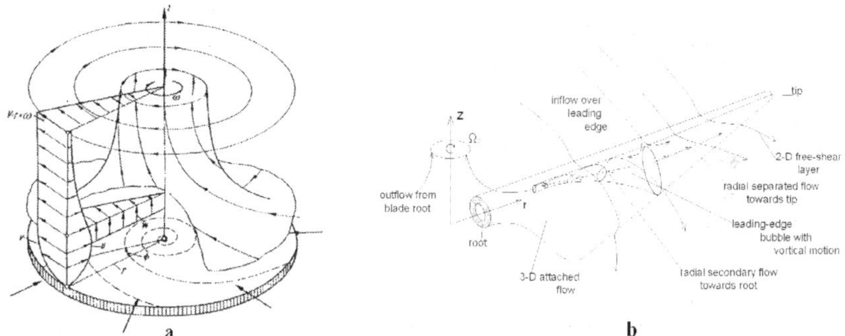

Fig. 10. Sucking-work mode of a wind turbine at high wind speeds (TSR≤3.0): a) concept of rotating flow over a stationary disk; b) model of the separated flow (Dumitrescu & Cardoş, 2009)

$$U_r = \Omega_b \left(\frac{V_w^2 (1-a)^2}{\Omega_b^2 r^2} + (1+d)^2 \right)^{\frac{1}{2}} r = \Omega_v \left(\frac{r}{c}, \frac{V_w}{\Omega_b r} \right) r \, , \tag{23}$$

a form that suggest an outer rotational flow with angular velocity higher than that of the constant-speed rotor, $\Omega_v > \Omega_b$ and $\Omega_v / \Omega_v = -\Omega_b / \Omega_b$.

In contrast with the Glauert weak wake (Glauert, 1963), the strong potential vortex-like wake is responsible for all the 3-D rotational effects, affecting the boundary layer on the turbine blades once the phenomenon of stall is initiated. The enhanced rotational outer flow and the bottom boundary layer are the cause of the high-lift effect of the airfoils at high angle of attack, i.e. the stall delay.

3.1 Boundary layer beneath a Rankine-like vortex

The axisymmetric angular velocity (Ekman) boundary layer developing on the suction side of the rotor disk is used to illustrate the contribution of the suction effect of a strong wake on the chordwise surface pressure distribution and volume flux in the radial direction. Thus, consider an axisymmetric rotating flow of a viscous incompressible fluid over a disk of radius R_0 (Fig. 11).

The angular velocity of the disk Ω_b is constant and that of the fluid far away from the disk Ω_v may be an arbitrary function of the radius r, provided it is stable, i.e. $\dfrac{d\left(\Omega_v r^2\right)}{dr} \geq 0$ and near the axis the outer flow behaves locally like a rigid body rotation, i.e. $r \to 0 : \dfrac{d\Omega_v}{dr} \to 0, \Omega_v \to \Omega_b$. The Rankine-like vortex assumed for the outer rotational flow satisfies both conditions and can be described by the vortex circulation $\Gamma_v \equiv \Omega_v r^2$ given by the empirical formula (Vatistas et al., 1991)

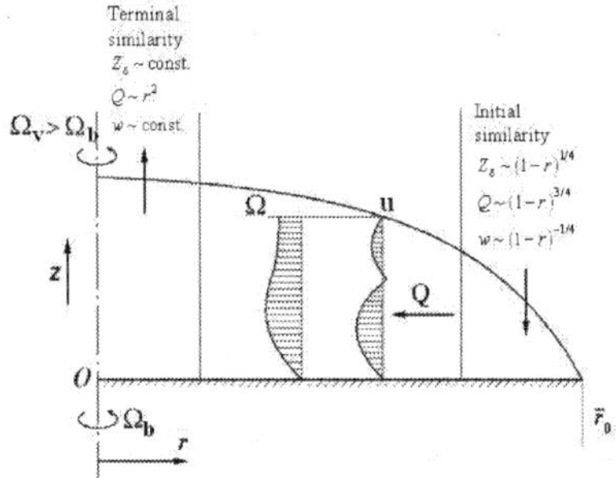

Fig. 11. Boundary layer on a rotating disk slower than the outer flow.

$$\Gamma_v = \Omega_b R_0^2 \sqrt{\frac{1+\dfrac{r_h^4}{R_0^4}}{1+\left(\dfrac{r}{r_h}\right)^4}}\left(\frac{r}{r_h}\right)^2 , \tag{24}$$

where r_h is the radius of forced vortex core.

This vortex model avoids the discontinuity of the velocity derivatives at the point of transition from free to forced modes and gives the possibility of the description of the flow near the hub.

For the inertial system of cylindrical coordinates (r,θ,z) considered in Fig. 10, the laminar boundary layer equations write as (Rott & Lewellen, 1966)

$$\frac{\partial(ru)}{\partial r}+\frac{\partial(rw)}{\partial z}=0 , \tag{25}$$

$$\frac{\partial(u^2)}{\partial r}+\frac{u^2-v^2}{r}+\frac{\partial(uw)}{\partial z}=-\frac{1}{\rho}\frac{\partial p}{\partial r}+\nu\frac{\partial^2 u}{\partial z^2} , \tag{26}$$

$$\frac{\partial(uv)}{\partial r}+\frac{2uv}{r}+\frac{\partial(vw)}{\partial z}=\nu\frac{\partial^2 v}{\partial z^2} , \tag{27}$$

with $\dfrac{\partial p}{\partial z}=0$, where (u,v,w) are the velocity components, p is the static pressure, ρ is the fluid density and $v=\dfrac{\mu}{\rho}$ is the kinematic viscosity.

The boundary conditions at the surface of the disk and the edge of boundary layer are

$$z=0: \quad u=w=0, \, v=\Omega_b r=\frac{\Gamma_b}{r} ,$$

$$z=Z_\delta: \quad u=0, \, v=V=\Omega_v r=\frac{\Gamma_v}{r} . \tag{28}$$

The outer flow radial momentum equation becomes then $\dfrac{V^2}{r} = \dfrac{1}{\rho}\dfrac{dp}{dr}$ and the pressure in the outer flow is

$$C_{ps} = \frac{p - p_0}{p_0 - p_c} = \frac{\displaystyle\int_0^r \frac{\Gamma_v^2}{r^3}\,dr}{\displaystyle\int_0^1 \frac{\Gamma_v^2}{r^3}\,dr} - 1, \tag{29}$$

where C_{ps} is the suction pressure coefficient normalized by the pressure at the vortex center, and the subscripts c and 0 indicate properties at the vortex center and the edge of disk, respectively.

The tangential and radial momentum-integral equations may be derived from equations (26) and (27) for radial and angular velocity profiles which satisfy the boundary conditions on the disk surface and the smoothness requirements at the edge of the boundary layer

$$u = U(r)\,f'\!\left(\frac{z}{Z_\delta}\right), f'(0) = 0, f'(1) = 0, f''(1) = 0 \tag{30}$$

$$\frac{v - \Omega_b r}{V - \Omega_b r} = g\!\left(\frac{z}{Z_\delta}\right), g(0) = 0, g(1) = 1, g'(1) = 0 \tag{31}$$

Outside the boundary layer, the tangential velocity $V(r) = \dfrac{\Gamma_v}{r}$ is assumed to be some given function, Eq. (24), while the radial velocity vanishes. The variables are the boundary-layer thickness Z_δ and the radial volume flux $Q = \displaystyle\int_0^{Z_\delta} r u \, dz$. The momentum-integral equations under these conditions are

$$\frac{d}{dr}\left[Q\left(\Gamma_f - \Gamma_b\right)\right] - \lambda_1 Q \frac{d\Gamma_f}{dr} = \lambda_1 \frac{r^2 \tau_\theta}{\rho}, \tag{32}$$

$$\lambda_2^2 \frac{d}{dr}\left[\frac{Q^2}{rZ_\delta}\right] + \frac{Z_\delta}{r^2}\left[\lambda_3\left(\Gamma_f - \Gamma_b\right)^2 + \lambda_4 \Gamma_b\left(\Gamma_f - \Gamma_b\right)\right] = -\frac{r\tau_r}{\rho}, \tag{33}$$

where λ_1, λ_2, λ_3 and λ_4 are profile form parameters defined in Table 1 (Rott & Lewellen, 1966)

Flow	$f'(\eta)$	$g(\eta)$	λ_1	λ_2	λ_3	λ_4	C_1	C_2
Laminar	$6.75\,\eta(1-\eta)^2$	$\eta(2-\eta)$	2.5	1.372	0.467	0.667	2	12
Turbulent	$1.69\,\eta^{1/7}(1-\eta)^2$	$\eta^{1/7}$	4.93	1.63	0.222	0.250	0.0225	0.0513

Table 1. Various parameters used in the axisymmetric boundary layer.

The Blasius shear laws for three-dimensional flow (Schlichting, 1979) cover both the laminar and turbulent case

$$\left|\tau\right| = C\rho V_\infty^2 \left(1 + \frac{U_m^2}{V_\infty^2}\right)^{1-\frac{\mu}{2}} \left(\frac{v}{V_\infty z_\delta}\right)^\mu,$$
(34)

$$\tau_\theta = C_1\rho V_\infty^2 \left(1 + \frac{U_m^2}{V_\infty^2}\right)^{\frac{1-\mu}{2}} \left(\frac{v}{V_\infty z_\delta}\right)^\mu , \tau_r = C_2\rho V_\infty U_m \left(1 + \frac{U_m^2}{V_\infty^2}\right)^{\frac{1-\mu}{2}} \left(\frac{v}{V_\infty z_\delta}\right)^\mu,$$
(35)

where U_m is equal to the maximum value of u and following Blasius, $\mu = 1$ (laminar flow), $\mu = 1/4$ and $C = 0.0225$ (turbulent flow).

This is the final result of the analytical formulation and the equations (32) and (33) must be integrated numerically for a particular circulation function Γ_v (Eq. (24)) with an appropriate choice of the boundary layer form parameters.

The onset of vortical flows like tornadoes can be visible on the constant-rotational speed rotor at low speed ratios ($\lambda < 3.0$). At the root area of the blade these flows generally behave like a vacuum pump which is featured by volume flow rate (Q) and suction pressure (C_{ps}).

Figures 12 and 13 show such characteristics induced by the wake modeled as a Rankine-like vortex with its axis normal to the rotor disk for different tip speed ratios. The suction characteristics show that the effects of the disk defined by $V_w / \Omega_b r \geq 1$ decrease rapidly with the radius towards the edge of the disk ($V_w / \Omega_b r = 1$) and have maximum values near the hub (here $r_h = 0.2R$).

On the other hand, large radial volume fluxes close to the hub along with the previous results (Fig. 7a) indicate a 3-D attached boundary layer with small leading-edge separation bubbles at high angles of attack.

Therefore, at low tip speed ratios and close to the hub, there is essentially a 3-D attached flow field and a theoretical description in this region can be performed by viscous-inviscid interactive flow models.

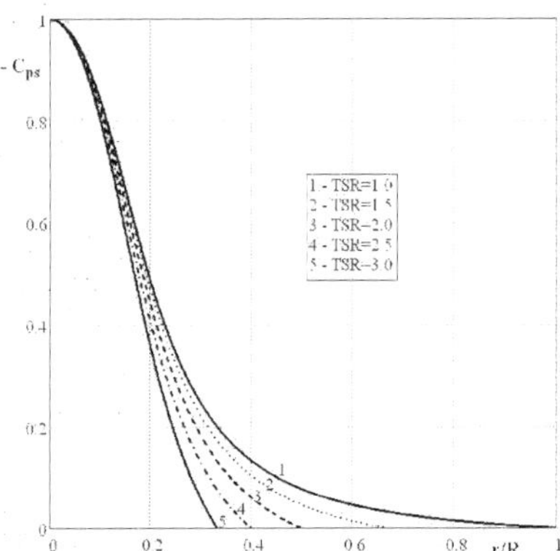

Fig. 12. Radial suction pressure induced by the wake at the $r_h / R = 0.2$.

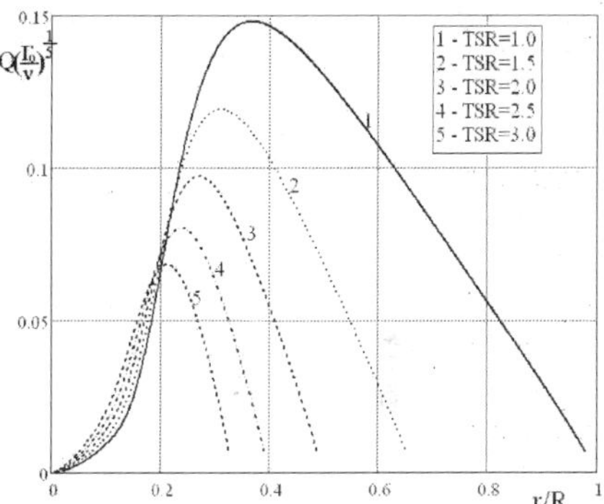

Fig. 13. Radial volume flow rate in the boundary layer beneath the wake at $r_h / R = 0.2$.

3.2 Blade-wake interaction

Flow at low tip speed ratios is dominated by a vortex-like wake which triggers the stall delay phenomenon on the inboard half-rotor disk. The dual working mode of the rotor blades as a stationary disk in a fast rotating flow near the hub-sucking mode, and as the rotating disk in the outboard separated flow-pumping mode, are clearly visible from their pressure distributions when the wind turbine operates at $\lambda \le 3.0$.

On the upper side of the inner blade sections there are two suction pressure fields: one is azimuthally uniform generated by circulation augmentation of wake and the other is the aerodynamic non uniform pressure field with a suction spike at the leading-edge (Fig. 4). These two pressure fields interact and give rise to the radial effects which contribute to stall delay, as well as to a higher lift coefficient at the root area of the blade. There are two main effects: superimposed pressure fields of wake (Fig. 12) and of blade (Fig. 4), which decrease the suction spikes at the leading edge until they cancel, and a radial flow effect induced by the volume flow merger in the boundary layers of wake and blade, resulting in velocity profile skewing and boundary-layer energizing (no separation). Assuming the conservation of circulation once the phenomenon of stall is initiated (here at $\lambda = 3.0$) the wake-blade pressure interaction for the increased wind speed ($\lambda < 3.0$) is a pressure redistribution process of pressures (C_{pU}, C_{ps}) along the chord at constant circulation (lift). The rule of pressure redistribution is sketched as is shown in Fig. 14.

As we will show in the next section, the phenomenon of stall-delay can be described as a three step process: generation of the strong wake by the clustering of shed vortices from the inboard blade segments at $\lambda = 3.0$, chordwise pressure redistribution at the constant value of circulation at the start regime ($\lambda = 3.0$), for $\lambda < 3.0$, followed by circulation decay for increasing radius. The pressure redistribution process is companied by the move of the pressure center to the midchord.

3.3 Application

The sucking suction effect is affected by two key non-dimensional parameters $\left(\dfrac{V_W}{\Omega_b r}, \dfrac{r}{c} \right)$.

Figures 15a and 15b illustrate the influence of these parameters on the chordwise pressure

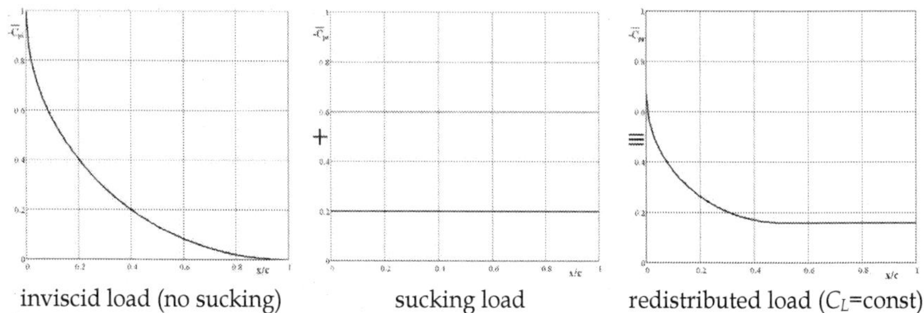

Fig. 14. The rule of the suction pressure with superimposed rotation effect in a normalized plan: $\bar{C}_p = C_p / |C_{pm}|$

distribution for the previous hypothetic blade, with constant post-stall incidence and chord along the span $\left(|C_{pm}| = 10\right)$. The figures also indicate the boundary layer state on the upper surface: S-separation and R- reattachment.

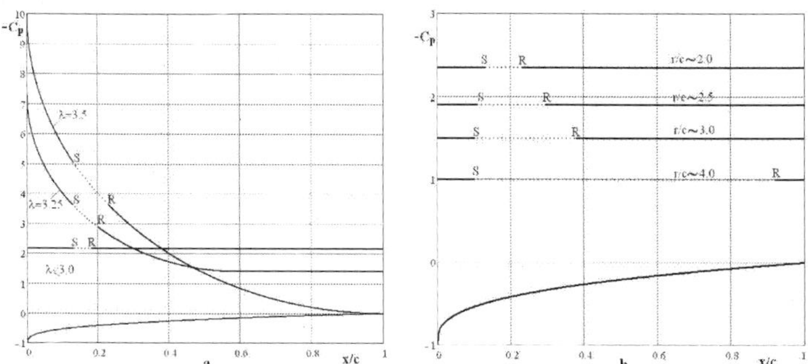

Fig. 15. Influence of TSR and r/c ratio on the suction pressure: a) $C_p(TSR)$ for $r/c = 2$; b) $C_p(r/c)$ for $TSR \leq 3$.

4. Model for delayed stall regime

Stall occurs at an airfoil when flow separates at high angles of attack, typically>15⁰, beyond this angle the 2-D lift coefficient drops significantly. It has long been known (Himmelskamp, 1947; Ronsten, 1992; Snel et al., 1993), however, that the observed power curve under conditions where most parts of the blade are stalled is consistent with significantly higher lift coefficients. The lack of a conceptual model for the complex 3-D flow field on the rotor blade, where stall begins, and how it progresses, has hindered the finding of an unanimously accepted solution. This section aims at giving a better understanding of the delayed stall events.

To interest an engineer in stall-delay one might cite the fact that the inboard regions of the blades are stalled for much of their operational time. Thus predicting blade loads and the power output during stall is very important in making good predictions of wind turbine performance. With a physicist one could reason that rotational stall admits interesting transient states from 3-D inboard delayed stall in partially separated flow to 2-D outboard massively separated flow. Here, the detailed physical understanding of the viscous flow surrounding the hub area is dominated by the interaction between the vorticity and pressure fields and their dynamics. However, all the existent models have been developed more or less intuitively without a basic fluid dynamics support. Six different models mostly used to correct the airfoil characteristics for stall delay, including a wide range of different assumptions, were recently analyzed (Breton et al., 2008). The conclusion was that none of the six models studied correctly represented the flow physics, and that this was ultimately responsible for their lack of generality.

4.1 New physics-based model

Flow Topology of Delayed Stall. There are two main flow regimes pertinent to wind turbine rotors: one is a flow close to rotating disk in axial flow at low velocity, practically at rest, ($\lambda > 3$, Fig. 2), and another is a rotating flow over a stationary or slowly rotating disk (Fig. 9). The last can produce delayed stall on the wind turbine blade sections operating at low tip speed ratios ($\lambda \leq 3$).

The key to the stall-delay phenomenon is the rotational flow surrounding the root area of the blade, i.e. the wake rotation. At low tip speed ratios (or high wind speeds), just as a vortex is shed from each tip blade, a vortex is also shed beginning from the section $\left\{ \dfrac{V_w}{\Omega_b r} = 1 \right\}$ of each blade. Then the blade vortices from $\left\{ \dfrac{V_w}{\Omega_b r} \geq 1 \right\}$ span will each be a line

vortex running axially to the center of the rotor. The direction of rotation of all $\left\{ \dfrac{V_w}{\Omega_b r} \geq 1 \right\}$

span vortices will be the same, forming a core (or root) vortex of total strength $\Gamma_v \gg \Gamma_b = \Omega_b r^2$. The root vortex is primarily responsible for inducing the rotational 3-D effects on blades. The acquisition of the tangential component of air velocity is compensated for by a fall in the static pressure (suction) in the wake.

There are two common characteristics which most strong potential vortices exhibit, like those produced at the blade root; the first is that the vortex velocity field above the surface boundary layer is always dominated by the tangential velocity component, while the second shows that the balance of forces between the flow outside the boundary layer and vortex core is governed by a balance between the centrifugal force and radial pressure gradient $\left(\dfrac{1}{\rho} \dfrac{\partial p}{\partial r} = \dfrac{V^2}{r} \right)$. In this working mode the centrifugal force has a strong stabilizing effect on

the boundary layer through the favorable radial pressure gradient which occurs. In contrast with the pumping - working mode at high tip speed ratios, now the centrifugal forces produce a centerwise suction effect, resulting mainly in boundary layer stabilizing against the separation. On the other hand, in the presence of a solid surface (blade), this balance of forces is disrupted by the friction retardation of the tangential flow close to the boundary. Under such a condition, the radial pressure gradient proceeds to drive the retarded boundary flow along the surface towards the center, resulting in large radial velocities residing close to the surface. Above the boundary layer where the cyclostrophic balance is still intact, the radial velocity inevitably falls to zero. This gives rise to an inflexion point in the radial velocity profile. At some location close to the axis, this inward flowing air moves away from the solid boundary and effuses into a vortex core. The radial pressure gradient has a very strong stabilizing effect on the boundary layer and acts to revert it to its reattaching, and even more to its laminar state upstream of the effusing core.

Modeling of Stall-Delay. The acquisition of the enhanced rotational flow (tangential component of velocity) by the increase in the kinetic energy of wind, associated with a strong suction of the air at the hub is governed by conservation of angular momentum, like the potential vortex, and conservation of energy:

$$\Gamma_v = const. , \tag{36}$$

$$\frac{p}{\rho} + \frac{\Gamma_v}{2r^2} = const. \tag{37}$$

Then, as the blade section moves through the air a circulation $\Gamma_{airfoil}$ develops around it. In order to comply with Kelvin's theorem,

$$\frac{D\Gamma}{Dt} = \frac{\Gamma_{airfoil} + \Gamma_{wake}}{\Delta t} = 0 , \tag{38}$$

a starting vortex Γ_{wake} must exist such that the total circulation around a line that surrounds both the airfoil and the wake remains unchanged. Since the circulation of the sectional airfoil, where the phenomenon of stall was initiated at $\lambda = 3.0$, is practically the blade circulation due to rotation, $\Gamma_{airfoil} = \Gamma_b = V_w^2 / \Omega_b = \Omega_b R^2 / \lambda^2$ (at $r / R = \lambda^{-1}$ the sectional airfoil is stalled for $\lambda = 3.0$), then $\Gamma_{wake,0} = -\Gamma_b \left(r = R / 3.0 \right)$.

The wake's circulation, which is constant up to the hub induces a suction pressure along the upper side of blade (Fig. 16) stabilizing the boundary layer against the separation at post-stall angles of attack (angles of 30 degrees exist currently at the blade root), followed by forming of a leading-edge separation bubble. The bubble at the innermost sectional airfoil has almost no effect on integrated loads, because it is never more than a few percent of the chord in length, and the airfoil acts as in an ideal fluid flow.

The phenomenon of stall-delay can be described as a three step process: rise of strong wake by clustering of vortices shed from the inboard blade span at $\lambda = 3.0$, inviscid (pressure) load redistribution along the airfoil chord at constant circulation or $C_{L,INV}$ of the start regime ($\lambda = 3.0$) for increased suction pressure at $\lambda < 3.0$, followed by spanwise circulation or lift decay involving the stretching of separation bubble surface all the way to the trailing edge for increasing radius; beyond that blade section ($r / R = \lambda^{-1}$) the flow is separated over the whole airfoil and the leading-edge stall occurs.

The specific mechanism for circulation (or lift) decay is not presently known. However, a measure of the degree of radial instability of the boundary layer and implicitly the lift decay can be obtained from the sucked volume flow rate (Fig. 13) into the boundary layer beneath the Rankine vortex-like wake. Thus, the maximum sucked flow rate would correspond to the minimum volume of the separation bubble and $C_{L,INV}$. Consequently C_L decay can be supposed to follow the flow rate decrease, $\dfrac{dC_L}{dr} \sim \dfrac{dQ}{dr}$, where the zero flow rate indicates the full stall occurring. More accurately, the curves of the volume flow rate show the state of

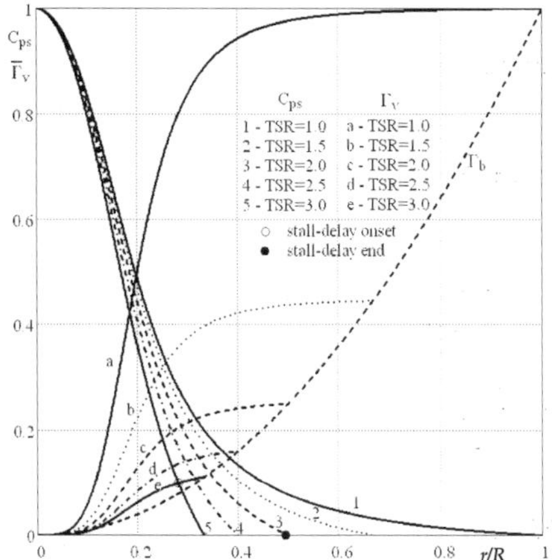

Fig. 16. Circulation $\Gamma / \Omega_b R^2$ and pressure $\left(C_{ps}\right)$ correlation for $TSR \leq 3.0$.

boundary layer, namely larger values of the flow rate signify smaller bubble volumes and thinner boundary layers for reducing tip speed ratios. Further, it can be easily found that a thicker boundary layer with distributed vorticity ($\lambda = 3.0$) is more stable, while a thinner boundary layer with more concentrated vorticity ($\lambda \to 1.0$) is less stable. Since the lift decay law depending on the stability of the boundary layer, is different from that suggested by Fig. 13, tending to be slower for higher λ and steeper for lower λ, an average decay corresponding to $\lambda = 2.0$ is assumed. Therefore, the decay is assumed to be terminated for all regimes at midspan, i.e. the end of the flow regime for $\lambda = 2.0$. This conclusion is consistent with the result found in (Dumitrescu & Cardoş, 2010), which indicates the full separation at $r / c = 4$.

Point	Flow Structure	Forces
1	The attached 3-D boundary layer with leading-edge separation bubble	Exceed 2-D C_{Lmax}, extrapolate linear regime
2	The separation bubble is sucked and its volume reduces	Maximum lift, $C_{L,INV}$ at $\lambda = 3.0$, followed by suction pressure redistribution and movement of pressure center to midchord; no correction to drag
3	Thick skewed 3-D boundary layer with stretching bubble	Gradual decay of lift for $r > r_0$; no correction to drag
4	The bubble breaks away from the trailing edge and forms a free shear layer	Readjust to stall regime

Table 2. Delayed stall events on a wind turbine blade.

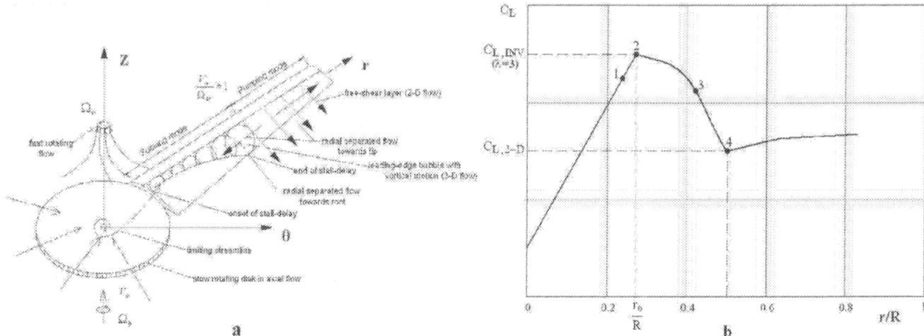

Fig. 17. Schematic showing the essential 3-D flow topology a) and delayed stall events b) on a wind turbine blade

The various stages of the delayed stall process are summarized schematically in Fig. 17. and Table 2. Stage 1 represents the delay in the onset of separation in response to reduction in adverse pressure gradients produced by the influence of the strong vortex wake at the root. The phenomenon is initiated at $\lambda = 3.0$, where the suction pressure equalizes to the kinetic energy of wake (Fig. 16) resulting mainly in boundary layer reattaching and the rise of a leading edge separation bubble, a stable focus as topological entity.

Stage 2 of the delayed stall process involves the suction pressure redistribution of the start pressure distribution with the conservation condition of its integrated load and the simultaneous move of pressure center to the midchord; as the suction/tip speed ratio increases/decreases the volume of the leading-edge separation bubble is reduced. The start pressure distribution is known from the inviscid aerodynamic loading solution over the chord; the pressure redistribution continues up to no suction peak exists at the leading-edge and even more no suction surface pressure gradient exists. On the other hand, the pressure redistribution is associated with a vorticity concentration process, which influences the stability of boundary layer.

Stage 3 is the sectional circulation (lift) decay which occurs at greater radii than the start location, where the separation bubble surface stretches all the way to the trailing edge. Beyond that zero flow rate location the flow is separated over the whole airfoil so that full stall occurs. By reason of the boundary layer stability it is inferred that the stall-delay effects are present up to a 50% of span for all the regimes.

Stage 4 is the relaxation to stall regime, which begins just as the integrated normal load of 2-D stall, C_{NS}, is reached (end of delayed stall) and continues as far as the recovery of the specific pressure distribution (suction peak at the leading edge) is achieved.

4.2 Application

The combined experimental rotor (Schepers et al. 1997) of the Renewable Energy Laboratory (NREL) is chosen to illustrate the stall-delay model proposed for a wind turbine blade. The rotor uses the NREL S 809 airfoil and a simpler BEM method. In this example, the one parameter pressure distribution on the upper surface of blade at the root with $\left|C_{pm}\right| = 11\left(\alpha = 22\,\text{deg}\right)$ is imposed and the hub is at approximately $0.20R$. The curves plotted in Fig. 19 are calculated with these conditions: the onset of stall-delay at $\lambda = 3.0$, $r_0 / R = 0.268$, $C_{L,INV(\lambda=3)} = 2.35$ and a linear approximate distribution of normal load decay (Fig. 18).

4

Verification of Lightning Protection Measures

Søren Find Madsen
Highvoltage.dk and Testinglab Denmark,
Denmark

1. Introduction

During the past decades, the size of wind turbines has increased constantly, which means that a 3MW turbine rising 150m above ground level has become today's standard. To optimize the overall efficiency these turbines are often located in clusters in mountainous regions or at off shore locations (wind farms). Such areas are selected as they are often favoured by constant high winds and don't interfere with public interests.

Building very large and heavy constructions in these less accessible areas makes the planning and completion of the project very difficult. The design of the turbine must reflect the situation where spare parts cannot be transported to the site overnight and it might be difficult or impossible to get heavy cranes etc. to the site to replace large components.

To match the output of electrical power, the blades have increased in size as well. In the early 90's blades had lengths of 15-20m and weighed less than a ton. Today blades of very large turbines have passed 60m of length and weigh more than 10 tons each. Besides the difficulties in manufacturing and transporting such three blades to the site, they also represent nearly 20% of the cost of the entire turbine. If one of the blades is damaged for mechanical reasons or due to lightning, the turbine is usually stopped and must be inspected before restart. The outage time of a 2-3MW turbine placed at an optimum location (in terms of wind) can easily cost several thousands of USD a day, adding an important sum to the overall cost of damage.

It is a well-known fact that wind turbines are frequently struck by direct lightning strikes. Especially air terminations must be designed as wearing components with the ability of replacement. The risk of getting struck has also proved to increase with the size of turbines. This fact cannot only be explained by the increase in collection area, but apparently the amount of upward initiated flashes from these relatively high structures has increased drastically as well.

Based on such technical and economic considerations, a serious effort has been made to improve the lightning protection of wind turbines during the past 5-10 years (Holboll et al., 2006; Larsen & Sorensen, 2003; Madsen, 2006). The contribution has resulted in an improved understanding of the issue of protecting composite blades from lightning discharges, followed by revised manufacturing procedures. Along with implementation of the findings on new blade designs, retrofit solutions for old blades complying with the research are also developed.

After performing extensive research within the area and designing optimum protection measures, the effectiveness of the final solutions must be verified. In June 2010, the revised lightning protection standard for wind turbines was published, (IEC 61400-24). Compared to the previous technical report from 2002 (IEC TR 61400-24), this document is stricter and much more demanding. The document includes several requirements for verification of the installed lightning protection systems on sub components. Due to the sizes of modern wind turbines and the difficulties of performing full scale tests, the revised IEC 61400-24 now describes three different methods of verifying the installed lightning protection system:

1. High voltage and high current tests.
2. Demonstration of similarity of the blade type (design) with a previously certified blade type, or a blade type with documented successful lightning protection in service for a long period under lightning strike conditions.
3. Designs that have had successful service experience, by using analysis tools previously verified by comparison with test results or with blade protection.

Bullet No. 2 is the usual choice for industries where the structural design does not change significantly over time, and where a large number of similar objects are in service in different regions of the world. Since the wind turbine industry has grown heavily during the past two decades and new materials and design principles are seen every day, the demonstration of similarity is seldom applicable. This leaves us with the two other methods, testing and modelling. The present chapter describes these two principles of verification, used extensively during the past ten years.

2. Full scale lightning verification tests

As seen from the above, verification has become a requirement, and the first proposed action was high voltage and high current testing. Traditionally, high voltage tests have been used to determine the exact location of initial leader attachment points, whereas high current tests are used to foresee the degree of damage once the lightning current appears (Madsen et al., 2006).

The lifetime aspect has become an issue of increasing interest, as turbines are often placed in regions of the world that are difficult or expensive to access. At the air termination system (often referred to as receptors on wind turbine blades) the charge associated with the lightning discharge will erode the metal surface and the surrounding laminate. The effect is cumulative, which means that regions in the world where lightning strikes contain a relatively large amount of charge also experience the largest degree of receptor erosion. Consequently, these sites have to be inspected more frequently, or the receptor design must reflect these wearing issues.

The specific energy in the lightning discharge is more related to the mechanical design and the cross sectional areas and fixations of down conductors, but the receptor design and the blade surface in vicinity of the receptor are relevant, too.

When designing full scale tests for verification of the lightning protection effectiveness, the tests should affect the structure (e.g. the blade) with an impact similar to the one experienced during a direct lightning strike to the structure. To get an understanding of the logical steps in developing such a full scale test, the discharge mechanisms on wind turbine blades and the associated verification are described.

2.1 Lightning discharge to a wind turbine

The process of lightning attachment and the following lightning current are described various places in literature. In general, a lightning strike to a wind turbine blade begins with a high electric field surrounding the blade tip. This field appears when a lightning leader from the cloud approaches the blade (downward initiated lightning strike) or it could be initiated by the field enhancement around sharp and grounded structures when the blade is subjected to the high static field conditions during a thunderstorm (possibly leading to an upward initiated lightning strike if the field is high enough for the inception of leaders).

Following the leader inception from the blade, is the leader interception where the leader from the cloud meets the leader from the blade (at which point it is defined where the lightning will strike), and finally conduction of the lightning current. Initially, the charge present in the leader tip and the leader channel itself will be neutralised by the ground potential from the blade. This charge neutralisation appears at the first return stroke, which is a high amplitude, long duration and highly energetic impulse current. After the first return stroke, additional charges in the thunder cloud have to be neutralized which results in subsequent strokes or continuing currents depending on the charge location in the cloud relative to the leader channel. Subsequent strokes are of short duration with less amplitude and energy content than by the first return stroke. However, due to the high current gradient (dI/dt), usually the subsequent strokes give high inductive voltage drops and the risk of side flashes within the blade. The continuing or intermediate current is responsible for the main charge transfer. During this sequence several hundreds of coulombs can be transported through the lightning channel to the air termination system, resulting in pitting and erosion on the surface of the air termination system.

2.2 Initial leader attachment test

The initial leader attachment test aims at identifying possible lightning attachment points on the turbine construction, typically the blade. The test is performed by elevating the entire blade or the part of the blade in concern (often the blade tip) above a grounded test plane in a high voltage laboratory. The ground plane simulates an equipotential plane of the electric field between an approaching lightning leader and the blade tip (downward initiated flash) or the static background field present prior to an upward initiated flash, Fig. 1.

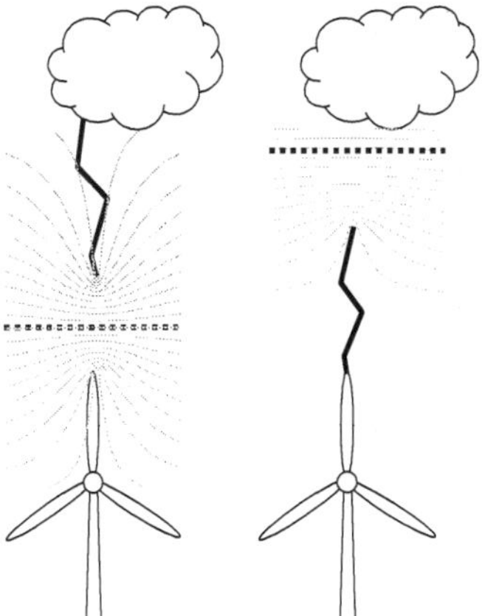

Fig. 1. The ground plane in the test setup simulates the equipotential plane (dotted line) some distance from the blade. Left: A downward initiated strike, Right: An upward initiated strike.

Subsequently, the elevated blade is connected to the output of a Marx generator injecting high voltage impulses into the lightning protection system of the blade. The voltage waveform is a switching type surge with a rise time of 250µs +/-20% and a decay time of 2500µs +/-60%. The voltage amplitude is selected so that flashover from the blade to the ground plane occurs at the rising flange of the pulse. Such a test simulates the situation in real life where the electric field surrounding the blade suddenly increases to the threshold value for ionization, and thereby initiates the following streamer and leader processes from the blade. The test setup used is shown on Fig. 2, having the blade elevated above a ground plane in the high voltage laboratory.

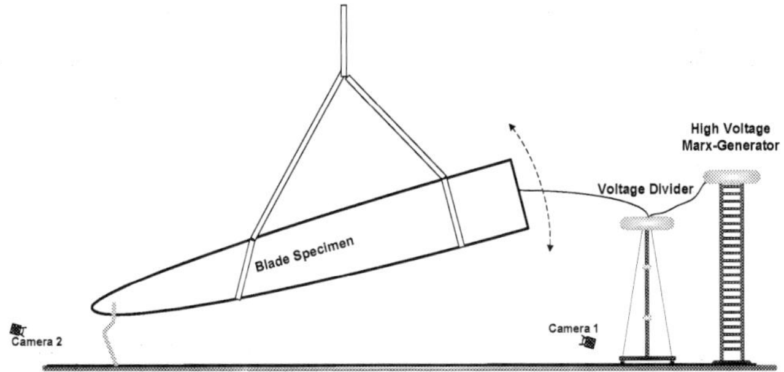

Fig. 2. Test setup of initial leader attachment test, on a blade tip section, 30°.

On Fig. 3 it is seen how the blade tip is oriented in three different angles to the test plane (90°, 60° and 30°). For each angle, the blade is further more pitched in four different pitch angles to represent all possible orientations, and by injecting three discharges at each position and each polarity, the blade has to experience at least fifty four flashovers during the initial leader attachment test.

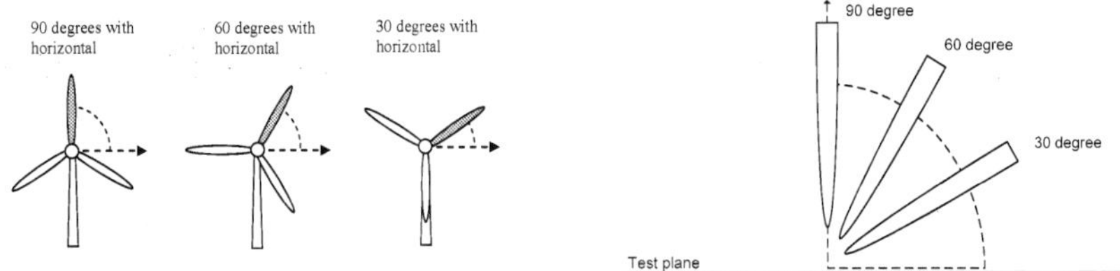

Fig. 3. The blade tip is oriented in three different angles to horizontal level and pitched in four different pitch angles for each orientation.

To pass the test, all discharges have to attach to the intended air termination (lightning receptor). If some of the discharges attach elsewhere by puncturing the blade laminate, the design must be reviewed, and the test must be repeated. After the initial leader attachment test is passed, the blade tip can continue the overall test procedure in the high current physical damage test.

2.3 High current physical damage test

Once the blade has passed the high voltage initial leader attachment test, it has been verified that lightning discharges most probably will attach to the intended air termination system.

Then the second task is to determine whether the design can handle the impact from the lightning current.

The following test method is based on an assessment of the annual number of direct lightning strikes to the turbine, the design lifetime and the lightning parameter distribution in IEC 62305.

2.3.1 Current components in IEC 62305-1

When a typical blade tip (a passive component with air termination system and down conductor) experiences a direct lightning strike to one of the intended air terminations, the different parameters of the discharge affect the component differently.

Air termination system

For the air termination system – also called receptors – the charge transfer in the entire flash is important, which means that these components should be tested with 300C of charge transfer in a single long duration stroke (for example 600 A in 0.5s). For lifetime testing, these values can be adjusted to represent a realistic lifetime. If the receptor cannot withstand the estimated lifetime it must be replaced during maintenance.

Down conductor

Considering the current path without arcing involved, the resistive heating and magnetic forces are of importance. This is tested by injecting an impulse current representing the first return stroke. The pulse must be of the type 10/350µs with peak amplitude of 200kA and a specific energy of 10MJ/Ω.

2.3.2 Classes of blade components

Different parts of the blade bear different risks of getting struck by lightning or conduct parts of the lightning current. Therefore, it is suggested to consider three classes of components which should be tested differently: The current amplitude, charge and specific energy listed in this section refer to the "Single shot test principle" as indicated in the standards, whereas a different approach considering the life time aspect is described in this section.

Class I:

Class I covers the attachment points on the blade found by the high voltage attachment test. It is believed that this high voltage test method reveals most of the possible locations for direct lightning attachment, meaning that these places are the only ones where current injection simulating a direct lighting attachment should take place. Tests in these areas comprise the following:

1. Injection of a 200kA 10/350µs pulse with a bolted connection to the attachment points (air terminations) considered, aiming at reaching the specific energy of 10MJ/Ω. This will reveal whether the various down conductor parts are dimensioned properly (thermally speaking), and if fastening of the down conductor can handle the magnetic forces associated with high amplitude currents.
2. Arc attachment of a combined impulse current (30kA or more) and a long duration stroke aiming at transferring the total charge in a flash of 300C. It must be emphasised that the charge only has an impact on the structure if it is injected through an open arc, so that the impulse current ignites the arc. Subsequently, the continuing current should deliver the charge of app. 300C.

Class II:

Class II defines the parts of the blade which are not subject to a direct lightning strike. However, by analysis they are seen to carry the entire lightning current. Such parts could be internal joints or connections on the down conductor. The test environment of Class II should contain injection of a 200kA 10/350µs pulse with a bolted connection to the air terminations considered, aiming at reaching the specific energy of 10MJ/Ω.

Structures of Class II, which have been tested indirectly during the test of a Class I component (e.g. a connection between the receptor and the down conductor), do not have to be tested again as a Class II component. Such a "receptor – down conductor interface" has achieved the same impact during the Class I test as defined by the Class II test, and therefore it is sufficiently tested by the bolted connection tests of Class I.

Class III:

Class III defines the parts of the blade which are not subject to a direct lightning strike. By analysis they are seen to carry smaller fractions of the lightning current. If it is possible to identify paths of the down conductor, which in worst case only conducts parts of the lightning currents, it is suggested to test these components with current pulses of less amplitude. If the specific part is believed to carry a fraction X of the entire current, the test conditions for Class III would be Injection of a X·200kA 10/350µs pulse with a bolted connection to the air terminations considered, aiming at reaching the specific energy of X^2·10MJ/Ω.

Scaling of the current values means, that if a certain parallel part of the down conductor is believed to carry half the current in worst case, the amplitude of the 10/350µs pulse should be ½·200kA = 100kA. The specific energy would then be reduced to $(½)^2$·10MJ/Ω = 2.5MJ/Ω.

2.3.3 Test method

For impulse current testing, the size and geometry of the test sample often plays a limited role in reaching the desired peak current amplitudes and current gradients. Therefore, it is advised to test as small specimens as possible only containing the relevant parts (receptors, connection components etc.).

The DC current transferring the charge of 300C is either generated by shorting a large battery bank through the specimen, or by rectifying the output of an AC set and connecting it across the specimen. In either case the charge only applies damage if it is injected through an open arc, hence the voltage of the source must be high enough to maintain the arc for the specified period of time. Ignition of the arc is done by an impulse current from a generator with a higher output voltage.

In principle, an impulse current generator responsible for applying the specific energy consists of a capacitor bank, which have been charged to a certain voltage and afterwards discharged through the test specimen. If the long tale of the 10/350µs is to be obtained with a minimum of capacitance in the bank, the crowbar principle should be used. This principle couples an inductance in series with the test specimen and shorts out the capacitor bank when the current amplitude is at its maximum. The two types of generators are often combined, which means that a 30-50kA oscillating impulse current can initiate the continuous current resulting in an overall charge transfer of 300C.

2.3.3.1 Bolted connection

By identification of structures, which will experience only the conduction of lightning current and not the impact of a direct strike (Class II), the specific energy of the impulse current is of interest. In these cases, terminations from the generator to the test specimen should be bolted connections, aiming at minimizing the overall impedance of the test circuit, and thereby increasing the specific energy of the current pulse, Fig. 4.

2.3.3.2 Arc injection

At lightning attachment points or other places where arcing in connection with the lightning current occurs (Class I), a test using open arcs must be conducted. Open arc attachment represents a relatively high impedance of the test circuit. The specific energy is limited, but the charge transfer will remain unaffected. Since the charging voltage is limited for current generators, the arc must be ignited by a thin ignition wire from the generator electrode to the specimen. Furthermore, the generator electrode must be of a jet diversion type, so that melted metal and plasma from the electrode material is blown away from the test specimen, Fig. 4.

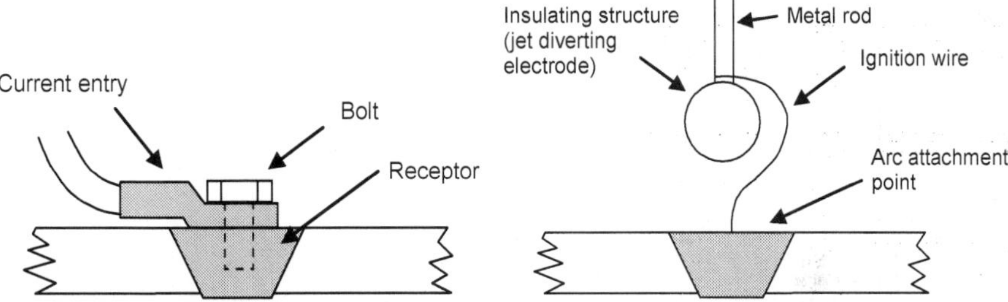

Fig. 4. Left: Bolted connection to an air termination system, Right: A thin ignition wire and a jet diverting electrode are used when arc injection tests are conducted.

2.3.4 Life time impact

The frequency of lightning attachments to wind turbines varies considerably from one site to another. In situations where wind turbines are struck several times a year, or even more often, the life time aspect becomes important. The issue is to determine how many strikes the turbine can withstand before it needs service or replacement of wearing components. In the IEC 61400-24, it is even a requirement that the turbine can remain operational between scheduled service/maintenance. This is ensured by verifying that the impact due to lightning is based on the expected amount of strikes and the severity of those strikes. Based on the probability density functions given in the IEC 62305-1, suggestions can be defined on how to simulate 20 years of lifetime for a blade on a typical site (Bertelsen et al. 2007). The methodology used suggests the following number of strikes and amplitudes for the different types of tests, Table 1.

Bolted connection		Arc attachment	
Number of impacts	Amplitude [kA]	Number of impacts	Charge [C]
2	200	4	300
3	150	4	200
8	100	6	150
7	50	6	100

Table 1. Peak current values for bolted connection with 10/350µs waveform and charge content in the arc attachment test, simulating 20 years of lifetime (Bertelsen et al. 2007).

By following these principles, each air termination system is affected with a total of 3500C in 20 discharges, and each part of the down conductor is exposed to 20 current pulses between 50kA and 200kA with specific energies up to $10MJ/\Omega$.

2.4 Test evaluation

The new standard IEC 61400-24 requires that wind turbine lightning protection systems are verified by either testing or numerical simulation. If simulation is used, the models used in the computation must be validated by testing.

The initial leader attachment test is nearly a pass/fail test, since the finite number of discharges either attaches correctly to air termination system or punctures the blade. Considering the high current test, it is a little more diffuse, since all impulse currents will introduce some degree of damage or wearing on the receptor or other components in the path of the lightning current. The current waveforms, peak amplitude, specific energy, and charge transfer should be recorded for each discharge. Besides this factual information, each discharge should be followed by a subjective analysis of whether the damage is acceptable.

Components as lightning receptors might suffer considerably from the charge transfer in an arc attachment test, why the possibility of replacing such components as part of scheduled maintenance should be discussed. In general the results and the evaluation pass/fail depend on the predefined success criteria.

2.5 Full scale lightning verification tests on other equipment, Nacelle, HUB, etc

Besides testing blade and blade tip sections which is the most frequently used test, full scale tests of other components are also conducted. When the blade or the nacelle is struck directly, the lightning current will always flow in the nacelle structure or close to the equipment located within or adjacent to the nacelle. For this reason, the immunity of the equipment must be verified, which has traditionally been done using the generic EMC standards, the IEC 61000-4-5 among others. To use these standards, the impact (in terms of current and voltage pulses) of the equipment at its actual position must be defined. To estimate the current level on a shielded cable within a complex nacelle structure is very difficult, and consequently the results of the estimation may be rather unreliable. Then the most straight forward solution is to conduct full scale tests and let the laws of physics decide the impact.

For this purpose, some test facilities offer to costumers full scale tests, in which the entire structure (Nacelle, HUB, etc.) is exposed to the lightning environment, i.e. impulse voltage and impulse currents as the ones used for blade verification. The entire system is operated under normal conditions, meaning that the blade bearings in a HUB is moving, the HUB on a nacelle is rotating, ensuring that all equipment is operational and that all sensors are active. By injecting realistic lightning currents in the range of 50kA to 200kA, and evaluating the output of all sensors and control systems, the immunity of the equipment can be verified.

Since inductive and capacitive couplings are included per default, there is no risk of over testing, and after passing the full scale equipment test, the manufacturer can definitely claim that their products meet the requirements. At the time of publication, this type of testing principle is quite new. For a couple of years, tests on aviation lights, wind sensors, GPS antennas and similar items have been conducted on a regular basis. The equipment is mounted on a mock-up of the actual installation, creating a similar impact in terms of the magnetic and electric field. An image of such a test is seen on Fig. 5. Tests on larger structures like wind turbine HUBs and nacelles have only been conducted a couple of times, and the results and images of such tests have not yet been published. However, it is expected that full scale tests on large equipment will be standard of near future, since the complexity of these systems makes it easier and more reliable to test on entire structures including all couplings.

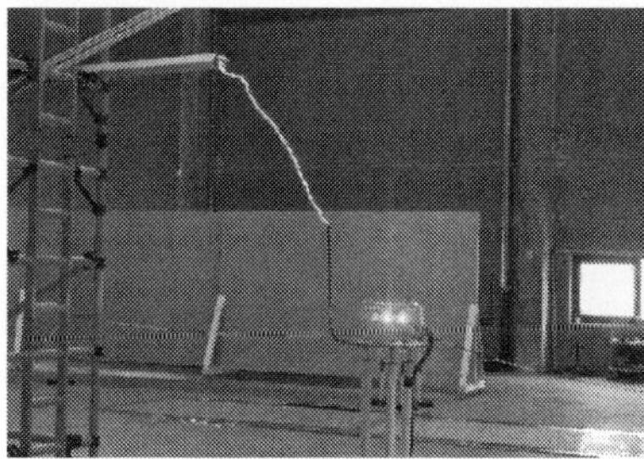

Fig. 5. A test of aircraft warning light, conducted using both impulse voltages and impulse currents, BTI Light systems A/S and Testinglab Denmark ApS.

3. Numerical modelling

Numerical modelling covers a wide range of mathematical tools to describe the physics of a certain structure or phenomenon. Concerning verification of wind turbine designs, the two main issues of lightning protection, *attachment process* and *current conduction* are simulated individually.

3.1 Modelling of lightning attachment points

Traditionally, the lightning attachment points on wind turbines have been defined by the Electro Geometrical Methods (Rolling Sphere, Protective Angle and Protective Mesh). Considering the blades, it implies that air terminations (receptors) are typically distributed along the length from radius 20m and outwards, although service experience shows that inboard receptors are only struck occasionally. Data supporting the fact that mainly the blade tips are struck was presented initially in 2006 (Madsen et al. 2006; Madsen et al. 2010). Based on the requirements in the IEC 61400-24 of mandatory verification, numerical tools capable of evaluating and verifying wind turbine blade protection designs have been invented.

3.1.1 Lightning research

In the quest of defining an appropriate model for determining the risk of lightning attachment points on wind turbines, it is valuable to investigate the tendencies in the lightning research community. Several research teams have worked with analytical and numerical methods to describe the attachment process. Results have been presented in scientific journals and lightning conferences.

The leader progression model developed by Dr. Marley Becerra and Professor Vernon Cooray considers all the physical relations and criteria governing the inception and stable propagation of the upward or connecting leader (Becerra, M. 2008). Their research is based on work by several other researchers during the past decades, and by comparing and combining these diverse theories, they have developed a unique model that considers the potential distribution surrounding grounded objects and the criteria for the inception of a stable positive upward leader.

Two cases apply for the positive upward leader. The first part considers the inception of a stable upward leader in the case of a static background field. Such a situation precedes the formation of an upward initiated lightning strike of negative polarity. In the second case, the electric field around the grounded structure is affected by a descending negative stepped leader, which means that the inception processes involve some dynamics.

The static version of the leader progression model has been verified using rocket triggered and altitude triggered lightning, and now it is used to evaluate the different characteristics of positive upward leaders. Only cloud-to-ground lightning flashes of negative polarity are treated by the model, corresponding to 90% of all naturally occurring lightning flashes. The remaining 10% are not covered by the present algorithms.

The engineering application intend to apply the highly scientific theories onto a physical structure like a wind turbine, and thereby develop an engineering tool that can be used to verify future wind turbine designs according to the IEC 61400-24.

3.1.2 Physical model

The recently used model to determine the attachment point distribution refers to the set of equations and algorithms presented in 2005 (Becerra, et al. 2005), and implemented in 2007, 2008 and 2009 (Bertelsen et al. 2007; Madsen et al. 2008, Madsen et al. 2009). Compared to the initial algorithms, the present set of equations now considers a static model where a downward leader is incorporated covering both upward and downward initiated lightning of negative polarity. The main physical parameters and conditions for incepting the positive leader from the grounded structure are the same for the early and the present models. The only difference is how the algorithm defines the striking inception distance and the associated attachment distribution.

3.1.3 Physics

Before a negative lightning strike occurs to a specific point on a structure, a leader is incepted from the point and propagates towards the charged region in the cloud (upward lightning) or the tip of the approaching stepped leader (downward lightning). It is generally believed that the point, which incepts the self-propagating leader, initially is also the point that will intercept the lightning strike (Golde, R.H. 1977). This leads in the direction that the point with the lowest stabilisation field (the least electric field sufficient for the inception of a stable and successful upward leader) will also incept an upward leader first and hence be the point struck by lightning.

However, since this is not the case that one single point receives all lightning attachments, the research team at Uppsala recommends that 'static leader inception zones' for each point in concern must be evaluated, and that the lateral coverage of these zones can be related to the probability of each point being struck.

Static leader inception is said to occur when the conditions for incepting a stable upward leader for a certain point is fulfilled, by the presence of a downward leader when the electric fields are calculated without considering the dynamics of the downward leader. Although introduction of a downward leader contradicts the definition of an upward initiated strike, the static leader inception zones are still used for evaluating both upward and downward initiated lightning (Becerra, M. 2008).

The static leader inception zone of a single point for a specific prospective peak current is defined as the horizontal area within which the tip of the downward leader at a certain but not constant height will incept a self-propagating upward leader. The principle is illustrated in Fig. 6: If a downward negative leader intrudes the horizontal area defined as the static leader inception zone, the corner of the building (green point) will incept a stable upward leader. If the vertical downward leader approaches outside this area, the conditions for incepting a stable upward leader at the green point is not met, and the lightning will strike elsewhere. Since the downward leader approaches at different lateral distances from the structure, the upward leader inception will occur when the downward leader tip is at different heights. Especially when the downward leader approaches outside of the structure, the leader will get closer to ground before inception of the upward leader from the point on the structure.

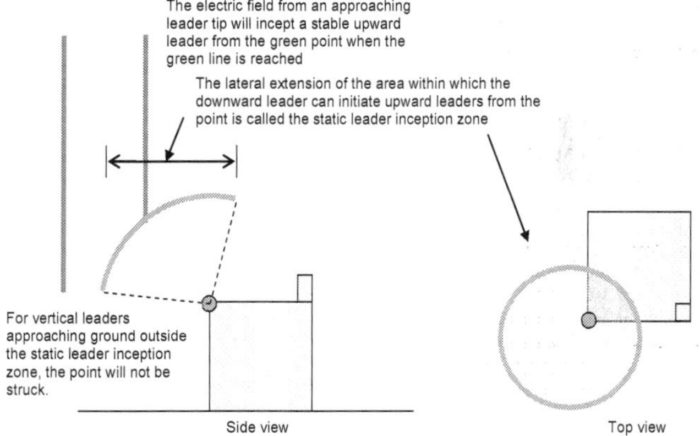

Fig. 6. The static leader inception zone used to evaluate the probability of each point incepting an upward leader.

If more than one point on a structure is considered, by independant treatment the static leader inception zones might coincide, so that the conditions for incepting an upward leader will be met at several points, if the downward leader approaches along the same vertical line. However, when this occurs it is unlikely that the different points will incept upward leaders for the same height of the downward leader, meaning that one of the points will incept the upward leader before the others, i.e. at a larger height of the downward leader.

On Fig. 7, a second point (blue) is considered exhibiting a larger inception zone at higher altitudes relative to the green point. Because the blue point will incept upward leaders at higher altitudes of the downward leader tip, when the downward leader approaches as shown, the blue inception zone covers the green inception zone. When correlating the inception zones with the probabilities, the blue area will attract more lightning strikes than the green area, meaning that the blue point is more exposed.

Consequences of following these principles: For each coordinate in the horizontal X-Y plane there exists a height of a vertical downward leader which corresponds to the inception of an upward leader from one single point on the structure considered. This single point on the structure will be the one struck by lightning if a vertical leader approaches along the specific X-Y coordinate.

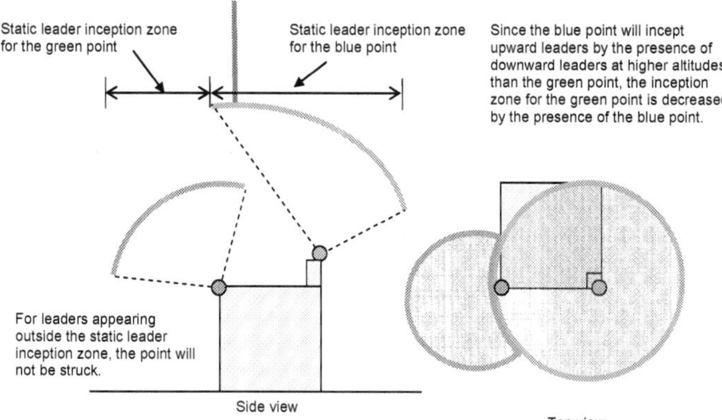

Fig. 7. If static leader inception zones coincide, the "area" will be given to the point that incepts the upward leader first.

The analysis must be conducted for several prospective peak currents, since the prospective peak current defines the charge distribution along the downward leader and hence the electric field surrounding each point on the structure. In the following section an algorithm is described which incorporates the principle of finding the single point on the structure that fulfils the upward leader inception criteria first, and thereby a method of assigning the static leader inception zones uniquely.

3.1.3 Algorithm

Two corners of a building are shown in a 2D plot on Fig. 8. The green and blue circle segments show the distances from the two corners (green and blue), at which a tip of a downward leader with a prospective peak return stroke current as indicated will initiate the inception of a stable upward leader from the corner considered. A larger prospective peak current implies that the inception of upward leaders will occur at larger distances, whereas smaller peak currents mean that the tip of the downward leader has to get closer to the point in concern to incept upward leaders.

A vertical downward leader is introduced along a fixed number of X-Y coordinates (only X coordinates on the drawing of Fig. 8) and extended to different heights or Z coordinates. For each location of the downward leader tip (x, y, z) and each of several different prospective peak currents, the electric field, and hence the conditions for incepting upward leaders from each point on the structure, is evaluated. Once the conditions are fulfilled for one point, the specific values of leader tip coordinates (x, y, z), the point for which the conditions are fulfilled (P_i) and the prospective peak current (I_p) are noted. Such a data set is seen on Fig. 8 concerning a vertical downward leader approaching the structure along (x_{dl}, y_{dl}), defined by a prospective return stroke current of 20kA, and extended to a height (z_{dl}) intersecting the 20kA striking inception distance for the green point.

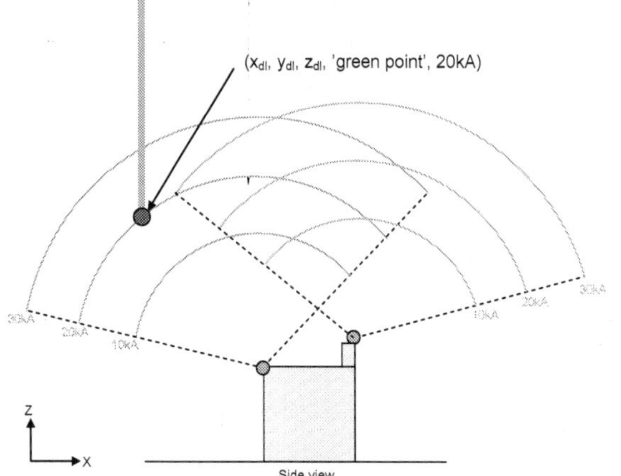

Fig. 8. Illustration of the algorithm for calculating static leader inception zones.

When a fine grid of coordinates for the downward leader tip are considered and the inception conditions for different prospective peak currents are evaluated in each coordinate, the acquired data sets will describe the coloured segments shown on Fig. 9. Here the area described by the lateral extension of each segment is defined as the stable leader inception zone for the corresponding peak return stroke current.

As mentioned, the evaluation of probabilities is done by considering the individual areas of the stable leader inception zones and relating them to the total stable leader inception zone of the structure.

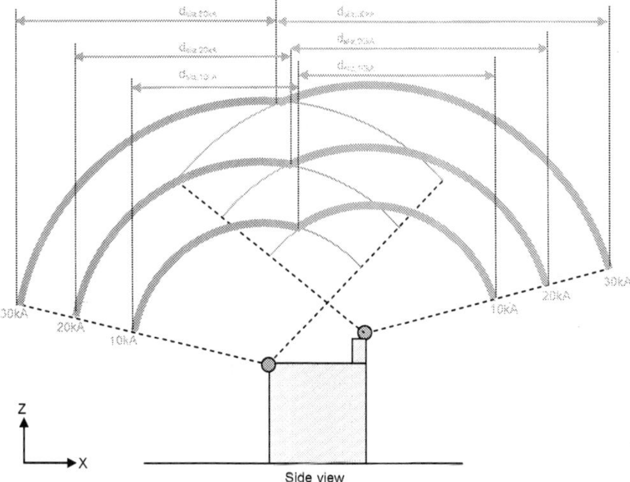

Fig. 9. The lateral extensions of the coloured segments for each peak current are the static leader inception zones.

On Fig. 10, a top view of the two points on the building considered previously is shown. The red dots correspond to each X-Y coordinate in which the vertical leader has approached the structure and the green and blue areas show the lateral extension of the stable leader inception zones considering a fixed prospective peak current.

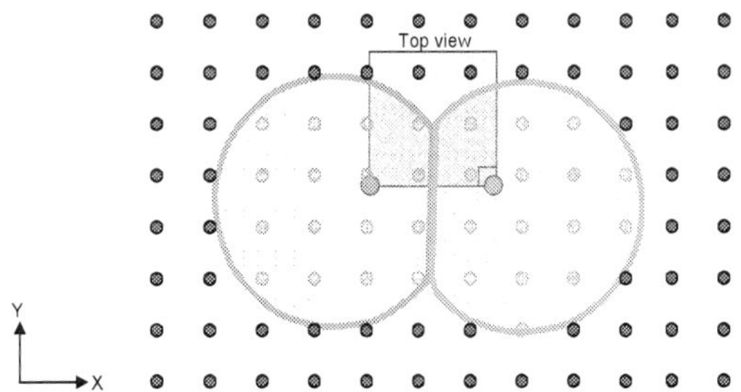

Fig. 10. The probabilities are evaluated based on the individual stable leader inception zones, for different fixed prospective peak currents.

The individual stable leader inception zones are found by determining which of the points, will incept a stable leader (in this case two points). The number of downward leader locations within the areas is proportional to the areas. Fig. 10. represents an example: The number of green points relative to the sum of green and blue points approximates the green area relative to the sum of the green and the blue area and hence the probability that the green point will be struck.

3.1.4 Case study

To verify the applicability of the simulation tool, a case study with a modern wind turbine is conducted. Several points along the blades and on the nacelle are considered. The final outcome is the probability of each point being struck. The calculations are performed with prospective peak currents of 60kA, 40kA, 20kA and 10kA. For higher or lower peak currents, the consequences are to be extrapolated based on these values.

Initially a 3D model of the wind turbine is defined within a box shaped analysis volume in a Finite Element Model (FEM) environment, Fig. 11. The size of the analysis volume is selected several times larger than the largest dimension of the turbine. Before assigning boundary conditions, the turbine model is subtracted from the analysis volume leaving only a single domain (between the turbine exterior and the boundaries of the analysis volume) for the analysis. The entire surface of the wind turbine as well as the ground at which it is standing is defined by a 0V boundary condition. The vertical boundaries of the analysis volume are assigned a boundary condition stating that the normal component of the electric displacement is zero (the potential is symmetric with respect to the boundary) and the upper horizontal boundary of the analysis volume is assigned a certain potential.

Poisson's equation is solved for the domain, and the vertical potential distribution above each point in concern is used as input to the following algorithms. The algorithms describe how this potential distribution is coupled with theoretical discharge physics, finally resulting in a quantity for each point known as the electric stabilisation field (Becerra, M. 2008). The electric field is calculated as the potential on the upper boundary necessary for

Fig. 11. A typical model of a turbine used in the research. The size of the analysis volume is considerably larger than the turbine, in this case 1km x 1km x 3km.

fulfilling the inception conditions divided by the height of the analysis volume, whereas the term successful upward leader refers to a leader that will propagate upwards self consistently.

Having the stabilisation field defined for each point on the structure, the second part of the process is to follow the procedure for assigning static leader inception zones, finally resulting in individual probabilities as seen above. To understand the basis of the discrete probabilities, 3D scatter plots of the successful leader inception points for the orientation 30° to horizontal level are shown in Fig. 12. Here it is clear how the upward leader inceptions occur with a larger distance to the downward leader tip for higher prospective peak currents (Left), whereas lower prospective peak currents allow the downward leader to approach closer to the turbine before upward leader inception (Right).

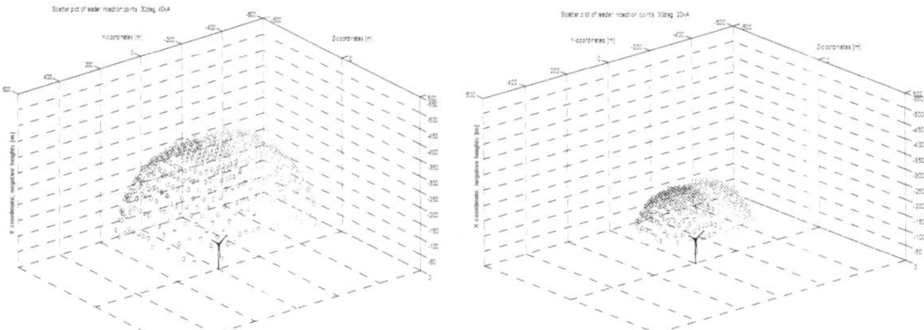

Fig. 12. Scatter plots showing origins of the downward leaders, leading to successful upward leader inception. Each colour corresponds to different attachment points. Left: 40kA, Right 20kA.

By evaluating the results presented graphically on Fig. 12 for three different rotor orientations and the four different peak current levels, an indication of the attachment point distribution for all possible situations is derived. In practice, it is done by counting the number of points with each individual colour and relating them to the total number of points (corresponding to the static leader inception zone defined previously). On Fig. 13, examples of the results considering two different rotor orientations and the 18 different points incepting lightning strikes are shown.

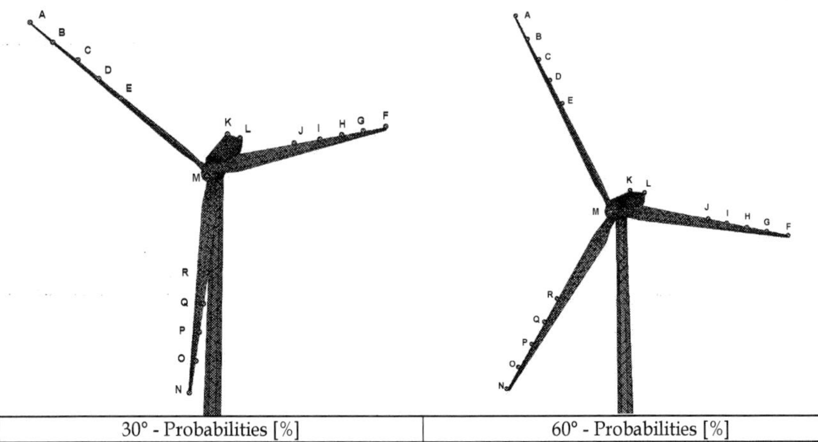

30° - Probabilities [%]				60° - Probabilities [%]			
Point	60kA	40kA	20kA	10kA	Point	60kA	40kA	20kA	10kA
A	50	50	50	50	A	100	100	99	98
F	50	50	50	50	F	0	0	1	2

Fig. 13. Attachment point distribution along five points for each blade, two points on the rear of the nacelle and the tip of the spinner.

In Fig 13. it is seen how the blade tips are the only exposed structures to peak currents down to 10kA and that the attachment distribution dictates equal probability for each of the upward pointing blades in the 30° orientation. For the second orientation (60° with horizontal), the probability of striking the upward pointing blade is by far larger than the probability of striking other parts of the turbine. However, as indicated in Fig. 13, the probability of striking the blade tip on the horizontal blade (point F) increases as the peak current is lowered.

Intuitively, the general conclusion based on the probabilities found above seems too simple. However, they depend strictly on the geometry and the algorithms derived. By investigating the situation having the rotor in the 30° orientation, the differences at the different peak return stroke currents are clarified. Fig. 14 shows three views of the turbine along with the points representing the leader tip positions at the successful inception of the upward leader. The blue points correspond to the situation where point A incepts upward leaders, whereas the green points represent the situations where point F receives the lightning strike.

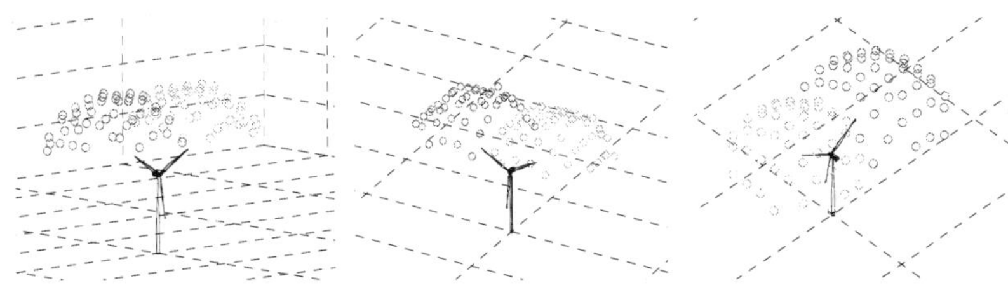

Fig. 14. Scatter plots visualising the 30° orientation considering 10kA prospective peak currents.

On Fig. 15 two plots of the same data are shown with a view parallel to the rotor axis and from directly above the turbine. In each case it is seen how the sphere caps drawn by the coloured points tend to wrap the turbine more smoothly at such low peak currents, so that the turbine geometry becomes more apparent to the leader tip. At high peak currents only the turbine extremities are exposed, whereas for low peak currents suddenly the less exposed structures on the turbine might incept lightning strikes.

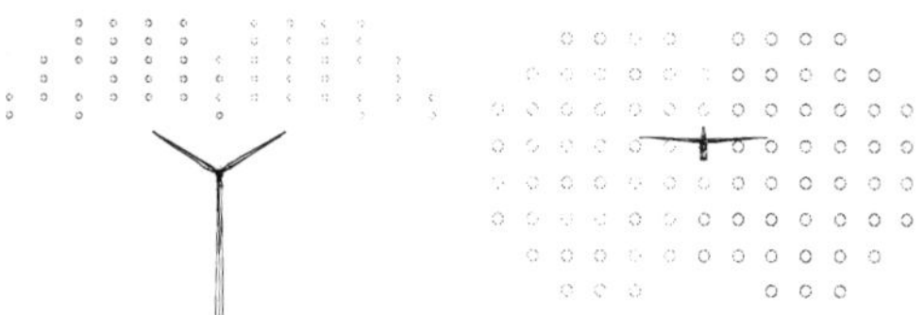

Fig. 15. Scatter plots shown parallel to the rotor axis (left) and from directly above the turbine (right), 10kA prospective peak current.

If full simulations were to be conducted at even smaller peak currents, the tendency would be that suddenly inboard receptors or the rear of the nacelle would be exposed enough to incept direct lightning strikes. However, at such low peak currents the associated damages are easier to control by suitable protection measures.

To prove this tendency, a simple situation is simulated manually by a vertical leader approaching directly above the turbine. Fig. 16 shows the leader tip height at connecting leader inception as well as the points from which the inception occurs for different peak currents. When lowering the current, the height of inception is lowered as well, meaning that the leader tip gets closer to the turbine before anything happens. Down to 8.5kA, the blade tip still incepts the connecting leaders first (A and F). At 8.25kA and 8kA, the fifth receptor pair (E and J) tends to incept leaders initially. The blade tips are not struck in this case. Lowering the current even further down to 7kA results in the exposure of the rear of the nacelle, since points K and L now incepts the initial leaders.

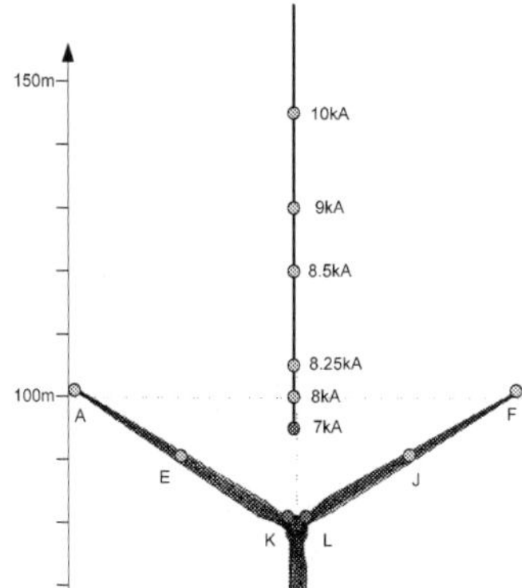

Fig. 16. By lowering the prospective peak return stroke current, attachment points elsewhere than the blade tips becomes possible.

Considering higher peak currents than the 60kA used in these simulations, the attachment distribution would be similar, as shown in the 60kA simulations, since the sphere caps will move further away from the turbine. The findings using vertical leaders therefore shows that inboard parts of the structure are only exposed to small amplitude lightning strikes, and that lightning strikes having peak amplitudes in excess of 10kA will attach to the blade tips.

3.1.4 Application of attachment point modelling

Modelling of the lightning attachment points on wind turbines is used to foresee where and with which amplitudes the lightning discharge will affect the structure. This enables the lightning protection engineer to place adequate protection measures at the right locations without over-engineering the solutions. To get the turbine designs certified by DNV, GL or similar, it requires that the protection principles applied are verified according to IEC 61400-24. Here either testing or modelling becomes necessary.

In the larger perspective, numerical modelling has also been used to address the issues of subdividing the wind turbine blades into lightning protection zones. The principle is known from the avionics industry, were the areas of an aircraft fuselage or a wing is divided into zones struck directly, experiencing a swept stroke, hang on zones and similar (SAE ARP 5414). The reason for considering zoning as an important part of the lightning protection design is that damages and attachment points inboard the blade tips - foreseen by the general EGM methods - are not experienced. Data from recent field surveys on modern wind turbines indicate that mainly attachments at the blade tips occur (Madsen et al. 2010).

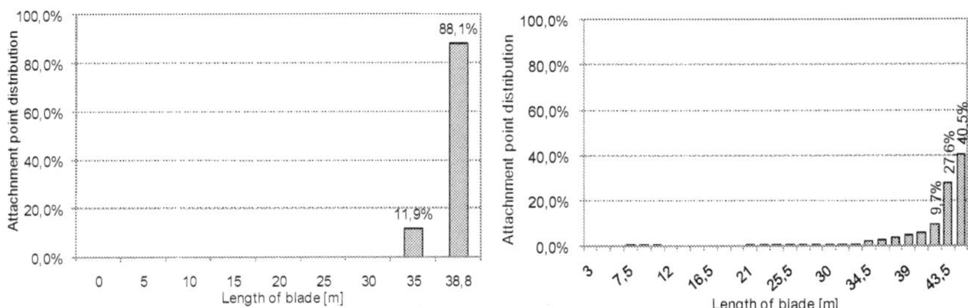

Fig. 17. Left: Attachment point distribution on 236 blades (39m) after two years of lightning exposure at the 'Horns Rev' wind farm, Right: Attachment point distribution of 2818 identified lightning attachment points on 45m blades.

From both graphs on Fig. 17 it is obvious how the blade tips are most favoured when it comes to lightning attachment. The main conclusion from the recent site inspection program is that the tip of the blades (within 1.5m) receives 70% of the lightning strikes, that 90% of the lightning strikes attaches within the outermost 6m of the blade, and that the remaining 10% attaches further inboard (6m from the tip). No correlations have been done so far considering the size of the erosion on receptors, and hence the current peak amplitude / specific energy / charge levels, but these are topics that will be addressed by the research team in future publications.

Based on the field surveys, and heavily supported by the numerical computations, it was therefore decided to define a zoning concept of wind turbine blades according to Fig. 18.

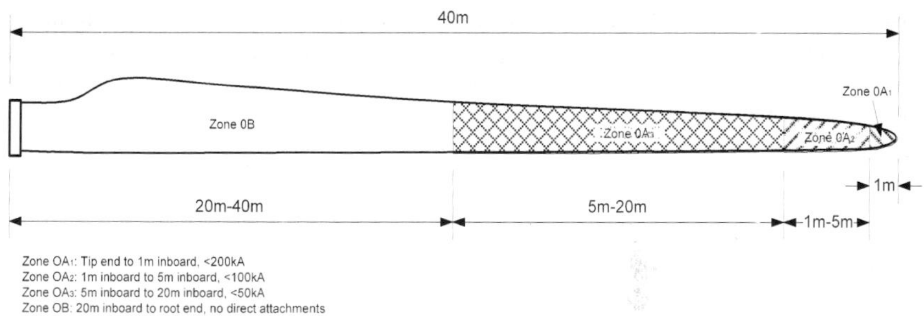

Fig. 18. New zoning concept based on the expected peak current amplitudes. (Madsen et al. 2010)

The zoning concept is regarded a possible upgrade for the test requirements in the next revision of the IEC 61400-24.

3.2 Modelling of magnetic fields

When a DC current is injected through a complex structure with several different paths, the current will be distributed according to the resistances of the different paths. There are no mutual couplings of neither inductive nor capacitive nature, since the currents or voltages are not time dependant. The solution of the current distribution is then straightforward, and can be performed using simple linear algebra.

If AC currents or transient currents are injected, the dI/dt of the AC current or the dU/dt of the AC voltage will introduce mutual couplings, which means that the current flowing in one conductor might induce a voltage on another conductor, or vice versa. In this case, the mutual couplings must be identified. It can be done analytically on very simple structures (two parallel wires, two wires of infinite length crossing at a fixed angle, etc.), but when it comes to real physical structures, numerical methods are required.

The numerical codes typically used are based on the FDTD (Finite Difference Time Domain) or the FEM (Finite Element Method). In both cases, the structure geometry is subdivided into a finite number of elements, and Maxwell Equations are then solved for each element respecting the mutual boundary conditions.

3.2.1 Current components

To model voltage drops during the interception of a lightning strike, the different components of the lightning strike must be considered individually. In the international standards for lightning protection, three characteristic current components for a Level 1 stroke are derived:

- The first return stroke, a 200kA current pulse with a rise time of 10μs and a decay time of 350μs. In the frequency domain, this waveform is simulated by an oscillating waveform exhibiting a frequency of 25kHz.
- The subsequent return stroke, a 50kA current pulse with a rise time of 0.25μs and a decay time of 100μs. In the frequency domain, this waveform is simulated by an oscillating waveform exhibiting a peak frequency of 1MHz.
- The continuing current, a DC current pulse of amplitude 200-800A and duration of up to a second.

In natural lightning, all possible combinations occur, but for verification of lightning protection systems (simulation and testing) these three individual components apply. In the case of determining the maximum magnetic fields within the nacelle, the first and the subsequent return stroke are of most concern.

Due to the frequencies of the lightning current and the permeability of the involved conductor materials for the nacelle structure, the skin effect becomes very important. Considering Iron with a relative permeability of 200 and conductivity in the range of 10^7 S/m, the skin depth for a 10kHz current component will be only 0.11mm, decreasing with increasing frequency. Therefore, the high frequency model (>10kHz) treats the solid structure of the nacelle as thin boundaries, since it can be assumed that all current flows at the structure extremity.

3.2.3 Modelling output

The simulations consider several different attachment points for the lightning strike, by injecting the lightning current into different places at the nacelle. A typical model of a wind turbine considering magnetic fields and current distribution when a blade is struck is seen on Fig. 19.

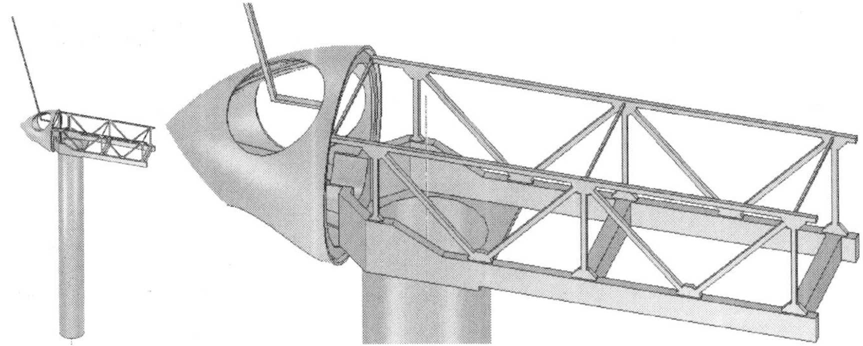

Fig. 19. The configuration where the turbine is struck on a blade pointing to the left with an angle of 45° with horizontal. The line extending from the HUB is simulating the blade down conductor.

In the case of a lightning strike to a blade located on the left side of the nacelle, the magnitude of the magnetic field during the first return stroke (200kA@25kHz) is illustrated on Fig. 20.

The magnetic field is visualized by drawing surfaces of equal magnitude. The red surface represents an area where the magnetic field attains a value of 30kA/m, the green surface represents a value of 20kA/m, the light blue surface represents a value of 10kA/m and the dark blue a value of 5kA/m.

The magnitude and distribution of the magnetic field around the geometry depends on the current path and current density on the surface of the structure. By evaluating the field distribution on Fig. 20, it is seen that the highest field strengths are obtained close to the main current paths where these are of limited size (the down conductor, lightning channel, etc.). At the rear of the nacelle and around the structural bars some metres away from the

HUB, the field is much lower. The current flowing in the nacelle construction works as the current in a faraday cage; hence the magnetic field in the centre of the nacelle is cancelled out to some degree.

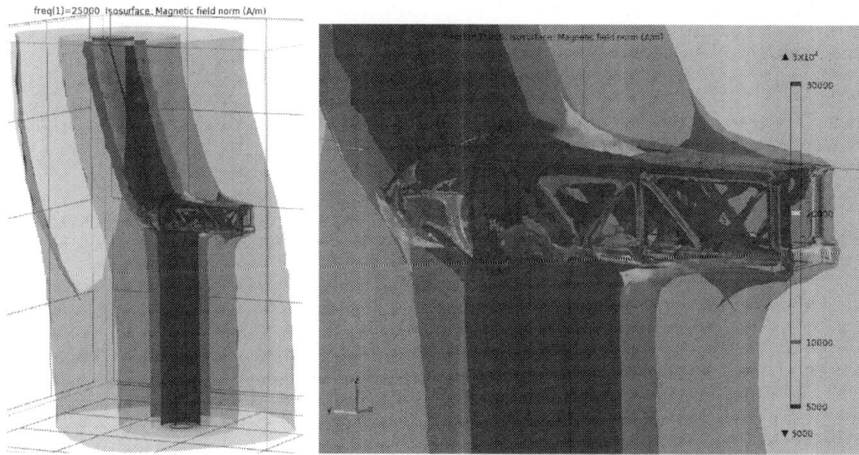

Fig. 20. Illustration of the magnetic field magnitude during a first return stroke represented by 200kA at 25kHz. The magnetic field is visualized by an iso-surface plot in which red represents a magnetic field strength of 30 kA/m, green represents 20 kA/m, light blue represents 10 kA/m and dark blue represents a field strength of 5 kA/m. Magnetic fields strength above 30 kA/m and below 5 kA/m has been omitted to simplify the illustration.

The current distribution within the different structural components also tends to minimize the magnetic field in the centre of the nacelle. This is seen more clearly at the following 2D slice plot, where the magnitude of the magnetic field is plotted for two different slice plots.

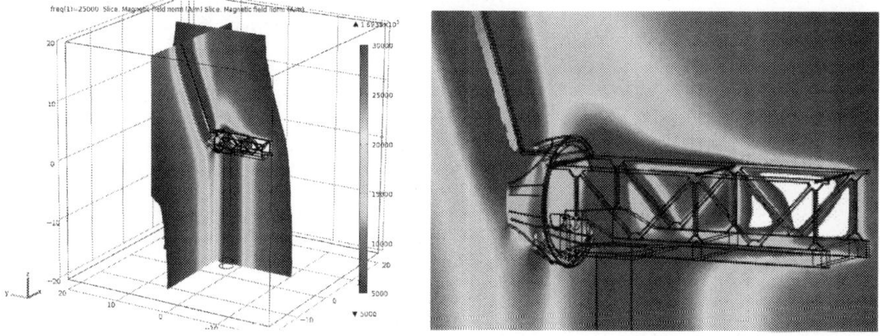

Fig. 21. 2D slice plot of the magnetic field when 200kA at 25kHz is conducted from one of the blades towards the tower base. The range for the plot is 5kA/m (blue) to 30kA/m (red).

The magnetic field is forced to the outside of the nacelle structure, due to the mutual coupling between the current flowing in the different structural components. The consequence is that the structural bars act as a Faraday cage for the interior of the nacelle, which is to be considered when placing panels and cables within the nacelle.

3.2.4 Application of results

Once the magnetic fields within the structure are known, panels and shielded cables can be selected to ensure certain compatibility between the control and sensor systems within the nacelle and the environment in terms of magnetic fields. Based on the current distribution also obtained by the numerical simulations, expressions can be derived, which couple lightning currents in the main structure with induced currents in cable shields.

Having the currents flowing on shielded cable or EMC enclosures with well-known transfer impedance, finally enables the designer to calculate the expected potential rise on conductors and hence select appropriate surge protection.

Along with testing, it is believed that future verification will benefit considerably by numerical modelling of lightning protection systems.

4. Conclusion

The present chapter presents general aspects of lightning protection to be considered when designing lightning protection systems for wind turbines. Since the release of the new standard IEC 61400-24, for lightning protection of wind turbines, verification of the protection measures has become mandatory. The verification can be done by either high voltage or high current testing, or by means of numerical modelling that has previously been verified against experimental findings or field surveys.

The test programme begins with an Initial Leader Attachment Test, defining where the turbine or in most cases the blade will most likely be struck. Hopefully, the blade will only be struck at places designed to handle the lightning current (lightning receptors) otherwise the design must be improved before passing the blade on to the high current test.

After defining these possible attachment points, the blade is tested in a high current laboratory to be subjected to the threat of the lightning current. The various lightning current waveforms are injected into the locations determined by the high voltage test, and the damage or wearing associated with these tests might require further design optimisation.

At an early stage of a design phase or in situations where testing is not an option, numerical modelling can be used as mean of verification. Basically the same two phenomena are modelled, the attachment process and the current conduction.

Attachment point modelling aims at identifying possible lightning attachment points on the wind turbine and defines the probabilities that certain areas will receive strikes of certain amplitudes. The methodology is used to foresee the most optimum placement of air termination systems on the nacelle and the blades, which is no longer applicable to the EGM methods according to IEC 61400-24.

Simulation of current distribution and magnetic fields in especially the nacelle structure is vital for design engineers to require a sufficient degree of shielding for their equipment. The magnetic environment within or adjacent to the nacelle structure during a lightning strike, is considerably higher than what the general EMC standards describe.

5. Acknowledgement

The research within lightning protection of wind turbines is carried out in a major community worldwide including representatives from the wind turbine manufacturers, wind turbine operators, test facilities, universities, public and private research institutes, etc. The knowledge accumulated within this group of researchers, and published at international conferences, in scientific journals and at commercial expos would not be possible without the involvement and professionalism of all participants.
A special acknowledgement is dedicated to friends and colleagues that have helped me and the wind turbine industry to gain a higher level of engineering expertise within lightning protection of wind turbines.

6. References

Madsen, S.F. (2006). Interaction between electrical discharges and materials for wind turbine blades particularly related to lightning protection, Ørsted-DTU, The Technical University of Denmark, Ph.D. Thesis, ISBN: 87-91184-60-6

Larsen, F.M & Sorensen, T. (2003). New lightning qualification test procedure for large wind turbine blades, *Proceedings of International Conference on Lightning and Static Electricity*, Blackpool, UK.

Madsen, S.F., Holboll, J., Henriksen, M., Bertelsen, K. & Erichsen, H.V. (2006) New test method for evaluating the lightning protection system on wind turbine blades. *Proceedings of the 28th International Conference on Lightning Protection*, Kanazawa, Japan.

Holboll, J., Madsen, S.F., Henriksen, M., Bertelsen, K. & Erichsen, H.V. (2006) Lightning discharge phenomena in the tip area of wind turbine blades and their dependency on material and environmental parameters. *Proceedings of the 28th International Conference on Lightning Protection*, Kanazawa, Japan.

Bertelsen, K., Erichsen, H.V. & Madsen, S.F. (2007) New high current test principle for wind turbine blades simulating the life time impact from lightning discharges. *Proceedings of the 30th International Conference on Lightning and Static Electricity*, Paris, France.

IEC 61400-24 Ed. 1.0. Wind turbines – Part 24: Lightning Protection, 2010.

IEC TR 61400-24. Wind turbine generator systems – Part 24: Lightning protection, 2002.

SAE ARP 5416. Aircraft Lightning Test Methods, Section 5: Direct Effects Test Methods, 2004.

IEC 62305-1 Ed. 1.0. Protection against lightning – Part 1: General principles, January 2006.

EN 50164-1 Lightning Protection Components (LPC) – Part 1: Requirements for connection components, September 1999.

Heater, J. and Ruei, R. (2003). A Comparison of Electrode Configurations for Simulation of Damage Caused by a Lightning Strike, *Proceedings of International Conference on Lightning and Static Electricity*, Blackpool, UK.

IEC 62305-2 Ed. 1.0, Protection against lightning – Part 2: Risk management, January 2006.

IEC 61000-4-5 Ed. 2.0, Electromagnetic compatibility (EMC) – Part 4-5: Testing and measurement techniques – Surge immunity test, November 2005.

Madsen, S.F., Bertelsen, K., Krogh, T.H., Erichsen, H.V., Hansen, A.N., Lønbæk, K.B. (2010) Proposal of new zoning concept considering lightning protection of wind turbine blades. *Proceedings of the 30th International Conference on Lightning Protection*, Calgari, Italy.

Becerra, M. (2008) On the Attachment of Lightning Flashes to Grounded Structures, Doctoral thesis, Uppsala University, ISBN :XXXX.

Becerra, M. and Cooray, V. (2005) A simplified model to represent the inception of upward leaders from grounded structures under the influence of lightning stepped leaders, *Procdings of the 29th International Conference on Lightning and Static Electricity*, Seattle Washington, USA.

Becerra, M., Cooray V. & Abidin H.Z. (2005) Location of the vulnerable points to be struck by lightning in complex structures, *Procdings of the 29th International Conference on Lightning and Static Electricity*, Seattle Washington, USA.

Bertelsen, K., Erichsen, H.V., Skov Jensen M.V.R. & Madsen, S.F. (2007) Application of numerical models to determine lightning attachment points on wind turbines, *Proceedings of the 30th International Conference on Lightning and Static Electricity*, Paris, France.

Madsen, S.F. & Erichsen, H.V. (2008) Improvements of numerical models to determine lightning attachment points on wind turbines, *Procdings of the 29th International Conference on Lightning Protection*, Uppsala, Sweden.

Cooray, V., Rakov, V. & Theethayi, N. (2004) The relationship between the leader charge and the return stroke current – Berger's data revisited, *Procdings of the 27th International Conference on Lightning Protection*, Avignon, France.

Madsen, S.F. & Erichsen, H.V. (2009) Numerical model to determine lightning attachment point distributions on wind turbines according to the revised IEC 61400-24, *Proceedings of the 31st International Conference on Lightning and Static Electricity*, Pittsfield, Massachussetts, USA.

Golde, R.H. (1977) Lightning Conductor, Chapter 17 in Golde, R.H. (Ed.), Lightning, vol. 2, Academic Press: London, UK.

SAE ARP 5414. Aircraft Lightning Zoning, 1999.

The value of γ, β and R^2 for all 16 different directions and for the combined data of all directions for KIA location is provided in Fig.3, 4 and 5 respectively.

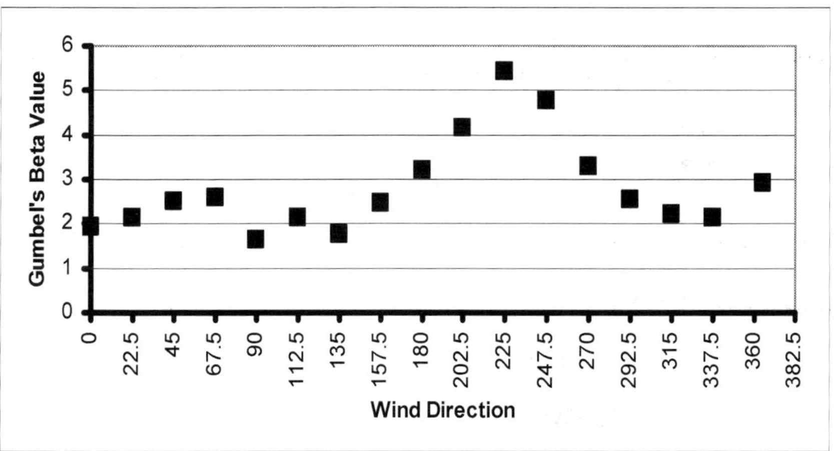

Fig. 4. Gumbel's β value for all 16 directions and for the data without considering the effect of wind direction for KIA location

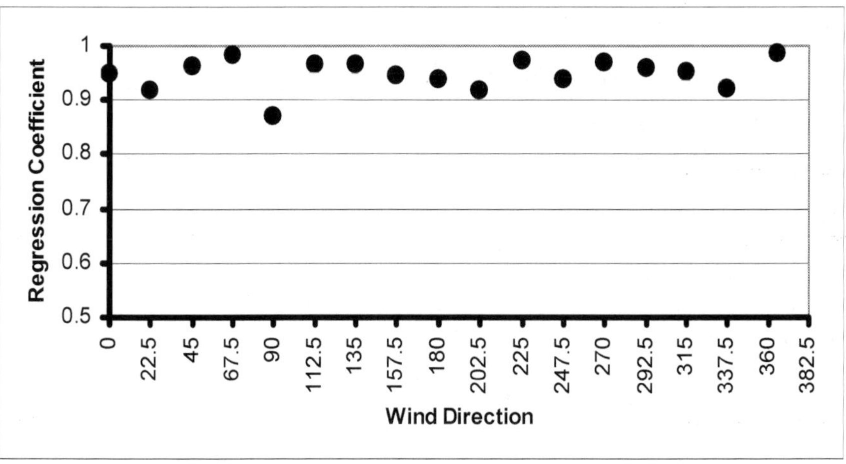

Fig. 5. Regression coefficient, R^2 for wind data from all 16 directions and for the data without considering the effect of wind direction for KIA location.

In the above three figures, the wind direction of 0, 22.5, 45, 67.5, 90, 112.5, 135, 157.5, 180, 202.5, 225, 247.5, 270, 292.5, 315 and 337.5 corresponds to wind blowing from N, NNE, NE, ENE, E, ESE, SE, SSE, S, SSW, SW, WSW, W, WNW, NW and NNW respectively. The γ, β

and R^2 values for the combined data, without considering the direction effect is provided for x axis at 365 degree (Though there is no 365 degree in reality), mainly for the sake of information and comparison. The γ value is found to be fluctuating from 5.3 to 16.2, β value is found to fluctuate from 1.65 to 5.4 and the value of R^2 is found fluctuating from 0.87 to 0.99. The value of γ and β can now be substituted in the formula $U_{TR} = \gamma - \beta \ln [-\ln\{(\lambda T_R-1)/(\lambda T_R)\}]$ in order to estimate the extreme wind speed for different return periods and for different directions. It can be seen that most of the values of regression coefficients are more than 0.9, which provides enough confidence in using the value of γ and β for the prediction of extreme wind speed.

The values of γ, β and R^2 for all the five locations in Kuwait can be obtained from Neelamani et al (2007). The wind speed for different return period, U_{TR} and for different locations and directions can then be estimated using the formula $U_{TR} = \gamma - \beta \ln[-\ln\{(\lambda T_R-1)/(\lambda T_R)\}]$. The predicted maximum wind speed for return periods of 10, 25, 50, 100 and 200 years from different directions for KIA location is given in Fig. 6.

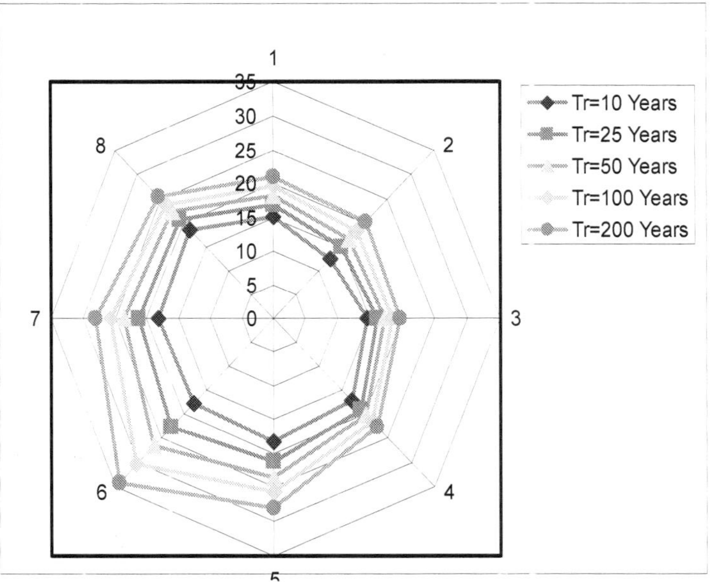

Fig. 6. The predicted maximum wind speed for different return periods and from different directions in KIA location in Kuwait

The following important information will be useful for the selection of suitable wind turbine, designing wind power plants and other tall structures in Kuwait:-

a.   The maximum wind speed for 100 year return period is expected to be of the order of 32.5 m/s and 30.5 m/s from SW direction and WSW directions respectively.

b.   The maximum wind speed from NW and SE direction (Most predominant wind direction in Kuwait) for a return period of 100 years is expected to be about 25 m/s and 22.3 m/s respectively.

c.   The smallest value of maximum wind speed for 100 year return period is expected to be about 19 m/s and is expected from the first quarter of the whole direction band (i.e. from N to E).

d.   From the present study it is found that the maximum wind speed without considering the directional effect is expected to be of the order of 31.24 m/s for a return period of 100 years in KIA location.

Similarly, from Fig. 7, one can see that the highest wind speed is from NW for any return period for the KISR location.

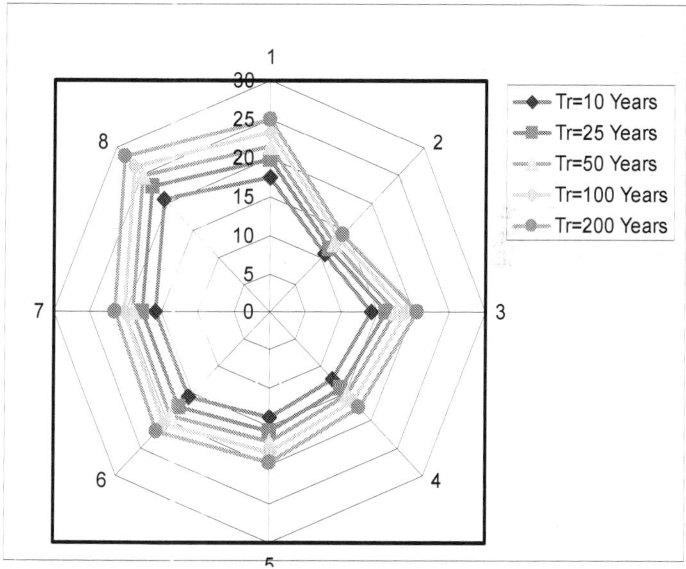

Fig. 7. The predicted maximum wind speed for different return periods and from different directions in KISR location in Kuwait

For Ras Al-Ardh location (Fig. 8), the highest wind speed for any return period is from N.
For Failaka island (Fig.9), the highest wind speed for any return period is from N, NW and SW.
Finally, for Al-Wafra (Fig.10), the highest wind speed for any return period is both from N and S.
In order to see the effect of change in spatial locations on the predicted 100 year return period wind speed, Fig.11 is provided. It can be seen that the change in spatial location to an extent of about 50 km from the main station (KIA) has reflected significant change in the predicted 100 year probable extreme wind speed.

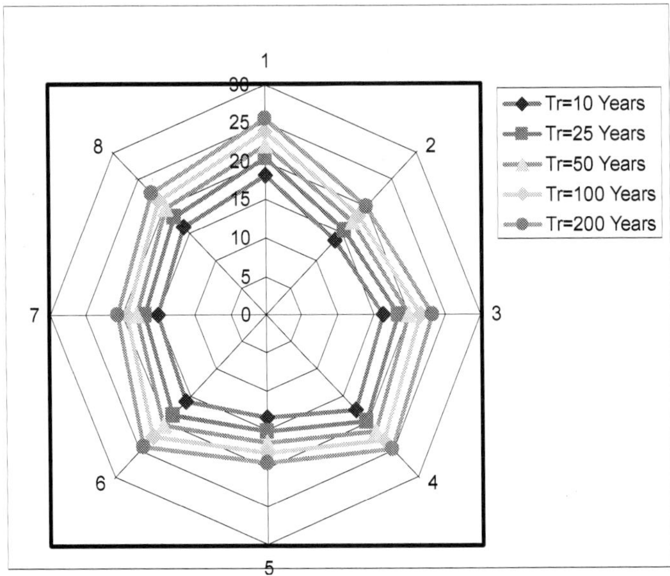

Fig. 8. The predicted maximum wind speed for different return periods and from different directions in Ras Al-Ardh location in Kuwait

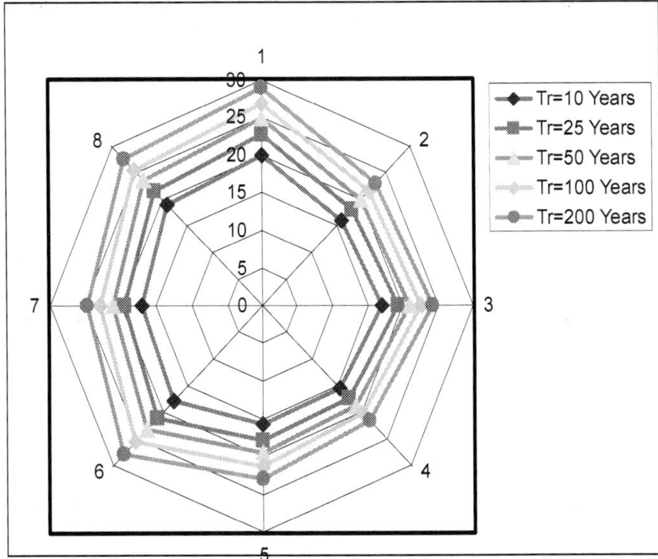

Fig. 9. The predicted maximum wind speed for different return periods and from different directions in Failaka Island location in Kuwait

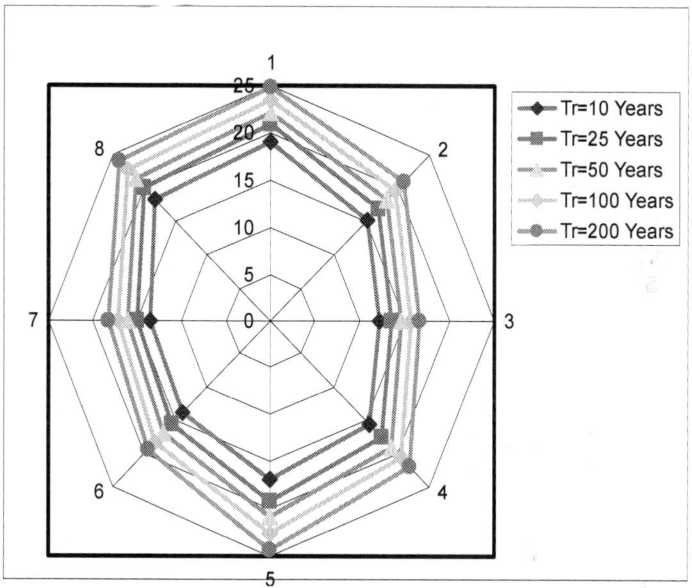

Fig. 10. The predicted maximum wind speed for different return periods and from different directions in Al-Wafra location in Kuwait

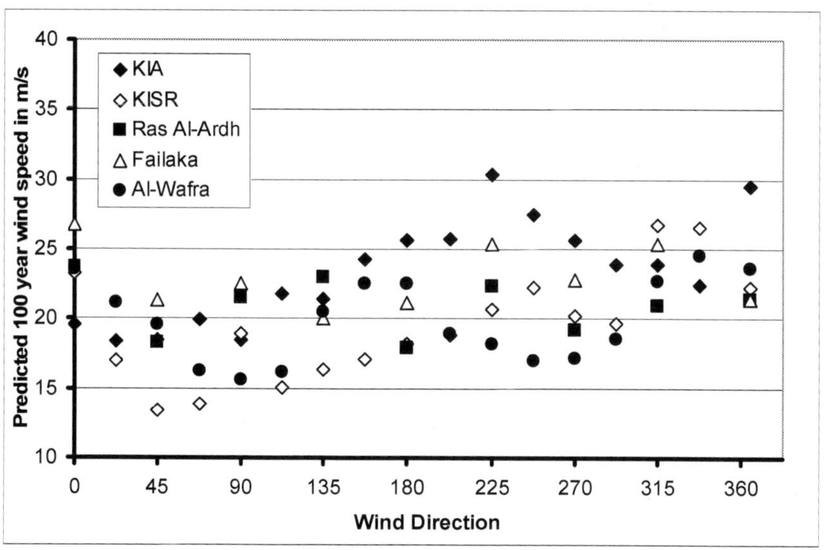

Fig. 11. Comparison of the predicted maximum wind speed for 100 year return periods and from different directions in KIA, KISR, Ras Al-Ardh, Failaka Island and Al-Wafra area

For example, the wind speed for 100 year return period from SW direction at KIA is about 30.3 m/s, whereas it is only about 18 m/s in Al-Wafra area. Al-Wafra area is located at the Southern boundary of Kuwait and is the main farming area in Kuwait. The plants and trees in the farm houses may dissipate significant amount of wind energy blowing from SW. Fig.11 can be used to select the appropriate expected extreme wind speed for 100 year return period at these 5 different locations and from different directions in Kuwait. This will help in appropriate design orientation of tall buildings in order to reduce the wind loading.

The extreme wind speed value and the associated direction will also be useful for the estimation of extreme sand and dust movements in Kuwait. This is because the extreme wind blowing from Iraq (North and North-West) and from Saudi Arabia (South and South-West) brings a large quantity of sand from the desert whereas the wind blowing from the Arabian Gulf side (North-East, East and South-East) moves significant amount of sand from Kuwait to the bordering countries.

4. Part 2: The effect of climate change on the extreme 10 minute average wind and gust speed

Climate change is already beginning to transform the life on Earth (http://www.nature.org/initiatives/climatechange/issues). Around the globe, seasons are shifting, temperatures are increasing and sea levels are rising. If proper actions are not taken now, the climate change will permanently alter the lands and waters we all depend upon for our survival. Some of the most dangerous consequences of climate change are Higher temperatures, changing landscapes, wildlife at risk, rising seas, increased risk of drought, fire and floods, stronger storms and increased storm damage, more heat-related illness and disease and significant global economic losses. In general, it is believed that global warming and climate change has more negative impacts than positive impacts. There are no proven findings on the effect of climate change on the extreme wind speeds at different locations on the earth. However an attempt can be made to visualize the impact, if measured wind data is available for the past many years, say at least for the past 50 years. How does one know the effect of global warming and climate change on the extreme winds and gusts. The only way is to get measured quality data for the past many years, divide them into few data segments (like 20 years of oldest data, 20 years of intermediate years and the latest 20 years), carry out the analysis on these data segments and analyze the trend of the predicted extreme values. For the present work, 45 years of measured data at KIA location (From 1957 to 2009) are available. The measured yearly maximum 10 minute average wind speed and the yearly maximum gust speed are as shown in Fig.12. It is seen that the gust speed has reached to 38 m/s during the past 53 years and the 10 minute average wind speed has reached to 30 m/s. The raw data obtained from Meteorology office of KIA for the present study is the maximum value of the 10 minute average wind speed and the gust speed for every day. Gumbel's extreme value distribution, as discussed in part I is adopted for obtaining the extreme 10 minute average wind and gust speed. The input data for extreme wind analysis (yearly maximum 10 minute average wind speed and the yearly maximum gust speed) is updated and extracted from the measured daily maximum 10 minute average wind and the gust speed data for the period 1957 to 2009. These data are separated into three sets of data groups covering the year 1957-1974, 1975-1992, 1993-2009, each of 18 years duration. Extreme wind analysis is carried out on each data group for both 10 minute average wind speed and the gust speed.

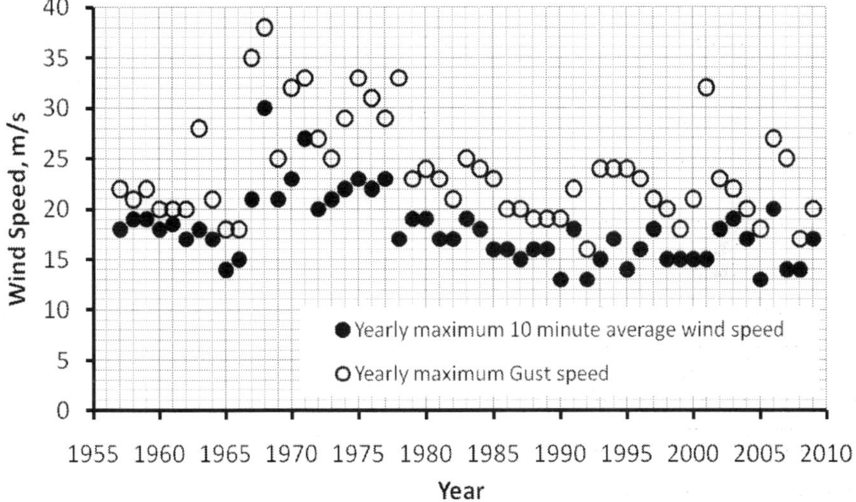

Fig. 12. The measured yearly maximum 10 minute average wind speed and the yearly maximum gust speed in Kuwait International Airport

Year	Max. Wind speed (m/s)	Max. Wind speed in descending order (m/s)	Rank, i	$Q=(i-0.44)/(N+0.12)$	$P=1-Q$	$T_R =1/(\lambda Q)$	$-\ln[-\ln(P)]$
1993	24	32	1	0.012411348	0.987589	80.57143	4.3829061
1994	24	27	2	0.034574468	0.965426	28.92308	3.34709822
1995	24	25	3	0.056737589	0.943262	17.625	2.84025512
1996	23	24	4	0.078900709	0.921099	12.67416	2.49875277
1997	21	24	5	0.10106383	0.898936	9.894737	2.23920429
1998	20	24	6	0.12322695	0.876773	8.115108	2.02869443
1999	18	23	7	0.145390071	0.85461	6.878049	1.85080821
2000	21	23	8	0.167553191	0.832447	5.968254	1.69616232
2001	32	22	9	0.189716312	0.810284	5.271028	1.5588833
2002	23	21	10	0.211879433	0.788121	4.719665	1.4350469
2003	22	21	11	0.234042553	0.765957	4.272727	1.32189836
2004	20	20	12	0.256205674	0.743794	3.903114	1.21742716
2005	18	20	13	0.278368794	0.721631	3.592357	1.1201187
2006	27	20	14	0.300531915	0.699468	3.327434	1.02880144
2007	25	18	15	0.322695035	0.677305	3.098901	0.94254836
2008	17	18	16	0.344858156	0.655142	2.899743	0.86061123
2009	20	17	17	0.367021277	0.632979	2.724638	0.78237526

Table 3. A Sample Table for the Extreme Gust Speed Analysis for the Wind Data during the year 1993 to 2009

A sample table of extreme wind analysis for the gust data during the year 1993 to 2009 is shown in Table 3.

Similar tables are prepared for all the other 5 data sets.

Gumbel distribution plot for the three data sets of 10 minute average wind speed and three data sets of gust speed is given in Fig.13 to 18. The equation of the best line fit and the correlation coefficient, R^2 are also provided in each figures.

The value of γ, β and R^2 for all the six different data sets is provided in table 4.

Type of data	Year range	γ	β	R2
10 minute average wind speed data	1957-1974	4.047	12.754	0.960
	1975-1992	3.123	12.086	0.895
	1993-2009	2.087	12.176	0.885
Gust speed data	1957-1974	6.325	14.033	0.960
	1975-1992	5.244	14.274	0.895
	1993-2009	3.858	15.227	0.885

Table 4. The value of γ, β and R^2 for all the six different data sets

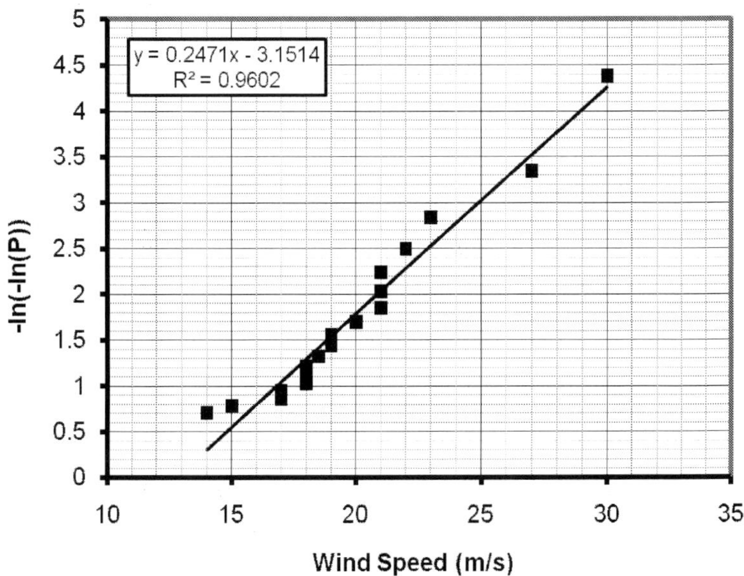

Fig. 13. Gumbel distribution plot for the maximum yearly 10 minute average wind speed data for the year 1957-1974

Part 2

Structural and Electromechanical Elements of Wind Power Conversion

6

Efficient Modelling of Wind Turbine Foundations

Lars Andersen and Johan Clausen
Aalborg University, Department of Civil Engineering
Denmark

1. Introduction

Recently, wind turbines have increased significantly in size, and optimization has led to very slender and flexible structures. Hence, the Eigenfrequencies of the structure are close to the excitation frequencies related to environmental loads from wind and waves. To obtain a reliable estimate of the fatigue life of a wind turbine, the dynamic response of the structure must be analysed. For this purpose, aeroelastic codes have been developed. Existing codes, e.g. FLEX by Øye (1996), HAWC by Larsen & Hansen (2004) and FAST by Jonkman & Buhl (2005), have about 30 degrees of freedom for the structure including tower, nacelle, hub and rotor; but they do not account for dynamic soil–structure interaction. Thus, the forces on the structure may be over or underestimated, and the natural frequencies may be determined inaccurately.

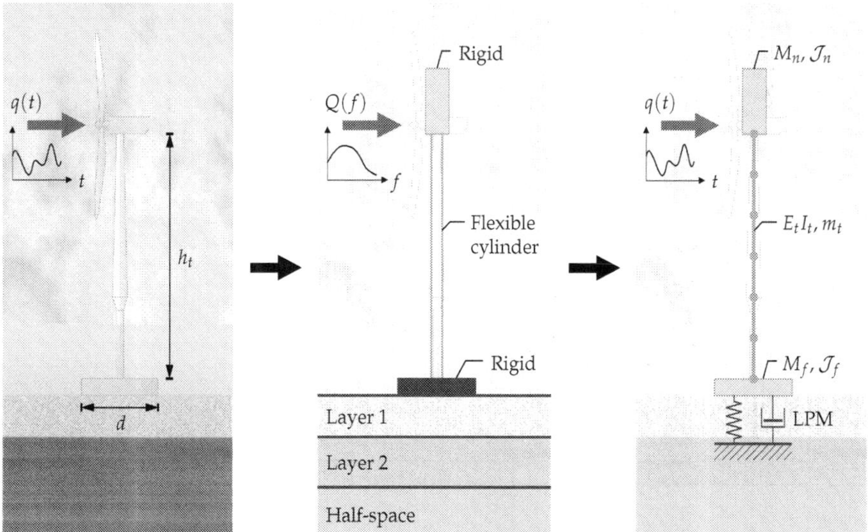

Fig. 1. From prototype to computational model: Wind turbine on a footing over a soil stratum (left); rigorous model of the layered half-space (centre); lumped-parameter model of the soil and foundation coupled with finite-element model of the structure (right).

Andersen & Clausen (2008) concluded that soil stratification has a significant impact on the dynamic stiffness, or impedance, of surface footings—even at the very low frequencies relevant to the first few modes of vibration of a wind turbine. Liingaard et al. (2007) employed a coupled finite-element/boundary-element model for the analysis of a flexible bucket foundation, finding a similar variation of the dynamic stiffness in the frequency range relevant for wind turbines. This illustrated the necessity of implementing a model of the turbine foundation into the aeroelasic codes that are utilized for design and analysis of the structure. However, since computation speed is of paramount importance, the model of the foundation should only add few degrees of freedom to the model of the structure. As proposed by Andersen (2010) and illustrated in Fig. 1, this may be achieved by fitting a lumped-parameter model (LPM) to the results of a rigorous analysis, following the concepts outline by Wolf (1994).

This chapter outlines the methodology for calibration and implementation of an LPM of a wind turbine foundation. Firstly, the formulation of rigorous computational models of foundations is discussed with emphasis on rigid footings, i.e. monolithic gravity-based foundations. A brief introduction to other types of foundations is given with focus on their dynamic stiffness properties. Secondly, Sections 2 and 3 provide an in-depth description of an efficient method for the evaluation of the dynamics stiffness of surface footings of arbitrary shapes. Thirdly, in Section 4 the concept of consistent lumped-parameter models is presented and the formulation of a fitting algorithm is discussed. Finally, Section 5 includes a number of example results that illustrate the performance of lumped-parameter models.

1.1 Types of foundations and their properties

The gravity footing is the only logical choice of foundation for land-based wind turbines on residual soils, whereas a direct anchoring may be applied on intact rock. However, for offshore wind turbines a greater variety of possibilities exist. As illustrated in Fig. 2, when the turbines are taken to greater water depths, the gravity footing may be replaced by a monopile, a bucket foundation or a jacket structure. Another alternative is the tripod which, like the jacket structure, can be placed on piles, gravity footings or spud cans (suction anchors). The latter case was studied by Senders (2005). In any case, the choice of foundation type is site dependent and strongly influenced by the soil properties and the environmental conditions, i.e. wind, waves, current and ice. Especially, current may involve sediment transport and scour on sandy and silty seabeds, which may lead to the necessity of scour protection around foundations with a large diameter or width.

Regarding the design of a wind turbine foundation, three limit states must be analysed in accordance with most codes of practice, e.g. the Eurocodes. For offshore foundations, design is usually based on the design guidelines provided by the API (2000) or DNV (2001). Firstly, the strength and stability of the foundation and subsoil must be high enough to support the structure in the ultimate limit state (ULS). Secondly, the stiffness of the foundation should ensure that the displacements of the structure are below a threshold value in the serviceability limit state (SLS). Finally, the wind turbine must be analysed regarding failure in the fatigue limit state (FLS), and this turns out to be critical for large modern offshore wind turbines.

The ULS is typically design giving for the foundations of smaller, land-based wind turbines. In the SLS and FLS the turbine may be regarded as fully fixed at the base, leading to a great simplification of the dynamic system to be analysed. However, as the size of the turbine increases, soil–structure interaction becomes stronger and due to the high flexibility of the structure, the first Eigenfrequencies are typically below 0.3 Hz.

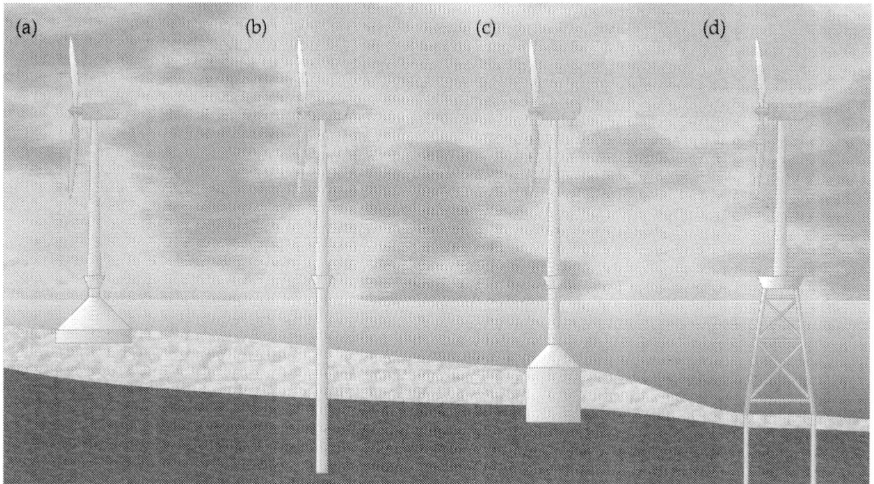

Fig. 2. Different types of wind turbine foundations used offshore a various water depths:
(a) gravity foundation; (b) monopile foundation; (c) monopod bucket foundation and
(d) jacket foundaiton.

An improper design may cause resonance due to the excitation from wind and waves, leading
to immature failure in the FLS. An accurate prediction of the fatigue life span of a wind turbine
requires a precise estimate of the Eigenfrequencies. This in turn necessitates an adequate
model for the dynamic stiffness of the foundation and subsoil. The formulation of such models
is the focus of such models. The reader is referred to standard text books on geotechnical
engineering for further reading about static behaviour of foundations.

1.2 Computational models of foundations for wind turbines

Several methods can be used to evaluate the dynamic stiffness of footings resting on
the surface of the ground or embedded within the soil. Examples include analytic,
semi-analytic or semi-empirical methods as proposed by Luco & Westmann (1971), Luco
(1976), Krenk & Schmidt (1981), Wong & Luco (1985), Mita & Luco (1989), Wolf (1994) and
Vrettos (1999) as well as Andersen & Clausen (2008). Especially, torsional motion of
footings was studied by Novak & Sachs (1973) and Veletsos & Damodaran Nair (1974) as
well as Avilés & Pérez-Rocha (1996). Rocking and horizontal sliding motion of footings was
analysed by Veletsos & Wei (1971) and Ahmad & Rupani (1999) as well as Bu & Lin (1999).
Alternatively, numerical analysis may be conducted using the finite-element method and the
boundary-element method. See, for example, the work by Emperador & Domínguez (1989)
and Liingaard et al. (2007).

For monopiles, analyses are usually performed by means of the Winkler approach in which
the pile is continuously supported by springs. The nonlinear soil stiffness in the axial direction
along the shaft is described by t–z curves, whereas the horizontal soil resistance along the shaft
is provided by p–y curves. Here, t and p is the resulting force per unit length in the vertical
and horizontal directions, respectively, whereas z and y are the corresponding displacements.
For a pile loaded vertically in compression, a similar model can be formulated for the tip

resistance. More information about these methods can be found in the design guidelines by API (2000) and DNV (2001).

Following this approach, El Naggar & Novak (1994a;b) formulated a model for vertical dynamic loading of pile foundations. Further studies regarding the axial response were conducted by Asgarian et al. (2008), who studied pile–soil interaction for an offshore jacket, and Manna & Baidya (2010), who compared computational and experimental results. In a similar manner, El Naggar & Novak (1995; 1996) studied monopiles subject to horizontal dynamic excitation. More work along this line is attributed to El Naggar & Bentley (2000), who formulated p–y curves for dynamic pile–soil interaction, and Kong et al. (2006), who presented a simplified method including the effect of separation between the pile and the soil. A further development of Winkler models for nonlinear dynamic soil behaviour was conducted by Allotey & El Naggar (2008). Alternatively, the performance of monopiles under cyclic lateral loading was studied by Achmus et al. (2009) using a finite-element model. Gerolymos & Gazetas (2006a;b;c) developed a Winkler model for static and dynamic analysis of caisson foundations fully embedded in linear or nonlinear soil. Further research regarding the formulation of simple models for dynamic response of bucket foundations was carried out by Varun et al. (2009). The concept of the monopod bucket foundation has been described by Houlsby et al. (2005; 2006) as well as Ibsen (2008). Dynamic analysis of such foundations were performed by Liingaard et al. (2007; 2005) and Liingaard (2006) as well as Andersen et al. (2009). The latter work will be further described by the end of this chapter.

2. Semi-analytic model of a layered ground

This section provides a thorough explanation of a semi-analytical model that may be applied to evaluate the response of a layered, or stratified, ground. The derivation follows the original work by Andersen & Clausen (2008). The fundamental assumption is that the ground may be analysed as a horizontally layered half-space with each soil layer consisting of a homogeneous linear viscoelastic material. In Section 3 the model of the ground will be used as a basis for the development of a numerical method providing the dynamic stiffness of a foundation over a stratum. Finally, in Section 5 this method will be applied to the analysis of gravity-based foundations for offshore wind turbines.

2.1 Response of a layered half-space

The surface displacement in time domain and in Cartesian space is denoted $u_i^{10}(x_1, x_2, t) = u_i(x_1, x_2, 0, t)$. Likewise the surface traction, or the load on the free surface, will be denoted $p_i^{10}(x_1, x_2, t) = p_i(x_1, x_2, 0, t)$. An explanation of the double superscript 10 is given in the next subsection. Here it is just noted that superscript 10 refers to the top of the half-space.

Further, let $g_{ij}(x_1 - y_1, x_2 - y_2, t - \tau)$ be the Green's function relating the displacement at the observation point $(x_1, x_2, 0)$ to the traction applied at the source point $(y_1, y_2, 0)$. Both points are situated on the surface of a stratified half-space with horizontal interfaces. The total displacement at the point $(x_1, x_2, 0)$ on the surface of the half-space is then found as

$$u_i^{10}(x_1, x_2, t) = \int_{-\infty}^{t} \int_{-\infty}^{\infty} \int_{-\infty}^{\infty} g_{ij}(x_1 - y_1, x_2 - y_2, t - \tau) p_j^{10}(y_1, y_2, \tau) \, \mathrm{d}y_1 \mathrm{d}y_2 d\tau. \qquad (1)$$

The displacement at any point on the surface of the half-space and at any instant of time may be evaluated by means of Eq. (1). However, this requires the existence of the Green's function $g_{ij}(x_1 - y_1, x_2 - y_2, t - \tau)$, which may be interpreted as the dynamic flexibility. Unfortunately, a closed-form solution cannot be established for a layered half-space, and in practice the temporal–spatial solution expressed by Eq. (1) is inapplicable.

Assuming that the response of the stratum is linear, the analysis may be carried out in the frequency domain. The Fourier transformation of the surface displacements with respect to time is defined as

$$U_i^{10}(x_1, x_2, \omega) = \int_{-\infty}^{\infty} u_i^{10}(x_1, x_2, t)e^{-i\omega t}dt \tag{2}$$

with the inverse Fourier transformation given as

$$u_i^{10}(x_1, x_2, t) = \frac{1}{2\pi} \int_{-\infty}^{\infty} U_i^{10}(x_1, x_2, \omega)e^{i\omega t}d\omega. \tag{3}$$

Likewise, a relationship can be established between the surface load $p_i^{10}(x_1, x_2, t)$ and its Fourier transform $P_i^{10}(x_1, x_2, \omega)$, and similar transformation rules apply to the Green's function, i.e. between $g_{ij}(x_1 - y_1, x_2 - y_2, t - \tau)$ and $G_{ij}(x_1 - y_1, x_2 - y_2, \omega)$. It then follows that

$$U_i^{10}(x_1, x_2, \omega) = \int_{-\infty}^{\infty} \int_{-\infty}^{\infty} G_{ij}(x_1 - y_1, x_2 - y_2, \omega)P_j^{10}(y_1, y_2, \omega)dy_1dy_2, \tag{4}$$

reducing the problem to a purely spatial convolution.

Further, assuming that all interfaces are horizontal, a transformation is carried out from the Cartesian space domain description into a horizontal wavenumber domain. This is done by a double Fourier transformation in the form

$$\overline{U}_i^{10}(k_1, k_2, \omega) = \int_{-\infty}^{\infty} \int_{-\infty}^{\infty} U_i^{10}(x_1, x_2, \omega)e^{-i(k_1x_1 + k_2x_2)}dx_1dx_2, \tag{5}$$

where the double inverse Fourier transformation is defined by

$$U_i^{10}(x_1, x_2, \omega) = \frac{1}{4\pi^2} \int_{-\infty}^{\infty} \int_{-\infty}^{\infty} \overline{U}_i^{10}(k_1, k_2, \omega)e^{i(k_1x_1 + k_2x_2)}dk_1dk_2. \tag{6}$$

By a similar transformation of the surface traction and the Green's function, Eq. (4) finally achieves the form

$$\overline{U}_i^{10}(k_1, k_2, \omega) = \overline{G}_{ij}(k_1, k_2, \omega)\overline{P}_j^{10}(k_1, k_2, \omega). \tag{7}$$

This equation has the advantage when compared to the previous formulation in space and time domain, that no convolution has to be carried out. Thus, the displacement amplitudes in the frequency–wavenumber domain are related directly to the traction amplitudes for a given set of the circular freqeuncy ω and the horizontal wavenumbers k_1 and k_2 via the Green's function tensor $\overline{G}_{ij}(k_1, k_2, \omega)$. When the load in the time domain varies harmonically in the form $p_i^{10}(x_1, x_2, t) = P_i(x_1, x_2)e^{i\omega t}$, the solution simplifies, since no inverse Fourier transformation over the frequency is necessary. $\overline{G}_{ij}(k_1, k_2, \omega)$ must only be evaluated at a single frequency.

The main advantage of the description in the frequency–horizontal wavenumber domain is that a solution for the stratum may be found analytically. In the following subsections, the derivation of $\overline{G}_{ij}(k_1, k_2, \omega)$ is described. As mentioned above, the derivation is based on the assumption that the material within each individual layer is linear elastic, homogeneous and isotropic. Further, material dissipation is confined to hysteretic damping, which has been found to be a reasonably accurate model for materials such as soil, even if the model is invalid from a physical point of view.

2.2 Flexibility matrix for a single soil layer

The stratum consists of J horizontally bounded layers, each defined by the Young's modulus E^j, the Poisson ratio ν^j, the mass density ρ^j and the loss factor η^j. Further, the layers have the depths h^j, $j = 1, 2, ..., J$. Thus, the equations of motion for each layer may advantageously be established in a coordinate system with the local x_3-coordinate x_3^j defined with the positive direction downwards so that $x_3^j \in [0, h^j]$, see Fig. 3.

2.2.1 Boundary conditions for displacements and stresses at an interface

In the frequency domain, and in terms of the horizontal wavenumbers, the displacements at the top and at the bottom of the jth layer are given, respectively, as

$$\overline{U}_i^{j0}(k_1, k_2, \omega) = \overline{U}_i(k_1, k_2, x_3^j = 0, \omega), \qquad \overline{U}_i^{j1}(k_1, k_2, \omega) = \overline{U}_i(k_1, k_2, x_3^j = h^j, \omega). \tag{8}$$

The meaning of the double superscript 10 applied in the definition of the flexibility or Green's function in the previous section now becomes somewhat clearer. Thus $\overline{U}_i^{10}$ are the displacement components at the top of the uppermost layer which coincides with the surface of the half-space. The remaining layers are counted downwards with $j = J$ referring to the bottommost layer. If an underlying half-space is present, its material properties are identified by index $j = J + 1$.

Similar to Eq. (8) for the displacements, the traction at the top and bottom of layer j are

$$\overline{P}_i^{j0}(k_1, k_2, \omega) = \overline{P}_i(k_1, k_2, x_3^j = 0, \omega), \qquad \overline{P}_i^{j1}(k_1, k_2, \omega) = \overline{P}_i(k_1, k_2, x_3^j = h^j, \omega). \tag{9}$$

The quantities defined in Eqs. (8) and (9) may advantageously be stored in vector form as

$$\overline{\mathbf{S}}^{j0} = \begin{bmatrix} \overline{\mathbf{U}}^{j0} \\ \overline{\mathbf{P}}^{j0} \end{bmatrix}, \qquad \overline{\mathbf{S}}^{j1} = \begin{bmatrix} \overline{\mathbf{U}}^{j1} \\ \overline{\mathbf{P}}^{j1} \end{bmatrix}, \tag{10}$$

where $\overline{\mathbf{U}}^{j0} = \overline{\mathbf{U}}^{j0}(k_1, k_2, \omega)$ is the column vector with the components $\overline{U}_i^{j0}$, $i = 1, 2, 3$, etcetera.

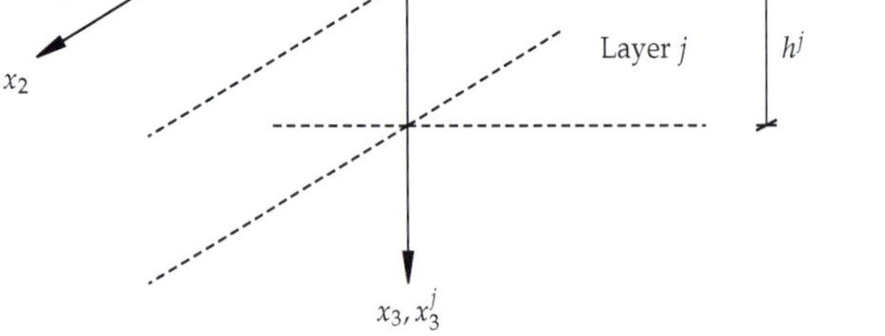

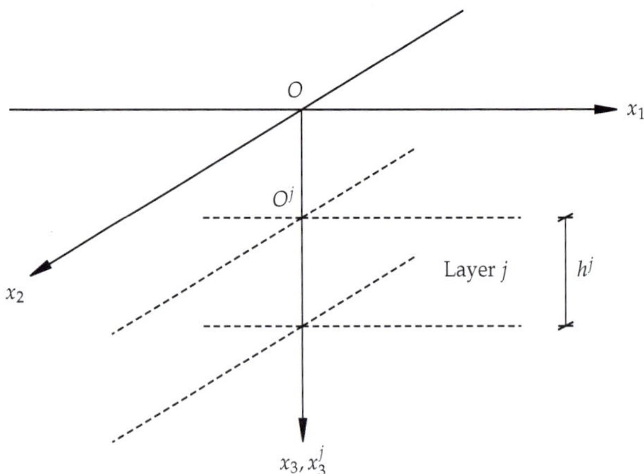

Fig. 3. Global and local coordinates for layer j with the depth h^j. The (x_1, x_2, x_3)-coordinate system has the origin O, whereas the local (x_1, x_2, x_3^j)-coordinate system has the origin O^j.

2.2.2 Governing equations for wave propagation in a soil layer

In the time domain, and in terms of Cartesian coordinates, the equations of motion for the layer are given in terms of the Cauchy equations, which in the absence of body forces read

$$\frac{\partial}{\partial x_k}\sigma_{ik}^j(x_1, x_2, x_3^j, t) = \rho^j \frac{\partial^2}{\partial t^2}u_i^j(x_1, x_2, x_3^j, t), \tag{11}$$

where $\sigma_{ik}^j(x_1, x_2, x_3^j, t)$ is the Cauchy stress tensor. On any part of the boundary, i.e. on the top and bottom of the layer, Dirichlet or Neumann conditions apply as defined by Eqs. (8) and (9), respectively. Initial conditions are of no interest in the present case, since the steady state solution is to be found.

Assuming hysteretic material dissipation defined by the loss factor η^j, the dynamic stiffness of the homogeneous and isotropic material may conveniently be described in terms of complex Lamé constants defined as

$$\lambda^j = \frac{\nu^j E^j \left(1 + i\,\mathrm{sign}(\omega)\eta^j\right)}{\left(1 + \nu^j\right)\left(1 - 2\nu^j\right)}, \qquad \mu^j = \frac{E^j\left(1 + i\,\mathrm{sign}(\omega)\eta^j\right)}{2\left(1 + \nu^j\right)}. \tag{12}$$

The sign function ensures that the material damping is positive in the entire frequency range $\omega \in [-\infty; \infty]$ involved in the inverse Fourier transformation (3).

Subsequently, the stress amplitudes $\hat{\sigma}_{ik}^j(x_1, x_2, x_3^j, \omega)$ may be expressed in terms of the dilation amplitudes $\hat{\Delta}^j(x_1, x_2, x_3^j, \omega)$, and the infinitesimal strain tensor amplitudes $\hat{\varepsilon}_{ik}^j(x_1, x_2, x_3^j, \omega)$,

$$\hat{\sigma}_{ik}^j(x_1, x_2, x_3^j, \omega) = \lambda^j \hat{\Delta}^j(x_1, x_2, x_3^j, \omega)\delta_{ik} + 2\mu^j \hat{\varepsilon}_{ik}^j(x_1, x_2, x_3^j, \omega), \tag{13}$$

where δ_{ij} is the Kronecker delta; $\delta_{ij} = 1$ for $i = j$ and $\delta_{ij} = 0$ for $i \neq j$. Further, the following definitions apply:

$$\hat{\Delta}^j(x_1, x_2, x_3^j, \omega) = \frac{\partial}{\partial x_k}U_k^j(x_1, x_2, x_3^j, \omega), \tag{14}$$

$$\hat{\varepsilon}_{ik}^j(x_1, x_2, x_3^j, \omega) = \frac{1}{2}\left(\frac{\partial}{\partial x_i}U_k^j(x_1, x_2, x_3^j, \omega) + \frac{\partial}{\partial x_k}U_i^j(x_1, x_2, x_3^j, \omega)\right). \tag{15}$$

It is noted that $\partial/\partial x_3^j = \partial/\partial x_3$, since the local x_3^j-axes have the same positive direction as the global x_3-axis.

Inserting Eqs. (12) to (15) into the Fourier transformation of the Cauchy equation given by Eq. (11), the Navier equations in the frequency domain are achieved:

$$\left(\lambda^j + \mu^j\right)\frac{\partial \hat{\Delta}^j}{\partial x_i} + \mu^j \frac{\partial^2 U_i^j}{\partial x_k \partial x_k} = -\omega^2 \rho^j U_i^j. \tag{16}$$

Applying the double Fourier transformation over the horizontal Cartesian coordinates as defined by Eq. (5), the Navier equations in the frequency–wavenumber domain become

$$\left(\lambda^j + \mu^j\right)ik_i\overline{\Delta}^j + \mu^j\left(\frac{d^2}{dx_3^2} - k_1^2 - k_2^2\right)\overline{U}_i^j = -\omega^2 \rho^j \overline{U}_i^j, \qquad i = 1,2, \tag{17a}$$

$$\left(\lambda^j + \mu^j\right)\frac{d\overline{\Delta}^j}{dx_3} + \mu^j\left(\frac{d^2}{dx_3^2} - k_1^2 - k_2^2\right)\overline{U}_3^j = -\omega^2 \rho^j \overline{U}_3^j, \tag{17b}$$

where $\overline{\Delta}^j = \overline{\Delta}^j(k_1, k_2, x_3^j, \omega)$ is the double Fourier transform of $\widehat{\Delta}^j(x_1, x_2, x_3^j, \omega)$ with respect to the horizontal Cartesian coordinates x_1 and x_2. Obviously,

$$\overline{\Delta}^j(k_1, k_2, x_3^j, \omega) = ik_1\overline{U}_1^j(k_1, k_2, x_3^j, \omega) + ik_2\overline{U}_2^j(k_1, k_2, x_3^j, \omega) + \frac{d\overline{U}_3^j(k_1, k_2, x_3^j, \omega)}{dx_3}. \tag{18}$$

Equations (17a) and (17b) are ordinary differential equations in x_3. When the boundary values at the top and the bottom of the layer expressed in Eqs. (8) and (9) are known, an analytical solution may be found as will be discussed below.

2.2.3 The solution for compression waves in a soil layer

The phase velocities of compression and shear waves, or P- and S waves, are identified as

$$c_P = \sqrt{\frac{\lambda^j + 2\mu^j}{\rho^j}}, \qquad c_S = \sqrt{\frac{\mu^j}{\rho^j}}, \tag{19}$$

respectively. It is noted that the phase velocities are complex when material damping is present. Further, in the frequency domain, the P- and S-waves in layer j are associated with the wavenumbers k_P^j and k_S^j,

$$\{k_P^j\}^2 = \frac{\omega^2}{\{c_P^j\}^2}, \qquad \{k_S^j\}^2 = \frac{\omega^2}{\{c_S^j\}^2}. \tag{20}$$

Introducing the parameters α_P^j and α_S^j as the larger of the roots to

$$\{\alpha_P^j\}^2 = k_1^2 + k_2^2 - \{k_P^j\}^2, \qquad \{\alpha_S^j\}^2 = k_1^2 + k_2^2 - \{k_S^j\}^2, \tag{21}$$

Eqs. (17a) and (17b) may conveniently be recast as

$$\left(\lambda^j + \mu^j\right) ik_i\overline{\Delta}^j + \mu^j\left(\frac{d^2 U_i^j}{dx_3^2} - \{\alpha_S^j\}^2 U_i^j\right) = 0, \qquad i = 1, 2, \tag{22a}$$

$$\left(\lambda^j + \mu^j\right)\frac{d\overline{\Delta}^j}{dx_3} + \mu^j\left(\frac{d^2 U_3^j}{dx_3^2} - \{\alpha_S^j\}^2 U_3^j\right) = 0. \tag{22b}$$

Equation (22a) is now multiplied with ik_i and Eq. (22b) is differentiated with respect to x_3. Adding the three resulting equations and making use of Eq. (18), an equation for the dilation is obtained in the form

$$\left(\lambda^j + \mu^j\right)\left(\frac{d^2}{dx_3^2} - k_1^2 - k_2^2\right)\overline{\Delta}^j + \mu^j\left(\frac{d^2}{dx_3^2} - \{\alpha_S^j\}^2\right)\overline{\Delta}^j = 0 \qquad \Rightarrow$$

$$\left(\lambda^j + 2\mu^j\right)\left(\frac{d^2}{dx_3^2} - k_1^2 - k_2^2\right)\overline{\Delta}^j + \mu^j\left(k_1^2 + k_2^2 - \{\alpha_S^j\}^2\right)\overline{\Delta}^j = 0 \qquad \Rightarrow$$

$$\left(\lambda^j + 2\mu^j\right)\left(\frac{d^2}{dx_3^2} - k_1^2 - k_2^2\right)\overline{\Delta}^j + \mu^j\{k_S^j\}^2\overline{\Delta}^j = 0. \tag{23}$$

The last derivation follows from Eq. (21). Further, Eqs. (19) and (20) involve that

$$\mu^j \{k_S^j\}^2 = \left(\lambda^j + 2\mu^j \right) \{k_P^j\}^2. \tag{24}$$

Inserting this result into Eq. (23), and once again making use of Eq. (21), we finally arrive at the ordinary homogenous differential equation

$$\frac{d^2 \overline{\Delta}^j}{dx_3^2} - \{\alpha_P^j\}^2 \overline{\Delta}^j = 0, \tag{25}$$

which has the full solution

$$\overline{\Delta}^j = a_1^j e^{\alpha_P^j x_3^j} + a_2^j e^{-\alpha_P^j x_3^j}. \tag{26}$$

Here a_1^j and a_2^j are integration constants that follow from the boundary conditions. Physically, the two parts of the solution (26) describe the decay of P-waves travelling in the negative and positive x_3-direction, respectively, i.e. P-waves moving up and down in the layer.

2.2.4 The solution for compression and shear waves in a soil layer
Insertion of the solution (26) into Eqs. (22a) and (22b) leads to three equations for the displacement amplitudes:

$$\frac{d^2 \overline{U}_i^j}{dx_3^2} - \{\alpha_S^j\}^2 \overline{U}_i^j = - \left(\frac{\lambda^j}{\mu^j} + 1 \right) ik_i \left(a_1^j e^{\alpha_P^j x_3^j} + a_2^j e^{-\alpha_P^j x_3^j} \right), \qquad i = 1, 2, \tag{27a}$$

$$\frac{d^2 \overline{U}_3^j}{dx_3^2} - \{\alpha_S^j\}^2 \overline{U}_3^j = - \left(\frac{\lambda^j}{\mu^j} + 1 \right) \alpha_P^j \left(a_1^j e^{\alpha_P^j x_3^j} - a_2^j e^{-\alpha_P^j x_3^j} \right). \tag{27b}$$

Solutions to Eqs. (27a) and (27b) are found in the form

$$\overline{U}_1^j = \overline{U}_{1,c}^j + \overline{U}_{1,p}^j = b_1^j e^{\alpha_S^j x_3^j} + b_2^j e^{-\alpha_S^j x_3^j} + b_3^j e^{\alpha_P^j x_3^j} + b_4^j e^{-\alpha_P^j x_3^j}, \tag{28a}$$

$$\overline{U}_2^j = \overline{U}_{2,c}^j + \overline{U}_{2,p}^j = c_1^j e^{\alpha_S^j x_3^j} + c_2^j e^{-\alpha_S^j x_3^j} + c_3^j e^{\alpha_P^j x_3^j} + c_4^j e^{-\alpha_P^j x_3^j}, \tag{28b}$$

$$\overline{U}_3^j = \overline{U}_{3,c}^j + \overline{U}_{3,p}^j = d_1^j e^{\alpha_S^j x_3^j} + d_2^j e^{-\alpha_S^j x_3^j} + d_3^j e^{\alpha_P^j x_3^j} + d_4^j e^{-\alpha_P^j x_3^j}, \tag{28c}$$

where the subscripts c and p denote the complimentary and the particular solutions, respectively. These include S- and P-wave terms, respectively. Like a_1^j and a_2^j, c_1^j, c_2^j, etc. are integration constants given by the boundary conditions at the top and the bottom of layer j. Apparently, the full solution has fourteen integration constants. However, a comparison of Eqs. (18) and (26) reveals that

$$\overline{\Delta}^j(k_1, k_2, x_3^j, \omega) = ik_1 \overline{U}_1^j + ik_2 \overline{U}_2^j + \frac{d\overline{U}_3^j}{dx_3} = a_1^j e^{\alpha_P^j x_3^j} + a_2^j e^{-\alpha_P^j x_3^j}. \tag{29}$$

By insertion of the complementary solutions, i.e. the first two terms in Eqs. (28a) to (28c), into Eq. (29) it immediately follows that

$$d_1^j = - \left(\frac{ik_1}{\alpha_S^j} b_1^j + \frac{ik_2}{\alpha_S^j} c_1^j \right), \qquad d_2^j = \frac{ik_1}{\alpha_S^j} b_2^j + \frac{ik_2}{\alpha_S^j} c_2^j. \tag{30}$$

124   Fundamental and Advanced Topics in Wind Power

functions of different powers are orthogonal. A further reduction of the number of integration constants is achieved by insertion of the particular solutions into the respective differential equations (27a) and (27b). Thus, after a few manipulations it may be shown that

$$b_3^j = -\frac{ik_1}{\{k_P^j\}^2}\,a_1^j, \qquad c_3^j = -\frac{ik_2}{\{k_P^j\}^2}\,a_1^j, \qquad d_3^j = -\frac{\alpha_P^j}{\{k_P^j\}^2}\,a_1^j, \tag{31a}$$

$$b_4^j = -\frac{ik_1}{\{k_P^j\}^2}\,a_2^j, \qquad c_4^j = -\frac{ik_2}{\{k_P^j\}^2}\,a_2^j, \qquad d_4^j = +\frac{\alpha_P^j}{\{k_P^j\}^2}\,a_2^j, \tag{31b}$$

where use has been made of the fact that

$$\frac{\lambda^j + \mu^j}{\mu^j\left(\{\alpha_S^j\}^2 - \{\alpha_P^j\}^2\right)} - \frac{\{c_P^j\}^2 - \{c_S^j\}^2}{\{c_S^j\}^2\left(\{k_P^j\}^2 - \{k_S^j\}^2\right)} = \frac{\{k_S^j\}^2 - \{k_P^j\}^2}{\{k_P^j\}^2\left(\{k_P^j\}^2 - \{k_S^j\}^2\right)} = -\frac{1}{\{k_P^j\}^2},$$

which follows from the definitions given in Eqs. (19) to (21). Thus, eventually only six of the original fourteen integration constants are independent, namely a_1^j, a_2^j, b_1^j, b_2^j, c_1^j and c_2^j. As already mentioned, the terms including a_1^j and a_2^j represent P-waves moving up and down in layer j. Inspection of Eqs. (28a) to (28c) reveals that the b_1^j and b_2^j terms represent S-waves that are polarized in the x_1-direction and which are moving up and down in the layer, respectively. Similarly, the c_1^j and c_2^j terms describe the contributions from S-waves polarized in the x_2-direction and travelling up and down in the layer, respectively. It becomes evident that the previously defined quantities α_P^j and α_S^j may be interpreted as exponential decay coefficients of P- and S-waves, respectively. When k_1 and k_2 are both small, α_P^j and α_S^j turn into "wavenumbers", as they become imaginary, cf. Eq. (21).

Once the displacement field is known, the stress components on any plane orthogonal to the x_3^j-axis may be found from Eq. (13) by letting index $k = 3$. The full solution for displacements, $\overline{\mathbf{U}}^j$, and traction, $\overline{\mathbf{P}}^j$, may then be written in matrix form as

$$\overline{\mathbf{S}}^j = \begin{bmatrix} \overline{\mathbf{U}}^j \\ \overline{\mathbf{P}}^j \end{bmatrix} = \mathbf{A}^j \mathbf{E}^j \mathbf{b}^j, \qquad \mathbf{b}^j = \begin{bmatrix} a_1^j & b_1^j & c_1^j & a_2^j & b_2^j & c_2^j \end{bmatrix}^T, \tag{32}$$

where $\mathbf{E}^j$ is a matrix of dimension (6×6). Only the diagonal terms

$$E_{11}^j = e^{\alpha_P^j x_3^j}, \quad E_{22}^j = E_{33}^j = e^{\alpha_S^j x_3^j}, \quad E_{44}^j = e^{-\alpha_P^j x_3^j}, \quad E_{55}^j = E_{66}^j = e^{-\alpha_S^j x_3^j}, \tag{33}$$

are nonzero. $\mathbf{A}^j$ is a matrix of dimension (6×6), the components of which follow from Eqs. (28) to (31) and (13). The computation of matrix $\mathbf{A}^j$ is further discussed below. Finally, the displacements and the traction at the two boundaries of layer j may be expressed as

$$\overline{\mathbf{S}}^{j0} = \mathbf{A}^{j0} \mathbf{b}^j, \qquad\qquad \mathbf{A}^{j0} = \mathbf{A}^j, \tag{34a}$$

$$\overline{\mathbf{S}}^{j1} = e^{\alpha_P^j h^j} \mathbf{A}^{j1} \mathbf{b}^j, \qquad \mathbf{A}^{j1} = \mathbf{A}^{j0} \mathbf{D}^j. \tag{34b}$$

Here $\mathbf{D}^j$ a (6×6) matrix with the nonzero components

$$D_{11}^j = 1, \quad D_{22}^j = D_{33}^j = e^{(\alpha_S^j - \alpha_P^j)h^j}, \quad D_{44}^j = e^{-2\alpha_P^j h^j}, \quad D_{55}^j = D_{66}^j = e^{-(\alpha_P^j + \alpha_S^j)h^j}, \tag{35}$$

found by evaluation of the matrix $\mathrm{e}^{-\alpha_p^j x_3^j}\mathbf{E}^j$ at $x_3^j = h^j$. Equations (34a) and (34b) may be combined in order to eliminate vector $\mathbf{b}^j$ which contains unknown integration constants. This provides a transfer matrix for the layer as proposed by Thomson (1950) and Haskell (1953),

$$\overline{\mathbf{S}}^{j1} = \mathrm{e}^{\alpha_p^j h^j}\mathbf{A}^{j1}[\mathbf{A}^{j0}]^{-1}\overline{\mathbf{S}}^{j0}, \tag{36}$$

forming a relationship between the displacements and the traction at the top and the bottom of a single layer.

The derivation of Eq. (36) has been based on the assumption that $\omega > 0$. When a static load is applied, the circular frequency is $\omega = 0$, whereby the wavenumbers of the P- and S-waves, i.e. k_P^j and k_S^j defined by Eq. (20), become zero and the integration constants b_3^j etc. given in Eq. (31) are undefined. Hence, the solution given in the previous section does not apply in the static case. However, for any practical purposes a useful approximation can be established for the static case by employing a low value of ω in the evaluation of $\overline{\mathbf{S}}^{j1}$.

2.3 Assembly of multiple layers

At an interface between two layers, the displacements should be continuous and there should be equilibrium of the traction. This may be expressed as $\overline{\mathbf{S}}^{j0} = \overline{\mathbf{S}}^{j-1,1}$, $j = 2, 3, ..., J$, i.e. the quantities at the top of layer j are equal to those at the bottom of layer $j - 1$. Proceeding in this manner, Eq. (36) for the single layer may be rewritten for a system of J layers,

$$\overline{\mathbf{S}}^{J1} = \mathrm{e}^{\Sigma\alpha}\mathbf{A}^{J1}[\mathbf{A}^{J0}]^{-1}\mathbf{A}^{J-1,1}[\mathbf{A}^{J-1,0}]^{-1}\cdots\mathbf{A}^{11}[\mathbf{A}^{10}]^{-1}\overline{\mathbf{S}}^{10}, \qquad \Sigma\alpha = \sum_{j=1}^{J}\alpha_p^j h^j. \tag{37}$$

Introducing the transfer matrix $\mathbf{T}$ defined as

$$\mathbf{T} = \begin{bmatrix} \mathbf{T}_{11} & \mathbf{T}_{12} \\ \mathbf{T}_{21} & \mathbf{T}_{22} \end{bmatrix} = \mathbf{A}^{J1}[\mathbf{A}^{J0}]^{-1}\mathbf{A}^{J-1,1}[\mathbf{A}^{J-1,0}]^{-1}\cdots\mathbf{A}^{11}[\mathbf{A}^{10}]^{-1}, \tag{38}$$

Equation (37) may in turn be written as $\overline{\mathbf{S}}^{J1} = \mathrm{e}^{\Sigma\alpha}\mathbf{T}\overline{\mathbf{S}}^{10}$, or

$$\begin{bmatrix} \overline{\mathbf{U}}^{J1} \\ \overline{\mathbf{P}}^{J1} \end{bmatrix} = \mathrm{e}^{\Sigma\alpha}\begin{bmatrix} \mathbf{T}_{11} & \mathbf{T}_{12} \\ \mathbf{T}_{21} & \mathbf{T}_{22} \end{bmatrix}\begin{bmatrix} \overline{\mathbf{U}}^{10} \\ \overline{\mathbf{P}}^{10} \end{bmatrix}, \qquad \Sigma\alpha = \sum_{j=1}^{J}\alpha_p^j h^j. \tag{39}$$

This establishes a relationship between the traction and the displacements at the free surface of the half-space and the equivalent quantities at the bottom of the stratum as originally proposed by Thomson (1950) and Haskell (1953).

2.4 Flexibility of a homogeneous or stratified ground

A stratified ground consisting of multiple soil layers may overlay bedrock. On the surface of the bedrock, the displacements are identically equal to zero and thus, by insertion into Eq. (39),

$$\begin{bmatrix} \overline{\mathbf{U}}^{J1} \\ \overline{\mathbf{P}}^{J1} \end{bmatrix} = \begin{bmatrix} \mathbf{0} \\ \overline{\mathbf{P}}^{J1} \end{bmatrix} = \mathrm{e}^{\Sigma\alpha}\begin{bmatrix} \mathbf{T}_{11} & \mathbf{T}_{12} \\ \mathbf{T}_{21} & \mathbf{T}_{22} \end{bmatrix}\begin{bmatrix} \overline{\mathbf{U}}^{10} \\ \overline{\mathbf{P}}^{10} \end{bmatrix}. \tag{40}$$

The first three rows of this matrix equation provide the identity

$$\overline{\mathbf{U}}^{10} = \overline{\mathbf{G}}_{\mathrm{rf}}\overline{\mathbf{P}}^{10}, \qquad \overline{\mathbf{G}}_{\mathrm{rf}} = -\mathbf{T}_{11}^{-1}\mathbf{T}_{12}. \tag{41}$$

$\overline{\mathbf{G}}_{\mathrm{rf}} = \overline{\mathbf{G}}_{\mathrm{rf}}(k_1, k_2, \omega)$ is the flexibility matrix for a stratum over a rigid bedrock. It is observed that the exponential function of the power $\Sigma\alpha$, defined in Eq. (39), vanishes in the formulation provided by Eq. (41). This is a great advantage from a computational point of view, since $e^{\Sigma\alpha}$ becomes very large for strata of great depths, which may lead to problems on a computer—even when double precision complex variables are employed.

Alternatively to a rigid bedrock, a half-space may be present underneath the stratum consisting of J layers. In this context, the material properties etc. of the half-space will be assigned the superscript $J+1$. The main difference between a semi-infinite half-space and a layer of finite depth is that only an upper boundary is present, i.e. the boundary situated at $x_3^{J+1} = 0$. Since the material is assumed to be homogeneous, no reflection of waves will take place inside the half-space. Further assuming that no sources are present in the interior of the half-space, only outgoing, i.e. downwards propagating, waves can be present. Dividing the matrices $\mathbf{A}^j$ and $\mathbf{E}^j$ for a layer of finite depth, cf. Eq. (32), into four quadrants, and the column vector $\mathbf{b}^j$ into two sub-vectors,

$$\mathbf{A}^j = \begin{bmatrix} \mathbf{A}^j_{11} & \mathbf{A}^j_{12} \\ \mathbf{A}^j_{21} & \mathbf{A}^j_{22} \end{bmatrix}, \qquad \mathbf{E}^j = \begin{bmatrix} \mathbf{E}^j_{11} & \mathbf{E}^j_{12} \\ \mathbf{E}^j_{21} & \mathbf{E}^j_{22} \end{bmatrix}, \qquad \mathbf{b}^j = \begin{bmatrix} \mathbf{b}^j_1 \\ \mathbf{b}^j_2 \end{bmatrix}, \tag{42}$$

it is evident that only half of the solution applies to the half-space, i.e.

$$\overline{\mathbf{s}}^{J+1} = \begin{bmatrix} \overline{\mathbf{U}}^{J+1} \\ \overline{\mathbf{P}}^{J+1} \end{bmatrix} = \begin{bmatrix} \mathbf{A}^{J+1}_{12} \\ \mathbf{A}^{J+1}_{22} \end{bmatrix} \mathbf{E}^{J+1}_{22} \mathbf{b}^{J+1}_2, \qquad \mathbf{b}^{J+1}_2 = \begin{bmatrix} a^{J+1}_2 & b^{J+1}_2 & c^{J+1}_2 \end{bmatrix}^T. \tag{43}$$

The terms including the integration constants a^{J+1}_1, b^{J+1}_1 and c^{J+1}_1 are physically invalid as they correspond to waves incoming from $x_3^{J+1} = \infty$, i.e. from infinite depth.

From Eq. (43), the traction on the interface between the bottommost layer and the half-space may be expressed in terms of the corresponding displacements by solution of

$$\overline{\mathbf{U}}^{J+1} = \mathbf{A}^{J+1}_{12} [\mathbf{A}^{J+1}_{22}]^{-1} \overline{\mathbf{P}}^{J+1}. \tag{44}$$

The matrix $\mathbf{E}^{J+1}_{22}$ reduces to the identity matrix of order 3, since all the exponential terms are equal to 1 for $x_3^{J+1} = 0$.

Firstly, if no layers are present in the model of the stratum, $J = 0$ and it immediately follows from Eq. (44) that Eq. (7), written in matrix form, becomes

$$\overline{\mathbf{U}}^{10} = \overline{\mathbf{G}}_{\mathrm{hh}} \overline{\mathbf{P}}^{10}, \qquad \overline{\mathbf{G}}_{\mathrm{hh}} = \mathbf{A}^{10}_{12} [\mathbf{A}^{10}_{22}]^{-1}, \tag{45}$$

where it is noted that the flexibility matrix for the homogeneous half-space $\overline{\mathbf{G}}_{\mathrm{hh}} = \overline{\mathbf{G}}_{\mathrm{hh}}(k_1, k_2, \omega)$ is given in the horizontal wavenumber–frequency domain.

Secondly, when J layers overlay a homogeneous half-space, continuity of the displacements, equilibrium of the traction and application of Eq. (44) provide

$$\overline{\mathbf{U}}^{J1} = \overline{\mathbf{U}}^{J+1,0} = \mathbf{A}^{J+1}_{12} [\mathbf{A}^{J+1}_{22}]^{-1} \overline{\mathbf{P}}^{J+1,0} = \mathbf{A}^{J+1}_{12} [\mathbf{A}^{J+1}_{22}]^{-1} \overline{\mathbf{P}}^{J1}. \tag{46}$$

Insertion of this result into Eq. (39) leads to the following system of equations:

$$\begin{bmatrix} \overline{\mathbf{U}}^{J1} \\ \overline{\mathbf{P}}^{J1} \end{bmatrix} = \begin{bmatrix} \mathbf{A}^{J+1}_{12} [\mathbf{A}^{J+1}_{22}]^{-1} \overline{\mathbf{P}}^{J1} \\ \overline{\mathbf{P}}^{J1} \end{bmatrix} = e^{\Sigma\alpha} \begin{bmatrix} \mathbf{T}_{11} & \mathbf{T}_{12} \\ \mathbf{T}_{21} & \mathbf{T}_{22} \end{bmatrix} \begin{bmatrix} \overline{\mathbf{U}}^{10} \\ \overline{\mathbf{P}}^{10} \end{bmatrix}. \tag{47}$$

From the bottommost three rows of the matrix equation, an expression of $\overline{\mathbf{P}}^{J1}$ is obtained which may be inserted into the first three equations. This leads to the solution

$$\overline{\mathbf{U}}^{10} = \overline{\mathbf{G}}_{1h}\overline{\mathbf{P}}^{10}, \tag{48}$$

where the flexibility matrix for the layered half-space $\overline{\mathbf{G}}_{1h} = \overline{\mathbf{G}}_{1h}(k_1, k_2, \omega)$ is given by

$$\overline{\mathbf{G}}_{1h} = \left(\mathbf{A}_{12}^{J+1}[\mathbf{A}_{22}^{J+1}]^{-1}\mathbf{T}_{21} - \mathbf{T}_{11}\right)^{-1}\left(\mathbf{T}_{12} - \mathbf{A}_{12}^{J+1}[\mathbf{A}_{22}^{J+1}]^{-1}\mathbf{T}_{22}\right). \tag{49}$$

Again the exponential function disappears. In the following, no distinction is made between $\overline{\mathbf{G}}_{rf}$, $\overline{\mathbf{G}}_{hh}$ and $\overline{\mathbf{G}}_{1h}$. The common notation $\overline{\mathbf{G}}$ will be employed, independent of the type of subsoil model.

2.5 Optimising the numerical evaluation of the Green's function

In order to obtain a solution in Cartesian space, a double inverse Fourier transformation over the horizontal wavenumbers is necessary as outlined by Eq. (6). A direct approach involves the evaluation of $\overline{\mathbf{G}}$ for numerous combinations of k_1 and k_2, leading to long computation times. However, as described in this subsection, a considerable reduction of the computation time can be achieved.

2.5.1 Computation of the matrices $\mathbf{A}^{j0}$ and $\mathbf{A}^{j1}$

The computation of the transfer matrix $\mathbf{T}$ involves inversion of the matrices $\mathbf{A}^{j0}$, $j = 1, 2, ..., J$. Further, the flexibility matrix $\overline{\mathbf{G}}(k_1, k_2, \omega)$ has to be evaluated for all combinations (k_1, k_2) before the transformation given by Eq. (6) may be applied. However, as pointed out by Sheng et al. (1999), the evaluation of $\mathbf{A}^j$, and therefore also the Green's function matrix $\overline{\mathbf{G}}$, is particularly simple along the line defined by $k_1 = 0$. To take advantage of this, a coordinate transformation is introduced in the form

$$\begin{bmatrix} k_1 \\ k_2 \\ x_3 \end{bmatrix} = \mathbf{R}(\varphi)\begin{bmatrix} \gamma \\ \alpha \\ x_3 \end{bmatrix}, \qquad \mathbf{R}(\varphi) = \begin{bmatrix} \sin\varphi & \cos\varphi & 0 \\ -\cos\varphi & \sin\varphi & 0 \\ 0 & 0 & 1 \end{bmatrix}. \tag{50}$$

This corresponds to a rotation of (k_1, k_2, x_3)-basis by the angle $\varphi - \pi/2$ around the x_3-axis as illustrated in Fig. 4. It follows from Eq. (50) that $R_{ij}(\varphi) = R_{ji}(\pi - \varphi)$, which in matrix–vector notation corresponds to $\{\mathbf{R}(\varphi)\}^T = \mathbf{R}(\pi - \varphi)$.

For any combination of k_1 and k_2, the angle φ is now defined so that $\gamma = 0$. The relationship between the coordinates in the two systems of reference is then given by

$$k_1 = \alpha\cos\varphi, \qquad k_2 = \alpha\sin\varphi, \qquad \alpha = \sqrt{k_1^2 + k_2^2}, \qquad \tan\varphi = \frac{k_2}{k_1}, \qquad \gamma = 0. \tag{51}$$

The computational advantage of this particular orientation of the (γ, α, x_3)-coordinate system is twofold. Firstly, the flexibility matrix may be evaluated along a line rather than over an area, and for any other combination of the wavenumbers, the Green's function matrix can be computed as

$$\overline{\mathbf{G}}(k_1, k_2, \omega) = \mathbf{R}(\varphi)\widehat{\overline{\mathbf{G}}}\{\mathbf{R}(\varphi)\}^T \qquad \text{or} \qquad \overline{G}_{ik}(k_1, k_2, \omega) = R_{il}(\varphi)\widehat{\overline{G}}_{lm}R_{km}(\varphi). \tag{52}$$

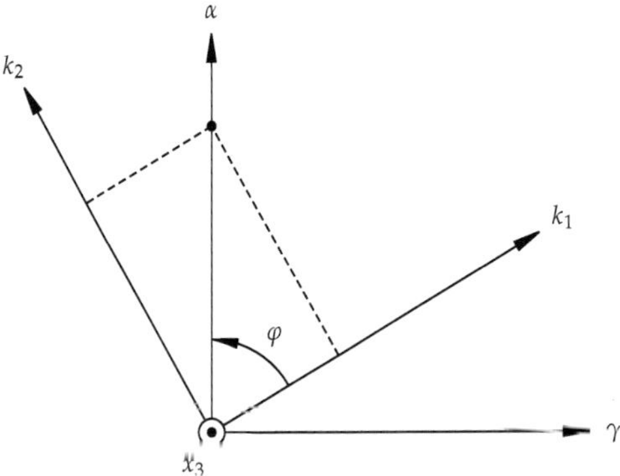

Fig. 4. Definition of the (k_1, k_2, x_3)- and (γ, α, x_3)-coordinate systems.

Here $\widehat{\overline{\mathbf{G}}} = \widehat{\overline{\mathbf{G}}}(\alpha, \omega) = \overline{\mathbf{G}}(0, \alpha, \omega)$. Secondly, the matrices $\mathbf{A}^{j0}$ and $\mathbf{A}^{j1}$—and therefore also $\mathbf{A}_{12}^{J+1}$ and $\mathbf{A}_{22}^{J+1}$—simplify significantly when one of the wavenumbers is equal to zero. Thus, when $k_1 = \gamma = 0$, $k_2 = \alpha$ and $\omega \neq 0$,

$$
\widehat{\mathbf{A}}^{j0} = \mathbf{A}^{j0}(0, \alpha, \omega) =
\begin{bmatrix}
0 & 1 & 0 & 0 & 1 & 0 \\
\widehat{A}_{21}^{j0} & 0 & 1 & \widehat{A}_{21}^{j0} & 0 & 1 \\
\widehat{A}_{31}^{j0} & 0 & \widehat{A}_{33}^{j0} & -\widehat{A}_{31}^{j0} & 0 & -\widehat{A}_{33}^{j0} \\
0 & \widehat{A}_{42}^{j0} & 0 & 0 & -\widehat{A}_{42}^{j0} & 0 \\
\widehat{A}_{51}^{j0} & 0 & \widehat{A}_{53}^{j0} & -\widehat{A}_{51}^{j0} & 0 & -\widehat{A}_{53}^{j0} \\
\widehat{A}_{61}^{j0} & 0 & \widehat{A}_{63}^{j0} & \widehat{A}_{61}^{j0} & 0 & \widehat{A}_{63}^{j0}
\end{bmatrix},
\tag{53a}
$$

where

$$
\widehat{A}_{21}^{j0} = -i\alpha/\{k_P^j\}^2, \quad \widehat{A}_{31}^{j0} = -\alpha_P^j/\{k_P^j\}^2, \quad \widehat{A}_{33}^{j0} = -i\alpha/\alpha_S^j,
\tag{53b}
$$

$$
\widehat{A}_{42}^{j0} = \alpha_S^j \mu^j, \quad \widehat{A}_{51}^{j0} = -2i\mu^j \alpha_P^j \alpha/\{k_P^j\}^2, \quad \widehat{A}_{53}^{j0} = \mu^j(\alpha^2/\alpha_S^j + \alpha_S^j),
\tag{53c}
$$

$$
\widehat{A}_{61}^{j0} = -\mu^j\left(\{k_S^j\}^2 + 2\{\alpha_S^j\}^2\right)/\{k_P^j\}^2, \quad \widehat{A}_{63}^{j0} = -2i\mu^j\alpha.
\tag{53d}
$$

At the bottom of the layer, the corresponding matrix is evaluated as $\widehat{\mathbf{A}}^{j1} = \widehat{\mathbf{A}}^{j0}\mathbf{D}^j$, where the components of the matrix $\mathbf{D}^j$ are given by Eq. (35). A result of the many zeros in $\widehat{\mathbf{A}}^{j0}$ and $\widehat{\mathbf{A}}_0^{j0}$ is that the matrices can be inverted analytically. This may reduce computation time significantly. The inversion of $\widehat{\mathbf{A}}^{j0}$ and $\widehat{\mathbf{A}}_0^{j0}$ is straightforward and will not be treated further.

Especially, for a homogeneous half-space, possibly underlying a stratum, the matrices $\widehat{\mathbf{A}}_{12}^{J+1,0}$ and $\widehat{\mathbf{A}}_{22}^{J+1,0}$ are readily obtained from the leftmost three columns of $\widehat{\mathbf{A}}^{j0}$, whereas $\widehat{\mathbf{A}}^{j1}$ is obtained as

$$
\widehat{\mathbf{A}}^{j1} = \widehat{\mathbf{A}}^{j0}\mathbf{D}^j
\tag{54}
$$

in accordance with Eq. (34b). Note that $\mathbf{D}^j$ is symmetric in the (k_1, k_2)-plane and that therefore $\widehat{\mathbf{D}}^j = \mathbf{D}^j$. This property follows from the definition of the exponential decay coefficients α_P^j and α_S^j given in Eq. (21), or the definition of α given by Eq. (51), along with the definition of $\mathbf{D}^j$, cf. Eq. (35). In other words it may be stated that α_P^j and α_S^j are invariant to rotation around

centre of the soil–foundation interface, the torsional and vertical displacements are completely decoupled from the remaining degrees of freedom. Thus, the impedance matrix simplifies to

$$
\mathbf{Z}(\omega) =
\begin{bmatrix}
Z_{11} & Z_{12} & 0 & Z_{14} & Z_{15} & 0 \\
Z_{12} & Z_{22} & 0 & Z_{24} & Z_{25} & 0 \\
0 & 0 & Z_{33} & 0 & 0 & 0 \\
Z_{14} & Z_{24} & 0 & Z_{44} & Z_{45} & 0 \\
Z_{15} & Z_{55} & 0 & Z_{45} & Z_{55} & 0 \\
0 & 0 & 0 & 0 & 0 & Z_{66}
\end{bmatrix}.
\tag{73}
$$

A further simplification of $\mathbf{Z}(\omega)$ is obtained if the moment of inertia around a given horizontal axis is invariant to a rotation of the footing around the z-axis. This is the case for the gravitation foundations that are typically utilised for wind turbines, i.e. circular, square, hexagonal and octagonal footings. With reference to Fig. 7, the moments of inertia are $I_{x_1} = I_{x_2} = I_\xi = I_\zeta$, where ζ is an arbitrary horizontal axis. As a result of this, $Z_{11} = Z_{22}$, $Z_{44} = Z_{55}$ and $Z_{15} = -Z_{24}$, and the coupling between sliding in the x_1-direction and rocking in the x_2-direction (and vice versa) vanishes, i.e.

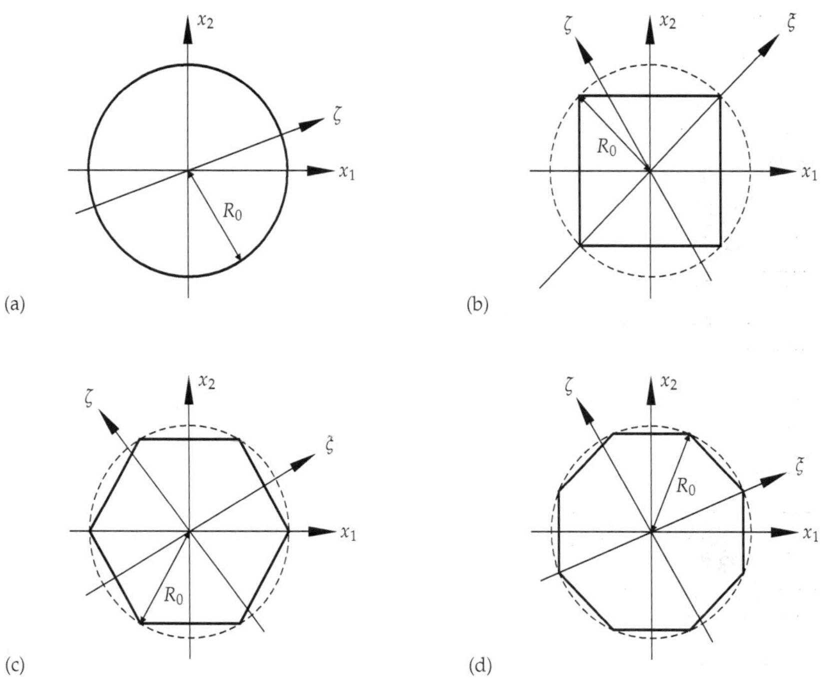

Fig. 7. Definition of axes for different geometries of a footing: (a) circular, (b) square, (c) hexagonal, and (d) octagonal footing. The horizontal plane is considered, and all the footings have the same characteristic length, R_0.

$$\mathbf{Z}(\omega) = \begin{bmatrix} Z_{11} & 0 & 0 & 0 & -Z_{24} & 0 \\ 0 & Z_{22} & 0 & Z_{24} & 0 & 0 \\ 0 & 0 & Z_{33} & 0 & 0 & 0 \\ 0 & Z_{24} & 0 & Z_{44} & 0 & 0 \\ -Z_{24} & 0 & 0 & 0 & Z_{55} & 0 \\ 0 & 0 & 0 & 0 & 0 & Z_{66} \end{bmatrix}. \tag{74}$$

3.1 Evaluation of the dynamic stiffness based on the flexibility of a layered ground

In order to compute the nonzero components of the impedance matrix $\mathbf{Z}(\omega)$, the distribution of the contact stresses at the interface between the footing and the ground due to given rigid body displacements has to be determined. However, Eq. (63) provides the displacement field for a known stress distribution. Generally this implies that the problem takes the form of an integral equation. For the particular case of a circular footing on a homogeneous half-space, Krenk & Schmidt (1981) derived a closed-form solution for the vertical impedance. Yong et al. (1997) proposed that the total contact stress be decomposed into a number of simple distributions obtained by a Fourier series with respect to the azimuthal angle and a polynomial in the radial direction, e.g.

$$P_r^{10}(r, \vartheta, \omega) = \sum_{m=1}^{M} \sum_{n=1}^{N} a_{mn} r^n \cos(m\vartheta) \tag{75}$$

for the component in the r-direction and a symmetric contact stress distribution. Similar expressions were given for the components in the q- (or ϑ-) and x_3-direction and for the antisymmetric case. The response to each of the contact stress distributions can be computed, and the coefficients a_{mn} are determined so that the prescribed rigid body displacements are obtained.

However, for arbitrary shapes of the footing it may be difficult to follow this idea. Hence, in this study a different approach is taken which has the following steps:

1. The displacement corresponding to each rigid body mode is prescribed at N points distributed uniformly at the interface between the footing and the ground.

2. The Green's function matrix is evaluated in the wavenumber domain along the α-axis, and Eq. (65b) is evaluated by application of Eq. (67).

3. The wavenumber spectrum for a simple distributed load with unit magnitude and rotational symmetry around a point on the ground surface is computed. As discussed in Example 2.6.3, a "bell-shaped" load based on a double Gaussian distribution has the advantage that the wavenumber spectrum is a monotonic decreasing function of α.

4. The response at point n to a load centred at point m is calculated for all combinations of $n, m = 1, 2, ..., N$. This provides a flexibility matrix for the footing.

5. The unknown magnitudes of the loads applied around each of the points are computed. Integration over the contact area provides the impedance.

The discretization of the soil–foundation interface into N points and the employment of the "bell-shaped" load distribution are visualized in Fig. 8 for a hexagonal footing.

In particular, if the surface traction vector in the wavenumber–frequency domain takes the form $\overline{P}_i^{10}(k_1, k_2, \omega) = \overline{D}(k_1, k_2)\, \widetilde{P}_i(\omega)$, $i = 1, 2, 3$, where $\overline{D}(k_1, k_2)$ is a stress distribution with unit magnitude and $\widetilde{P}_i(\omega)$ is an amplitude, Eq. (65a) may be computed as

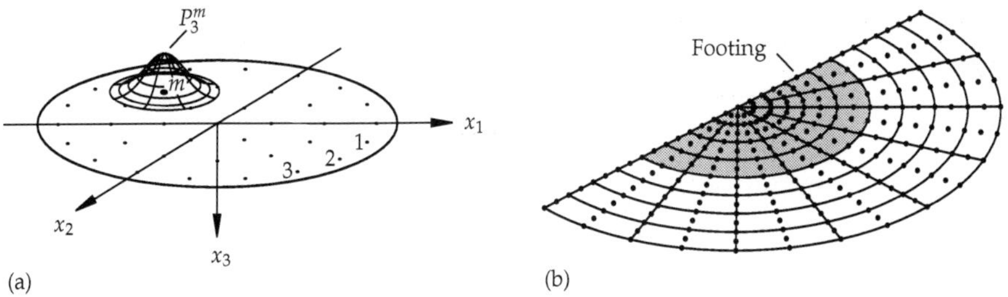

Fig. 8. Discretization of the soil–foundation interface for a hexagonal footing. The vertical component of the "bell-shaped" load at point m is shown.

$$\mathbf{U}^{10} = \mathbf{R}(\theta)\,\widehat{\mathbf{G}}\,[\mathbf{R}(\theta)]^T\,\widetilde{\mathbf{P}}, \qquad \widehat{\mathbf{G}} = \widehat{\mathbf{G}}(r,\omega) = \frac{1}{2\pi}\int_0^\infty \widetilde{\mathbf{G}}\,\widehat{\overline{D}}\,\alpha\,d\alpha. \tag{76}$$

Here it is noted that $\widehat{\overline{D}} = \widehat{\overline{D}}(\alpha) = \overline{D}(0,\alpha)$, since an axisymmetric distribution is assumed. The choice of contact stress distribution and various discretization aspects are discussed below. Alternatively, a boundary element model based on the Green's function for the layered half-space may be employed. However, this involves some additional work, since the Green's function for traction has to be evaluated.

3.2 Discretization considerations

In order to achieve an accurate and efficient computation of the impedance matrix for a footing with the present method, a number of issues need consideration:

1. Equation (76) has to be evaluated numerically. This requires a computation of $\widehat{\overline{\mathbf{G}}}(\alpha,\omega)$ for a number of discrete wavenumbers. All peaks in the wavenumber spectrum must be represented well, demanding a fine discretization in the low wavenumber range—in particular for a half-space with little material damping.

2. No significant contributions may exist from the products $\widehat{\overline{D}}(\alpha)\,\widehat{\overline{G}}_{ij}(\alpha,\omega)$, $i,j = 1,2,3$, for wavenumbers beyond the truncation point in the numerical evaluation of the integral in Eq. (76).

3. Enough points should be employed at the soil–structure interface in order to provide a good approximation of the contact stress distribution.

Concerning item 1 it is of paramount importance to determine the wavenumber below which the wavenumber spectrum may have narrow-banded peaks. Here use can be made of the fact that the longest wave present in a homogeneous half-space is the Rayleigh wave. An approximate upper limit for the Rayleigh wavenumber is provided by the inequality $\alpha_R = \omega/c_R < 1.2\,\omega/c_S$ for $\nu \in [0;0.5]$. For a stratum with J layers overlaying a homogeneous half-space, the idea is now to determine the quantity

$$\alpha_1 = 2\omega/\min\left\{c_S^1, c_S^2, ..., c_S^{J+1}\right\}. \tag{77}$$

where index $J+1$ refers to the underlying homogeneous half-space. In a stratum, waves with wavenumbers higher than α_1 are generally subject to strong material dissipation since they arise from P- or S-waves being reflected multiple times at the interfaces between layers. Only if the loss factor is $\eta^j = 0$ for all layers, undamped Love waves may exist; but this situation is not likely to appear in real soils where typical values are $\eta^j \approx 0.01$ to 0.1.

Concerning item 2 it has been found by numerical experiments that the integral of Eq. (76) may be truncated beyond the wavenumber α_2 determined as

$$\alpha_2 = \max\left\{5\,\alpha_1, 20\,\alpha_0\right\}, \quad \alpha_0 = 2\pi/R_0. \tag{78}$$

Here R_0 is a characteristic length of the foundation, e.g. the diameter of a circular footing. For strata with $\eta^j > 0.01$ for all layers it has been found that accurate results are typically obtained by Simpson integration with 2000 points in the wavenumber range $\alpha \in [0; \alpha_1]$ and 500 points in the range $\alpha \in [\alpha_1; \alpha_2]$. As discussed above, the numerical evaluation of the integral in the range $\alpha \in [\alpha_1, \alpha_2]$ is particularly efficient for the "bell-shaped" load distribution, since $D(\alpha)\,\overline{G}_{ij}(\alpha, \omega), i, j = 1, 2, 3$, are all monotone functions beyond α_1.

Finally, concerning item 3, $\mathbf{U}^{10}$ must be evaluated for all combinations of receiver and source points, which involves a high number of computations if $\widehat{\mathbf{G}}(r, \omega)$ is to be evaluated directly for each value of r. Instead, an alternative approach is suggested. Firstly, $\widehat{\mathbf{G}}(r, \omega)$ is determined at a number of points on the r-axis from $r = 0$ to $r = 2R_0$. Subsequently $\mathbf{U}^{10}$ is found by Eq. (76) using linear interpolation of $\widehat{\mathbf{G}}(r, \omega)$. It has been found that 250 points in this discretization provides a fast solution of satisfactory accuracy.

4. Consistent lumped-parameter models for wind-turbine foundations

Dynamic soil–structure interaction of wind turbines may be analysed by the finite-element method (FEM), the boundary-element method (BEM) or the domain-transformation method (DTM) described in the previous sections. These methods are highly adaptable and may be applied to the analysis of wave-propagation problems involving stratified soil, embedded foundations and inclusions or inhomogeneities in the ground. However, this comes at the cost of great computation times, in particular in the case of time-domain analysis of transient structural response over large periods of time. Thus, rigorous numerical models based on the FEM, the BEM, or the DTM, are not useful for real-time simulations or parametric studies in situations where only the structural response is of interest.

Alternatively, soil–structure interaction may be analysed by experimental methods. However, the models and equipment required for such analyses are expensive and this approach is not useful in a predesign phase. Hence, the need arises for a computationally efficient model which accounts for the interaction of a wind turbine foundation with the surrounding/underlying soil. A fairly general solution is the so-called *lumped-parameter model*, the development of which has been reported by Wolf (1991a), Wolf & Paronesso (1991), Wolf (1991b), Wolf & Paronesso (1992), Wolf (1994), Wolf (1997), Wu & Lee (2002), and Wu & Lee (2004). The present section is, to a great extent, based on this work.

The basic concept of a lumped-parameter model is to represent the original problem by a simple mechanical system consisting of a few so-called *discrete elements*, i.e. springs, dashpots and point masses which are easily implemented in standard finite-element models or aero-elastic codes for wind turbines. This is illustrated in Fig. 1 for a surface footing on a layered ground. The computational model consists of two parts: a model of the structure (e.g. a finite-element model) and a lumped-parameter model (LPM) of the foundation and the subsoil. The formulation of the model has three steps:

1. A rigorous frequency-domain model is applied for the foundation (in this case a footing on a soil stratum) and the frequency response is evaluated at a number of discrete frequencies.

2. A lumped-parameter model providing approximately the same frequency response is calibrated to the results of the rigorous model.

3. The structure itself (in this case the wind turbine) is represented by a finite-element model (or similar) and soil–structure interaction is accounted for by a coupling with the LPM of the foundation and subsoil.

Whereas the application of rigorous models like the BEM or DTM is often restricted to the analysis in the frequency domain—at least for any practical purposes—the LPM may be applied in the frequency domain as well as the time domain. This is ideal for problems involving linear response in the ground and nonlinear behaviour of a structure, which may typically be the situation for a wind turbine operating in the serviceability limit state (SLS).
It should be noted that the geometrical damping present in the original wave-propagation problem is represented as material damping in the discrete-element model. Thus, no distinction is made between material and geometrical dissipation in the final lumped-parameter model—they both contribute to the same parameters, i.e. damping coefficients.
Generally, if only few discrete elements are included in the lumped-parameter model, it can only reproduce a simple frequency response, i.e. a response with no resonance peaks. This is useful for rigid footings on homogeneous soil. However, inhomogeneous or flexible structures and stratified soil have a frequency response that can only be described by a lumped-parameter model with several discrete elements resulting in the presence of internal degrees of freedom. When the number of internal degrees of freedom is increased, so is the computation time. However, so is the quality of the fit to the original frequency response. This is the idea of the so-called *consistent lumped-parameter model* which is presented in this section.

4.1 Approximation of soil–foundation interaction by a rational filter

The relationship between a generalised force resultant, $f(t)$, acting at the foundation–soil interface and the corresponding generalised displacement component, $v(t)$, can be approximated by a differential equation in the form:

$$\sum_{i=0}^{k} A_i \frac{\mathrm{d}^i v(t)}{\mathrm{d}t^i} = \sum_{j=0}^{l} B_j \frac{\mathrm{d}^j f(t)}{\mathrm{d}t^j}. \tag{79}$$

Here, A_i, $i = 1, 2, \ldots, k$, and B_j, $j = 1, 2, \ldots, l$, are real coefficients found by curve fitting to the exact analytical solution or the results obtained by some numerical method or measurements. The rational approximation (79) suggests a model, in which higher-order temporal derivatives of both the forces and the displacements occur. This is undesired from a computational point of view. However, a much more elegant model only involving the zeroth, the first and the second temporal derivatives may be achieved by a rearrangement of the differential operators. This operation is simple to carry out in the frequency domain; hence, the first step in the formulation of a rational approximation is a Fourier transformation of Eq. (79), which provides:

$$\sum_{i=0}^{k} A_i (i\omega)^i V(\omega) = \sum_{j=0}^{l} B_j (i\omega)^j Q(\omega) \quad \Rightarrow$$

$$Q(\omega) = \widehat{Z}(i\omega) V(\omega), \quad \widehat{Z}(i\omega) = \frac{\sum_{i=0}^{k} A_i (i\omega)^i}{\sum_{j=0}^{l} B_j (i\omega)^j}, \tag{80}$$

where $V(\omega)$ and $F(\omega)$ denote the complex amplitudes of the generalized displacements and forces, respectively. It is noted that in Eq. (80) it has been assumed that the reaction force $F(\omega)$ stems from the response to a single displacement degree of freedom. This is generally not the case. For example, as discussed in Section 3, there is a coupling between the rocking moment–rotation and the horizontal force–translation of a rigid footing. However, the model (80) is easily generalised to account for such behaviour by an extension in the form $F_i(\omega) = \widehat{Z}_{ij}(i\omega)V_j(\omega)$, where summation is carried out over index j equal to the degrees of freedom contributing to the response. Each of the complex stiffness terms, $\widehat{Z}_{ij}(i\omega)$, is given by a polynomial fraction as illustrated by Eq. (80) for $\widehat{Z}(i\omega)$. This forms the basis for the derivation of so-called consistent lumped-parameter models.

4.2 Polynomial-fraction form of a rational filter

In the frequency domain, the dynamic stiffness related to a degree of freedom, or to the interaction between two degrees of freedom, i and j, is given by $\widetilde{Z}_{ij}(a_0) = Z^0_{ij}S_{ij}(a_0)$ (no sum on i,j). Here, $Z^0_{ij} = Z_{ij}(0)$ denotes the static stiffness related to the interaction of the two degrees of freedom, and $a_0 = \omega R_0/c_0$ is a dimensionless frequency with R_0 and c_0 denoting a characteristic length and wave velocity, respectively. For example, for a circular footing with the radius R_0 on an elastic half-space with the S-wave velocity c_S, $a_0 = \omega R_0/c_S$ may be chosen. With the given normalisation of the frequency it is noted that $\widetilde{Z}_{ij}(a_0) = Z_{ij}(c_0 a_0/R_0) = Z_{ij}(\omega)$.

For simplicity, any indices indicating the degrees of freedom in question are omitted in the following subsections, e.g. $\widetilde{Z}(a_0) \sim \widetilde{Z}_{ij}(a_0)$. The frequency-dependent stiffness coefficient $S(a_0)$ for a given degree of freedom is then decomposed into a singular part, $S_s(a_0)$, and a regular part, $S_r(a_0)$, i.e.

$$\widetilde{Z}(a_0) = Z^0 S(a_0), \qquad S(a_0) = S_s(a_0) + S_r(a_0), \tag{81}$$

where Z^0 is the static stiffness, and the singular part has the form

$$S_s(a_0) = k^\infty + i a_0 c^\infty. \tag{82}$$

In this expression, k^∞ and c^∞ are two real-valued constants which are selected so that $Z^0 S_s(a_0)$ provides the entire stiffness in the high-frequency limit $a_0 \to \infty$. Typically, the stiffness term $Z^0 k^\infty$ vanishes and the complex stiffness in the high-frequency range becomes a pure mechanical impedance, i.e. $S_s(a_0) = i a_0 c^\infty$. This is demonstrated in Section 5 for a two different types of wind turbine foundations interacting with soil.

The regular part $S_r(a_0)$ accounts for the remaining part of the stiffness. Generally, a closed-form solution for $S_r(a_0)$ is unavailable. Hence, the regular part of the complex stiffness is usually obtained by fitting of a rational filter to the results obtained with a numerical or semi-analytical model using, for example, the finite-element method (FEM), the boundary-element method (BEM) or the domain-transformation method (DTM). Examples are given in Section 5 for wind turbine foundations analysed by each of these methods.

Whether an analytical or a numerical solution is established, the output of a frequency-domain analysis is the complex dynamic stiffness $\widetilde{Z}(a_0)$. This is taken as the "target solution", and the regular part of the stiffness coefficient is found as $S_r(a_0) = \widetilde{Z}(a_0)/Z^0 - S_s(a_0)$. A rational approximation, or filter, is now introduced in the form

$$S_r(a_0) \approx \widehat{S}_r(i a_0) = \frac{P(i a_0)}{Q(i a_0)} = \frac{p_0 + p_1(i a_0) + p_2(i a_0)^2 + \ldots + p_N(i a_0)^N}{q_0 + q_1(i a_0) + q_2(i a_0)^2 + \ldots + q_M(i a_0)^M}. \tag{83}$$

The orders, N and M, and the coefficients, p_n ($n = 0, 1, \ldots, N$) and q_m ($m = 0, 1, \ldots, M$), of the numerator and denominator polynomials $P(ia_0)$ and $Q(ia_0)$ are chosen according to the following criteria:

1. To obtain a unique definition of the filter, one of the coefficients in either $P(ia_0)$ or $Q(ia_0)$ has to be given a fixed value. For convenience, $q_0 = 1$ is chosen.

2. Since part of the static stiffness is already represented by $S_s(0) = k^\infty$, this part of the stiffness should not be provided by $S_r(a_0)$ as well. Therefore, $p_0/q_0 = p_0 = 1 - k^\infty$.

3. In the high-frequency limit, $S(a_0) = S_s(a_0)$. Thus, the regular part must satisfy the condition that $\widehat{S}_r(ia_0) \to 0$ for $a_0 \to \infty$. Hence, $N < M$, i.e. the numerator polynomial $P(ia_0)$ is at least one order lower than the denominator polynomial, $Q(ia_0)$.

Based on these criteria, Eq. (84) may advantageously be reformulated as

$$S_r(a_0) \approx \widehat{S}_r(ia_0) = \frac{P(ia_0)}{Q(ia_0)} = \frac{1 - k^\infty + p_1(ia_0) + p_2(ia_0)^2 + \ldots + p_{M-1}(ia_0)^{M-1}}{1 + q_1(ia_0) + q_2(ia_0)^2 + \ldots + q_M(ia_0)^M}. \tag{84}$$

Evidently, the polynomial coefficients in Eq. (84) must provide a physically meaningful filter. By a comparison with Eqs. (79) and (80) it follows that p_n ($n = 1, 2, \ldots, M-1$) and q_j ($m = 1, 2, \ldots M$) must all be real. Furthermore, no poles should appear along the positive real axis as this will lead to an unstable solution in the time domain. This issue is discussed below.

The total approximation of $S(a_0)$ is found by an addition of Eqs. (82) and (84) as stated in Eq. (81). The approximation of $S(a_0)$ has two important characteristics:

- It is exact in the static limit, since $S(a_0) \approx \widehat{S}(ia_0) + S_s(a_0) \to 1$ for $a_0 \to 0$.

- It is exact in the high-frequency limit. Here, $S(a_0) \to S_s(a_0)$ for $a_0 \to \infty$, because $\widehat{S}_r(ia_0) \to 0$ for $a_0 \to \infty$.

Hence, the approximation is double-asymptotic. For intermediate frequencies, the quality of the fit depends on the order of the rational filter and the nature of the physical problem. Thus, in some situations a low-order filter may provide a very good fit to the exact solution, whereas other problems may require a high-order filter to ensure an adequate match—even over a short range of frequencies. As discussed in the examples given below in Section 5, a filter order of $M = 4$ will typically provide satisfactory results for a footing on a homogeneous half-space. However, for flexible, embedded foundations and layered soil, a higher order of the filter may be necessary—even in the low-frequency range relevant to dynamic response of wind turbines.

4.3 Partial-fraction form of a rational filter

Whereas the polynomial-fraction form is well-suited for curve fitting to measured or computed responses, it provides little insight into the physics of the problem. To a limited extent, such information is gained by a recasting of Eq. (84) into partial-fraction form,

$$\widehat{S}_r(ia_0) = \sum_{m=1}^{M} \frac{R_m}{ia_0 - s_m}, \tag{85}$$

where s_m, $m = 1, 2, \ldots, M$, are the poles of $\widehat{S}_r(ia_0)$ (i.e. the roots of $Q(ia_0)$), and R_j are the corresponding residues. The conversion of the original polynomial-fraction form into the partial-fraction expansion form may be carried out in MATLAB with the built-in function `residue`.

The poles s_m are generally complex. However, as discussed above, the coefficients q_m must be real in order to provide a rational approximation that is physically meaningful in the time domain. To ensure this, any complex poles, s_m, and the corresponding residues, R_m, must appear as conjugate pairs. When two such terms are added together, a second-order term with real coefficients appears. Thus, with N conjugate pairs, Eq. (85) can be rewritten as

$$\widehat{S}_r(ia_0) = \sum_{n=1}^{N} \frac{\beta_{0n} + \beta_{1n}ia_0}{\alpha_{0n} + \alpha_{1n}ia_0 + (ia_0)^2} + \sum_{n=N+1}^{M-N} \frac{R_n}{ia_0 - s_n}, \qquad 2N \leq M. \tag{86}$$

The coefficients α_{0n}, α_{1n}, β_{0n} and β_{1n}, $n = N+1, N+2, \ldots, M-N$, are given by

$$\alpha_{0n} = [s_n^{\Re}]^2 + [s_n^{\Im}]^2, \quad \alpha_{1n} = -2s_n^{\Re}, \quad \beta_{0n} = -2(R_n^{\Re}s_n^{\Re} + R_n^{\Im}s_n^{\Im}), \quad \beta_{1n} = 2R_n^{\Re}, \tag{87}$$

where $s_n^{\Re} = \Re(s_n)$ and $s_n^{\Im} = \Im(s_n)$ are the real and imaginary parts of the complex conjugate poles, respectively. Similarly, the real and imaginary parts of the complex conjugate residues are denoted by $R_n^{\Re} = \Re(R_n)$ and $R_n^{\Im} = \Im(R_n)$, respectively.

By adding the singular term in Eq. (82) to the expression in Eq. (85), the total approximation of the dynamic stiffness coefficient $S(a_0)$ can be written as

$$\widehat{S}(ia_0) = k^{\infty} + ia_0c^{\infty} + \sum_{n=1}^{N} \frac{\beta_{0n} + \beta_{1n}ia_0}{\alpha_{0n} + \alpha_{1n}ia_0 + (ia_0)^2} + \sum_{n=N+1}^{M-N} \frac{R_n}{ia_0 - s_n}. \tag{88}$$

The total approximation of the dynamic stiffness in Eq. (88) consists of three characteristic types of terms, namely a constant/linear term, $M - 2N$ first-order terms and N second-order terms. These terms are given as:

$$\text{Constant/linear term:} \qquad k^{\infty} + ia_0c^{\infty} \tag{89a}$$

$$\text{First-order term:} \qquad \frac{R}{ia_0 - s} \tag{89b}$$

$$\text{Second-order term:} \qquad \frac{\beta_1 ia_0 + \beta_0}{\alpha_0 + \alpha_1 ia_0 + (ia_0)^2}. \tag{89c}$$

4.4 Physical interpretation of a rational filter

Now, each term in Eq. (89) may be identified as the frequency-response function for a simple mechanical system consisting of springs, dashpots and point masses. Physically, the summation of terms (88) may then be interpreted as a parallel coupling of $M - N + 1$ of these so-called discrete-element models, and the resulting lumped-parameter model provides a frequency-response function similar to that of the original continuous system. In the subsections below, the calibration of the discrete-element models is discussed, and the physical interpretation of each kind of term in Eq. (89) is described in detail.

4.4.1 Constant/linear term

The constant/linear term given by Eq. (89a) consists of two known parameters, k^{∞} and c^{∞}, that represent the singular part of the dynamic stiffness. The discrete-element model for the constant/linear term is shown in Fig. 9.

The equilibrium formulation of Node 0 (for harmonic loading) is as follows:

$$\kappa U_0(\omega) + i\omega\gamma\frac{R_0}{c_0} U_0(\omega) = P_0(\omega) \tag{90}$$

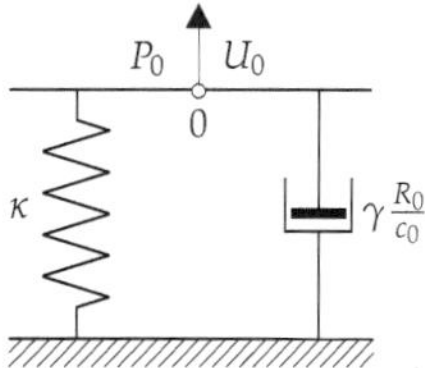

Fig. 9. The discrete-element model for the constant/linear term.

Recalling that the dimensionless frequency is introduced as $a_0 = \omega R_0/c_0$, the equilibrium formulation in Eq. (90) results in a force–displacement relation given by

$$P_0(a_0) = (\kappa + ia_0\gamma)\, U_0(a_0). \tag{91}$$

By a comparison of Eqs. (89a) and (91) it becomes evident that the non-dimensional coefficients κ and γ are equal to k^∞ and c^∞, respectively.

4.4.2 First-order terms with a single internal degree of freedom

The first-order term given by Equation (89b) has two parameters, R and s. The layout of the discrete-element model is shown in Fig. 10a. The model is constructed by a spring $(-\kappa)$ in parallel with another spring (κ) and dashpot $(\gamma\frac{R_0}{c_0})$ in series. The serial connection between the spring (κ) and the dashpot $(\gamma\frac{R_0}{c_0})$ results in an internal node (internal degree of freedom). The equilibrium formulations for Nodes 0 and 1 (for harmonic loading) are as follows:

$$\text{Node } 0: \quad \kappa\big(U_0(\omega) - U_1(\omega)\big) - \kappa U_0(\omega) = P_0(\omega) \tag{92a}$$

$$\text{Node } 1: \quad \kappa\big(U_1(\omega) - U_0(\omega)\big) + i\omega\gamma\frac{R_0}{c_0}U_1(\omega) = 0. \tag{92b}$$

After elimination of $U_1(\omega)$ in Eqs. (92a) and (92b), it becomes clear that the force–displacement relation of the first-order model is given as

$$P_0(a_0) = \frac{-\dfrac{\kappa^2}{\gamma}}{ia_0 + \dfrac{\kappa}{\gamma}}\, U_0(a_0). \tag{93}$$

By comparing Eqs. (89b) and (93), κ and γ are identified as

$$\kappa = \frac{R}{s}, \qquad \gamma = -\frac{R}{s^2}. \tag{94}$$

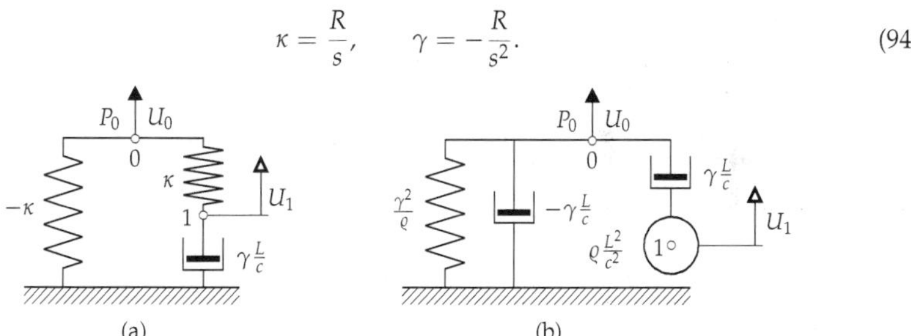

(a) (b)

Fig. 10. The discrete-element model for the first-order term: (a) Spring-dashpot model; (b) monkey-tail model.

It should be noted that the first-order term could also be represented by a so-called "monkey-tail" model, see Fig. 10b. This turns out to be advantageous in situations where κ and γ in Eq. (94) are negative, which may be the case when R is positive (s is negative). To avoid negative coefficients of springs and dashpots, the monkey-tail model is applied, and the resulting coefficients are positive. By inspecting the equilibrium formulations for Nodes 0 and 1, see Fig. 10b, the coefficients can be identified as

$$\gamma = \frac{R}{s^2}, \qquad \varrho = -\frac{R}{s^3}. \tag{95}$$

Evidently, the internal degree of freedom in the monkey-tail model has no direct physical meaning in relation to the original problem providing the target solution. Thus, at low frequencies the point mass may undergo extreme displacements, and in the static case the displacement is infinite. This lack of direct relationship with the original problem is a general property of the discrete-element models. They merely provide a mechanical system that leads to a similar frequency response.

4.4.3 Second-order terms with one or two internal degrees of fredom

The second-order term given by Eq. (89c) has four parameters: α_0, α_1, β_0 and β_1. An example of a second-order discrete-element model is shown in Fig. 11a. This particular model has two internal nodes. The equilibrium formulations for Nodes 0, 1 and 2 (for harmonic loading) are as follows:

$$\text{Node 0:} \qquad \kappa_1\left(U_0(\omega) - U_1(\omega)\right) - \kappa_1 U_0(\omega) = P_0(\omega) \tag{96a}$$

$$\text{Node 1:} \qquad \kappa_1\left(U_1(\omega) - U_0(\omega)\right) + i\omega\gamma_1\frac{R_0}{c_0}\left(U_1(\omega) - U_2(\omega)\right) = 0 \tag{96b}$$

$$\text{Node 2:} \qquad \kappa_2 U_2(\omega) + i\omega\gamma_2\frac{R_0}{c_0}U_2(\omega) + i\omega\gamma_1\frac{R_0}{c_0}\left(U_2(\omega) - U_1(\omega)\right) = 0. \tag{96c}$$

After some rearrangement and elimination of the internal degrees of freedom, the force-displacement relation of the second-order model is given by

$$P_0(a_0) = \frac{-\kappa_1^2\frac{\gamma_1+\gamma_2}{\gamma_1\gamma_2}ia_0 - \frac{\kappa_1^2\kappa_2}{\gamma_1\gamma_2}}{(ia_0)^2 + \left(\kappa_1\frac{\gamma_1+\gamma_2}{\gamma_1\gamma_2} + \frac{\kappa_2}{\gamma_2}\right)ia_0 + \frac{\kappa_1\kappa_2}{\gamma_1\gamma_2}}U_0(a_0). \tag{97}$$

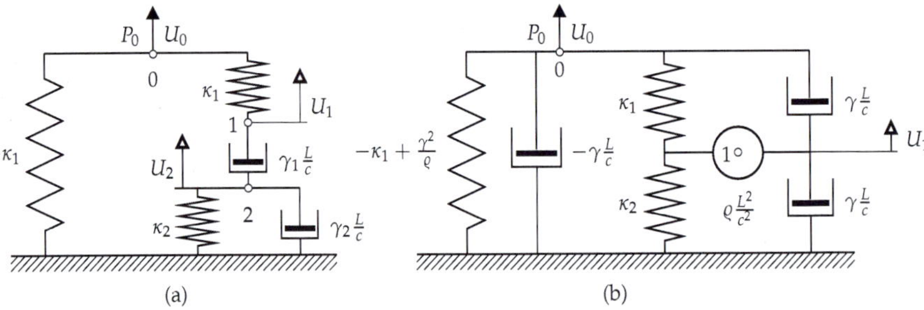

Fig. 11. The discrete-element model for the second-order term: (a) Spring-dashpot model with two internal degrees of freedom; (b) spring-dashpot-mass model with one internal degree of freedom.

By a comparison of Eqs. (89c) and (97), the four coefficients in Eq. (97) are identified as

$$\kappa_1 = -\frac{\beta_0}{\alpha_0}, \qquad \gamma_1 = -\frac{\alpha_0\beta_1 - \alpha_1\beta_0}{\alpha_0^2}, \tag{98a}$$

$$\kappa_2 = \frac{\beta_0}{\alpha_0^2}\frac{(-\alpha_0\beta_1 + \alpha_1\beta_0)^2}{\alpha_0\beta_1^2 - \alpha_1\beta_0\beta_1 + \beta_0^2}, \qquad \gamma_2 = \frac{\beta_0^2}{\alpha_0^2}\frac{-\alpha_0\beta_1 + \alpha_1\beta_0}{\alpha_0\beta_1^2 - \alpha_1\beta_0\beta_1 + \beta_0^2}. \tag{98b}$$

Alternatively, introducing a second-order model with springs, dampers and a point mass, it is possible to construct a second-order model with only one internal degree of freedom. The model is sketched in Fig. 11b. The force–displacement relation of the alternative second-order model is given by

$$P_0(a_0) = \frac{2\left(\frac{\kappa_1\gamma}{\varrho} + \frac{\gamma^3}{\varrho^2}\right)ia_0 - \frac{\kappa_1^2}{\varrho} + \frac{(\kappa_1+\kappa_2)\gamma^2}{\varrho^2}}{(ia_0)^2 + 2\frac{\gamma}{\varrho}ia_0 + \frac{\kappa_1+\kappa_2}{\varrho}}U_0(a_0). \tag{99}$$

By equating the coefficients in Eq. (99) to the terms of the second-order model in Eq. (89c), the four parameters κ_1, κ_2, γ and ϱ can be determined. In order to calculate ϱ, a quadratic equation has to be solved. The quadratic equation for ϱ is

$$a\varrho^2 + b\varrho + c = 0 \quad \text{where} \quad a = \alpha_1^4 - 4\alpha_0\alpha_1^2, \quad b = -8\alpha_1\beta_1 + 16\beta_0, \quad c = 16\frac{\beta_1^2}{\alpha_1^2}. \tag{100}$$

Equation (100) results in two solutions for ϱ. To ensure real values of ϱ, $b^2 - 4ac \geq 0$ or $\alpha_0\beta_1^2 - \alpha_1\beta_0\beta_1 + \beta_0^2 \geq 0$. When ϱ has been determined, the three remaining coefficients can be calculated by

$$\kappa_1 = \frac{\varrho\alpha_1^2}{4} - \frac{\beta_1}{\alpha_1}, \quad \kappa_2 = \varrho\alpha_0 - \kappa_1, \quad \gamma = \frac{\varrho\alpha_1}{2}. \tag{101}$$

4.5 Fitting of a rational filter

In order to get a stable solution in the time domain, the poles of $\widehat{S}_r(ia_0)$ should all reside in the second and third quadrant of the complex plane, i.e. the real parts of the poles must all be negative. Due to the fact that computers only have a finite precision, this requirement may have to be adjusted to $s_m < -\varepsilon$, $m = 1, 2, \ldots, M$, where ε is a small number, e.g. 0.01.

The rational approximation may now be obtained by curve-fitting of the rational filter $\widehat{S}_r(ia_0)$ to the regular part of the dynamic stiffness, $S_r(a_0)$, by a least-squares technique. In this process, it should be observed that:

1. The response should be accurately described by the lumped-parameter model in the frequency range that is important for the physical problem being investigated. For soil–structure interaction of wind turbines, this is typically the low-frequency range.

2. The "exact" values of $S_r(a_0)$ are only measured—or computed—over a finite range of frequencies, typically for $a_0 \in [0; a_{0max}]$ with $a_{0max} = 2 \sim 10$. Further, the values of $S_r(a_0)$ are typically only known at a number of discrete frequencies.

3. Outside the frequency range, in which $S_r(a_0)$ has been provided, the singular part of the dynamic stiffness, $S_s(a_0)$, should govern the response. Hence, no additional tips and dips should appear in the frequency response provided by the rational filter beyond the dimensionless frequency a_{0max}.

Firstly, this implies that the order of the filter, M, should not be too high. Experience shows that orders about $M = 2 \sim 8$ are adequate for most physical problems. Higher-order filters than this are not easily fitted, and lower-order filters provide a poor match to the "exact" results. Secondly, in order to ensure a good fit of $\widehat{S}_r(ia_0)$ to $S_r(a_0)$ in the low-frequency range, it is recommended to employ a higher weight on the squared errors in the low-frequency range, e.g. for $a_0 < 0.2 \sim 2$, compared with the weights in the medium-to-high-frequency range. Obviously, the definition of low, medium and high frequencies is strongly dependent on the problem in question. For example, frequencies that are considered high for an offshore wind turbine, may be considered low for a diesel power generator.

For soil-structure interaction of foundations, Wolf (1994) suggested to employ a weight of $w(a_0) = 10^3 \sim 10^5$ at low frequencies and unit weight at higher frequencies. This should lead to a good approximation in most cases. However, numerical experiment indicates that the fitting goodness of the rational filter is highly sensitive to the choice of the weight function $w(a_0)$, and the guidelines provided by Wolf (1994) are not useful in all situations. Hence, as an alternative, the following fairly general weight function is proposed:

$$w(a_0) = \frac{1}{\left(1 + (\varsigma_1 a_0)^{\varsigma_2}\right)^{\varsigma_3}}. \tag{102}$$

The coefficients ς_1, ς_2 and ς_3 are heuristic parameters. Experience shows that values of about $\varsigma_1 = \varsigma_2 = \varsigma_3 = 2$ provide an adequate solution for most foundations in the low-frequency range $a_0 \in [0; 2]$. This recommendation is justified by the examples given in the next section. For analyses involving high-frequency excitation, lower values of ς_1, ς_2 and ς_3 may have to be employed.

Hence, the optimisation problem defined in Table 1. However, the requirement of all poles lying in the second and third quadrant of the complex plane is not easily fulfilled when an optimisation is carried out by least-squares (or similar) curve fitting of $\widehat{S}_r(ia_0)$ to $S_r(a_0)$ as suggested in Table 1. Specifically, the choice of the polynomial coefficients $q_j, j = 1, 2, \ldots, m$, as the optimisation variables is unsuitable, since the constraint that all poles of $\widehat{S}_r(ia_0)$ must have negative real parts is not easily incorporated in the optimisation problem. Therefore, instead of the interpretation

$$Q(ia_0) = 1 + q_1(ia_0) + q_2(ia_0)^2 + \ldots + q_M(ia_0)^M, \tag{103}$$

an alternative approach is considered, in which the denominator is expressed as

$$Q(ia_0) = (ia_0 - s_1)(ia_0 - s_2) \cdots (ia_0 - s_M) = \prod_{m=1}^{M} (ia_0 - s_m). \tag{104}$$

In this representation, $s_m, m = 1, 2, \ldots, M$, are the roots of $Q(ia_0)$. In particular, if there are N complex conjugate pairs, the denominator polynomial may advantageously be expressed as

$$Q(ia_0) = \prod_{n=1}^{N} (ia_0 - s_n)(ia_0 - s_n^*) \cdot \prod_{n=N+1}^{M-N} (ia_0 - s_n). \tag{105}$$

where an asterisk ($*$) denotes the complex conjugate. Thus, instead of the polynomial coefficients, the roots s_n are identified as the optimisation variables.

A rational filter for the regular part of the dynamic stiffness is defined in the form:

$$S_r(a_0) \approx \widehat{S}_r(ia_0) = \frac{P(ia_0)}{Q(ia_0)} = \frac{1 - k^\infty + p_1(ia_0) + p_2(ia_0)^2 + \ldots + p_{M-1}(ia_0)^{M-1}}{1 + q_1(ia_0) + q_2(ia_0)^2 + \ldots + q_M(ia_0)^M}.$$

Find the optimal polynomial coefficients p_n and q_m which minimize the object function $F(p_n, q_m)$ in a weighted-least-squares sense subject to the constraints $G_1(p_n, q_m), G_2(p_n, q_m), \ldots, G_M(p_n, q_m)$.

Input:

$$\begin{aligned}
M &: \quad \text{order of the filter} \\
p_n^0, &\quad n = 1, 2, \ldots, M-1, \\
q_m^0, &\quad m = 1, 2, \ldots, M, \\
a_{0j}, &\quad j = 1, 2, \ldots, J, \\
S_r(a_{0j}), &\quad j = 1, 2, \ldots, J, \\
w(a_{0j}), &\quad j = 1, 2, \ldots, J.
\end{aligned}$$

Variables:

$$\begin{aligned}
p_n, &\quad n = 1, 2, \ldots, M-1, \\
q_m, &\quad m = 1, 2, \ldots, M.
\end{aligned}$$

Object function:

$$F(p_n, q_m) = \sum_{j=1}^{J} w(a_{0j}) \left(\widehat{S}_r(ia_{0j}) - S_r(a_{0j}) \right)^2.$$

Constraints:

$$\begin{aligned}
G_1(p_n, q_m) &= \Re(s_1) < -\varepsilon, \\
G_2(p_n, q_m) &= \Re(s_2) < -\varepsilon, \\
&\vdots \\
G_M(p_n, q_m) &= \Re(s_M) < -\varepsilon.
\end{aligned}$$

Output:

$$\begin{aligned}
p_n, &\quad n = 1, 2, \ldots, M-1, \\
q_m, &\quad m = 1, 2, \ldots, M.
\end{aligned}$$

Here, p_n^0 and q_m^0 are the initial values of the polynomial coefficients p_n and q_m, whereas $S_r(a_{0j})$ are the "exact" value of the dynamic stiffness evaluated at the J discrete dimensionless frequencies a_{0j}. These are either measured or calculated by rigourous numerical or analytical methods. Further, $\widehat{S}_r(ia_{0j})$ are the values of the rational filter at the same discrete frequencies, and $w(a_0)$ is a weight function, e.g. as defined by Eq. (102) with $\varsigma_1 = \varsigma_2 = \varsigma_3 = 2$. Finally, s_m are the poles of the rational filter $\widehat{S}_r(ia_0)$, i.e. the roots of the denominator polynomial $Q(ia_0)$, and ε is a small number, e.g. $\varepsilon = 0.01$.

Table 1. Fitting of rational filter by optimisation of polynomial coefficients.

Accordingly, in addition to the coefficients of the numerator polynomial $P(ia_0)$, the variables in the optimisation problem are the real and imaginary parts $s_n^{\Re} = \Re(s_n)$ and $s_n^{\Im} = \Im(s_n)$ of the complex roots $s_n, n = 1, 2, \ldots, N$, and the real roots $s_n, n = N+1, N+2, \ldots, M-N$.

The great advantage of the representation (105) is that the constraints on the poles are defined directly on each individual variable, whereas the constraints in the formulation with $Q(ia_0)$ defined by Eq. (103), the constraints are given on functionals of the variables. Hence, the solution is much more efficient and straightforward. However, Eq. (105) has two disadvantages when compared with Eq. (103):

- The number of complex conjugate pairs has to be estimated. However, experience shows that as many as possible of the roots should appear as complex conjugates—e.g. if M is even, $N = M/2$ should be utilized. This provides a good fit in most situations and may, at the same time, generate the lumped-parameter model with fewest possible internal degrees of freedom.

A rational filter for the regular part of the dynamic stiffness is defined in the form:

$$S_r(a_0) \approx \widehat{S}_r(ia_0) = \frac{P(ia_0)}{Q(ia_0)} = \frac{1 - k^\infty + p_1(ia_0) + p_2(ia_0)^2 + \ldots + p_{M-1}(ia_0)^{M-1}}{\prod_{m=1}^{N}(ia_0 - s_m)(ia_0 - s_m^*) \cdot \prod_{m=N+1}^{M-N}(ia_0 - s_m)}.$$

Find the optimal polynomial coefficients p_n and the poles s_m which minimise the object function $F(p_n, s_m)$ subject to the constraints $G_0(p_n, s_m), G_1(p_n, s_m), \ldots, G_N(p_n, s_m)$.

Input:

$$\begin{aligned}
M &: \quad \text{order of the filter} \\
N &: \quad \text{number of complex conjugate pairs,} \quad 2N \leq M \\
p_n^0, & \quad n = 1, 2, \ldots, M - 1, \\
s_m^{\Re 0}, & \quad m = 1, 2, \ldots, N, \\
s_m^{\Im 0}, & \quad m = 1, 2, \ldots, N, \\
s_m^0, & \quad m = 1, 2, \ldots, M - N, \\
a_{0j}, & \quad j = 1, 2, \ldots, J, \\
S_r(a_{0j}), & \quad j = 1, 2, \ldots, J, \\
w(a_{0j}), & \quad j = 1, 2, \ldots, J.
\end{aligned}$$

Variables:

$$\begin{aligned}
p_n, & \quad n = 1, 2, \ldots, M - 1, \\
s_m^{\Re}, & \quad m = 1, 2, \ldots, N, & \quad s_m^{\Re} < -\varepsilon, \\
s_m^{\Im}, & \quad m = 1, 2, \ldots, N, & \quad s_m^{\Im} > +\varepsilon, \\
s_m, & \quad m = N + 1, 2, \ldots, M - N, & \quad s_m < -\varepsilon.
\end{aligned}$$

Object function:

$$F(p_n, s_m) = \sum_{j=1}^{J} w(a_{0j})\left(\widehat{S}_r(ia_{0j}) - S_r(a_{0j})\right)^2.$$

Constraints:

$$\begin{aligned}
G_0(p_n, s_m) &= 1 - \prod_1^M (-s_m) = 0, \\
G_k(p_n, s_m) &= \zeta s_k^{\Re} + s_k^{\Im} < 0, & \quad k = 1, 2, \ldots, N.
\end{aligned}$$

Output:

$$\begin{aligned}
p_n, & \quad n = 1, 2, \ldots, M - 1, \\
s_m^{\Re}, & \quad m = 1, 2, \ldots, N, \\
s_m^{\Im}, & \quad m = 1, 2, \ldots, N, \\
s_m, & \quad m = N + 1, 2, \ldots, M - N.
\end{aligned}$$

Here, superscript 0 indicates initial values of the respective variables, and $\widehat{S}_r(ia_{0j})$ are the values of the rational filter at the same discrete frequencies. Further, $\zeta \approx 10 \sim 100$ and $\varepsilon \approx 0.01$ are two real parameters. Note that the initial values of the poles must conform with the constraint $G_0(p_n, s_m)$. For additional information, see Table 1.

Table 2. Fitting of rational filter by optimisation of the poles.

- In the representation provided by Eq. (103), the correct asymptotic behaviour is automatically ensured in the limit $ia_0 \to 0$, i.e. the static case, since $q_0 = 1$. Unfortunately, in the representation given by Eq. (105) an additional equality constraint has to be implemented to ensure this behaviour. However, this condition is much easier implemented than the constraints which are necessary in the case of Eq. (103) in order to prevent the real parts of the roots from being positive.

Eventually, instead of the problem defined in Table 1, it may be more efficient to solve the optimisation problem given in Table 2. It is noted that additional constraints are suggested, which prevent the imaginary parts of the complex poles to become much (e.g. 10 times) bigger than the real parts. This is due to the following reason: If the real part of the complex pole s_m vanishes, i.e. $s_m^{\Re} = 0$, this results in a second order pole, $\{s_m^{\Im}\}^2$, which is real and positive.

Evidently, this will lead to instability in the time domain. Since the computer precision is limited, a real part of a certain size compared to the imaginary part of the pole is necessary to ensure a stable solution.

Finally, as an alternative to the optimisation problems defined in Table 1 and Table 2, the function $S(ia_0)$ may be expressed by Eq. (86), i.e. in partial-fraction form. In this case, the variables in the optimization problem are the poles and residues of $S(ia_0)$. In the case of the second-order terms, these quantities are replaced with α_0, α_1, β_0 and β_1. At a first glance, this choice of optimisation variable seems more natural than p_n and s_m, as suggested in Table 2. However, from a computational point of view, the mathematical operations involved in the polynomial-fraction form are more efficient than those of the polynomial-fraction form. Hence, the scheme provided in Table 2 is recommended.

5. Time-domain analysis of soil–structure interaction

In this section, two examples are given in which consistent lumped-parameter models are applied to the analysis of foundations and soil–structure interaction. The first example concerns a rigid hexagonal footing on a homogeneous or layered ground and was first presented by Andersen (2010). The frequency-domain solution obtained by the domain-transformation method presented in Sections 2–3 is fitted by LPMs of different orders. Subsequently, the response of the original model and the LPMs are compared in frequency and time domain.

In the second example, originally proposed by Andersen et al. (2009), LPMs are fitted to the frequency-domain results of a coupled boundary-element/finite-element model of a flexible embedded foundation. As part of the examples, the complex stiffness of the foundation in the high-frequency limit is discussed, i.e. the coefficients k^∞ and c^∞ in Eq. (82) are determined for each component of translation and rotation of the foundation. Whereas no coupling exists between horizontal sliding and rocking of surface footings in the high-frequency limit, a significant coupling is present in the case of embedded foundations—even at high frequencies.

5.1 Example: A footing on a homogeneous or layered ground
The foundation is modelled as a regular hexagonal rigid footing with the side length r_0, height h_0 and mass density ρ_0. This geometry is typical for offshore wind turbine foundations.

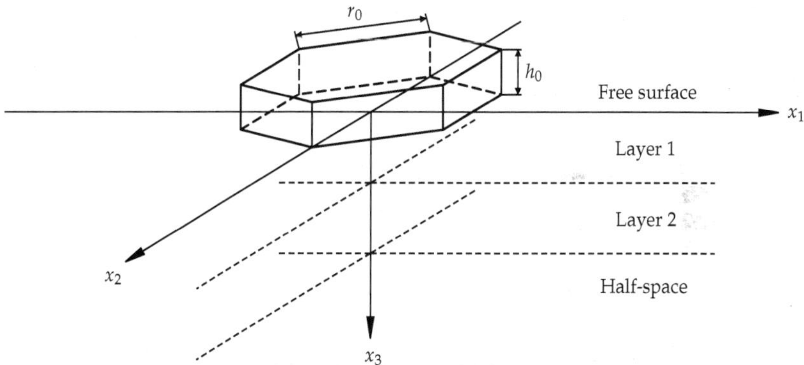

Fig. 12. Hexagonal footing on a stratum with three layers over a half-space.

As illustrated in Fig. 12, the centre of the soil–foundation interface coincides with the origin of the Cartesian coordinate system. The mass of the foundation and the corresponding mass moments of inertia with respect to the three coordinate axes then become:

$$M_0 = \rho_0 h_0 A_0, \quad \mathcal{J}_1 = \mathcal{J}_2 = \rho_0 h_0 \mathcal{I}_0 + \frac{1}{3}\rho_0 h_0^3 A_0, \quad \mathcal{J}_3 = 2\rho_0 h_0 \mathcal{I}_0, \tag{106a}$$

where A_0 is the area of the horizontal cross-section and $\mathcal{I}_0$ is the corresponding geometrical moment of inertia,

$$A_0 = \frac{3\sqrt{3}}{2}r_0^2, \quad \mathcal{I}_0 = \frac{5\sqrt{3}}{16}r_0^4. \tag{106b}$$

It is noted that $\mathcal{I}_0$ is invariant to rotation of the foundation around the x_3-axis. This property also applies to circular or quadratic foundations as discussed in Section 3.

5.1.1 A footing on a homogeneous ground

Firstly, we consider a hexagonal footing on a homogeneous visco-elastic half-space. The footing has the side length $r_0 = 10$ m, the height $h_0 = 10$ m and the mass density $\rho_0 = 2000$ kg/m^3, and the mass and mass moments of inertia are computed by Eq. (106). The properties of the soil are $\rho^1 = 2000$ kg/m^3, $E^1 = 10^4$ kPa, $\nu^1 = 0.25$ and $\eta^1 = 0.03$. However, in the static limit, i.e. for $\omega \to 0$, the hysteretic damping model leads to a complex impedance in the frequency domain. By contrast, the lumped-parameter model provides a real impedance, since it is based on viscous dashpots. This discrepancy leads to numerical difficulties in the fitting procedure and to overcome this, the hysteretic damping model for the soil is replaced by a linear viscous model at low frequencies, in this case below 1 Hz.

In principle, the time-domain solution for the displacements and rotations of the rigid footing is found by inverse Fourier transformation, i.e.

$$v_i(t) = \frac{1}{2\pi}\int_{-\infty}^{\infty} V_i(\omega)e^{i\omega t}\,d\omega, \quad \theta_i(t) = \frac{1}{2\pi}\int_{-\infty}^{\infty} \Theta_i(\omega)e^{i\omega t}\,d\omega, \quad i = 1,2,3. \tag{107}$$

The displacements, rotations, forces and moments in the time domain are visualised in Fig. 13. In the numerical computations, the frequency response spectrum is discretized and accordingly, the time-domain solution is found by a Fourier series.

According to Eqs. (72) and (74), the vertical motion $V_3(\omega)$ as well as the torsional motion $\Theta_3(\omega)$ (see Fig. 6) are decoupled from the remaining degrees of freedom of the hexagonal footing. Thus, $V_3(\omega)$ and $\Theta_3(\omega)$ may be fitted by independent lumped-parameter models.

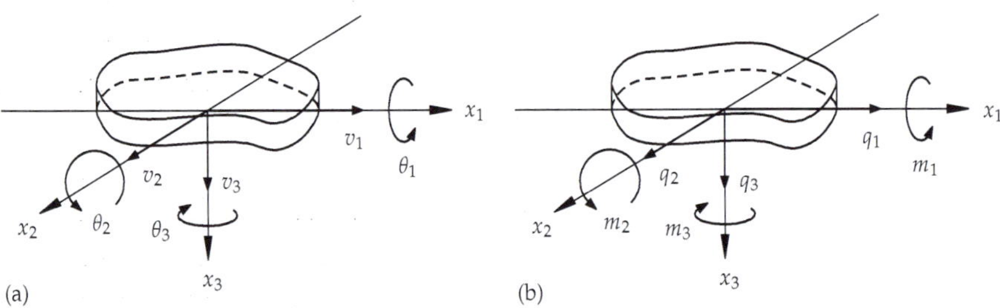

Fig. 13. Degrees of freedom for a rigid surface footing in the time domain: (a) displacements and rotations, and (b) forces and moments.

In the following, the quality of lumped-parameter models based on rational filters of different orders are tested for vertical and torsional excitation.

For the footing on the homogeneous half-space, rational filters of the order 2–6 are tested. Firstly, the impedance components are determined in the frequency-domain by the method presented in Section 2. The lumped-parameter models are then fitted by application of the procedure described in Section 4 and summarised in Table 2. The two components of the normalised impedance, S_{33} and S_{66}, are shown in Figs. 14 and 16 as functions of the physical frequency, f. It is noted that all the LPMs are based on second-order discrete-element models including a point mass, see Fig. 11b. Hence, the LPM for each individual component of the impedance matrix, $\mathbf{Z}(\omega)$, has 1, 2 or 3 internal degrees of freedom.

With reference to Fig. 14, a poor fit of the vertical impedance is obtained with $M = 2$ regarding the absolute value of S_{33} as well as the phase angle. A lumped-parameter model with $M = 4$ provides a much better fit in the low-frequency range. However, a sixth-order lumped-parameter model is required to obtain an accurate solution in the medium-frequency range, i.e. for frequencies between approximately 1.5 and 4 Hz. As expected, further analyses show that a slightly better match in the medium-frequency range is obtained with the weight-function coefficients $\varsigma_1 = 2$ and $\varsigma_2 = \varsigma_3 = 1$. However, this comes at the cost of a poorer match in the low-frequency range. Finally, it has been found that no improvement is achieved if first-order terms, e.g. the "monkey tail" illustrated in Fig. 10b, are allowed in the rational-filter approximation.

Figure 16 shows the rational-filter approximations of S_{66}, i.e. the non-dimensional torsional impedance. Compared with the results for the vertical impedance, the overall quality of the fit is relatively poor. In particular the LPM with $M = 2$ provides a phase angle which is negative in the low frequency range. Actually, this means that the geometrical damping provided by the second-order LPM becomes negative for low-frequency excitation. Furthermore, the stiffness is generally under-predicted and as a consequence of this an LPM with $M = 2$ cannot be used for torsional vibrations of the surface footing.

A significant improvement is achieved with $M = 4$, but even with $M = 6$ some discrepancies are observed between the results provide by the LPM and the rigorous model. Unfortunately, additional studies indicate that an LPM with $M = 8$ does not increase the accuracy beyond that of the sixth-order model.

Next, the dynamic soil–foundation interaction is studied in the time domain. In order to examine the transient response, a pulse load is applied in the form

$$p(t) = \begin{cases} \sin(2\pi f_c t)\sin(0.5\pi f_c t) & \text{for} \qquad 0 < t < 2/f_c \\ 0 & \text{otherwise.} \end{cases} \tag{108}$$

In this analysis, $f_c = 2$ Hz is utilised, and the responses obtained with the lumped-parameter models of different orders are computed by application of the Newmark β-scheme proposed by Newmark Newmark (1959). Figure 15 shows the results of the analysis with $q_3(t) = p(t)$, whereas the results for $m_3(t) = p(t)$ are given in Fig. 17.

In the case of vertical excitation, Fig. 15 shows that even the LPM with $M = 2$ provides an acceptable match to the "exact" results achieved by inverse Fourier transformation of the frequency-domain solution. In particular, the maximum response occurring during the excitation is well described. However, an improvement in the description of the damping is obtained with $M = 4$. For torsional motion, the second-order LPM is invalid since it provides negative damping. Hence, the models with $M = 4$ and $M = 6$ are compared in Fig. 17. It is clearly demonstrated that the fourth-order LPM provides a poor representation of the torsional impedance, whereas an accurate prediction of the response is achieved with the sixth-order model.

Subsequently, lumped-parameter models are fitted for the horizontal sliding and rocking motion of the surface footing, i.e. $V_2(\omega)$ and $\Theta_1(\omega)$ (see Fig. 6). As indicated by Eqs. (72) and (74), these degrees of freedom are coupled via the impedance component Z_{24}. Hence, two analyses are carried out. Firstly, the quality of lumped-parameter models based on rational filters of different orders are tested for horizontal and moment excitation. Secondly, the significance of coupling is investigated by a comparison of models with and without the coupling terms.

Similarly to the case for vertical and torsional motion, rational filters of the order 2–6 are tested. The three components of the normalised impedance, S_{22}, $S_{24} = S_{42}$ and S_{44}, are shown in Figs. 18, 20 and 22 as functions of the physical frequency, f. Again, the lumped-parameter models are based on discrete-element model shown in Fig. 11b, which reduces the number of internal degrees of freedom to a minimum. Clearly, the lumped-parameter models with $M = 2$ provide a poor fit for all the components S_{22}, S_{24} and S_{44}. However, Figs. 18 and 22 show that an accurate solution is obtained for S_{22} and S_{44} when a 4th model is applied, and the inclusion of an additional internal degree of freedom, i.e. raising the order from $M = 4$ to $M = 6$, does not increase the accuracy significantly. However, for S_{24} an LPM with $M = 6$ is much more accurate than an LPM with $M = 4$ for frequencies $f > 3$ Hz, see Fig. 20.

Subsequently, the transient response to the previously defined pulse load with centre frequency $f_c = 2$ Hz is studied. Figure 19 shows the results of the analysis with $q_2(t) = p(t)$, and the results for $m_1(t) = p(t)$ are given in Fig. 21. Further, the results from an alternative analysis with no coupling of sliding and rocking are presented in Fig. 23. In Fig. 19 it is observed that the LPM with $M = 2$ provides a poor match to the results of the rigorous model. The maximum response occurring during the excitation is well described by the low-order LPM. However the damping is significantly underestimated by the LPM. Since the loss factor is small, this leads to the conclusion that the geometrical damping is not predicted with adequate accuracy. On the other hand, for $M = 4$ a good approximation is obtained with regard to both the maximum response and the geometrical damping. As suggested by Fig. 18, almost no further improvement is gained with $M = 6$. For the rocking produced by a moment applied to the rigid footing, the lumped-parameter model with $M = 2$ is useless. Here, the geometrical damping is apparently negative. However, $M = 4$ provides an accurate solution (see Fig. 21) and little improvement is achieved by raising the order to $M = 6$ (this result is not included in the figure).

Alternatively, Fig. 23 shows the result of the time-domain solution for a lumped-parameter model in which the coupling between sliding and rocking is disregarded. This model is interesting because the two coupling components S_{24} and S_{42} must be described by separate lumped-parameter models. Thus, the model with $M = 4$ in Fig. 23 has four less internal degrees of freedom than the corresponding model with $M = 4$ in Fig. 21. However, the two results are almost identical, i.e. the coupling is not pronounced for the footing on the homogeneous half-space. Hence, the sliding–rocking coupling may be disregarded without significant loss of accuracy. Increasing the order of the LPMs for S_{22} and S_{44} from 4 to 8 results in a model with the same number of internal degrees of freedom as the fourth-order model with coupling; but as indicated by Fig. 23, this does not improve the overall accuracy. Finally, Fig. 20 suggests that the coupling is more pronounced when a load with, for example, $f_c = 1.5$ or 3.5 Hz is applied. However, further analyses, whose results are not presented in this paper, indicate that this is not the case.

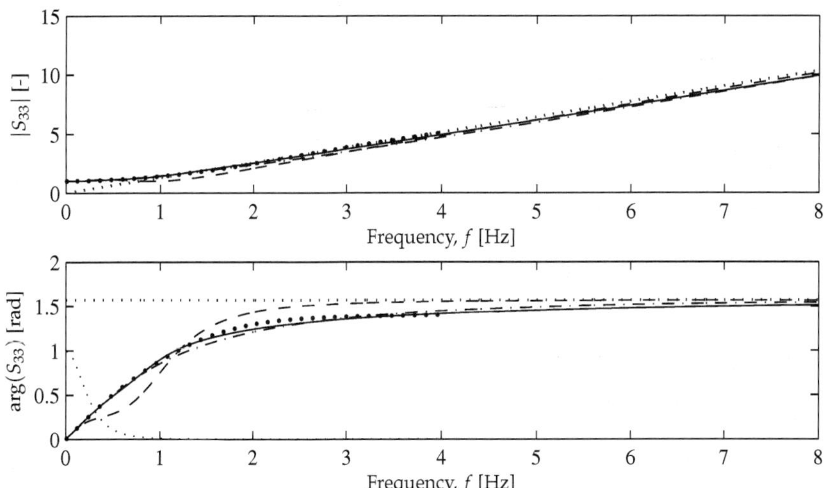

Fig. 14. Dynamic stiffness coefficient, S_{33}, obtained by the domain-transformation model (the large dots) and lumped-parameter models with $M = 2$ ($--$), $M = 4$ ($-\cdot-\cdot$), and $M = 6$ (——). The thin dotted line ($\cdots$) indicates the weight function w (not in radians), and the thick dotted line ($\cdots$) indicates the high-frequency solution, i.e. the singular part of S_{33}.

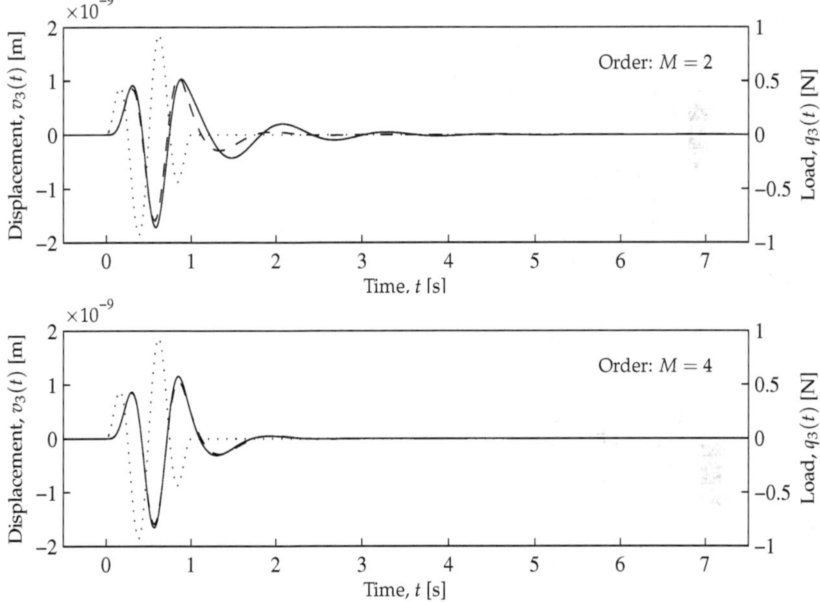

Fig. 15. Response $v_3(t)$ obtained by inverse Fourier transformation ($--$) and lumped-parameter model (——). The dots ($\cdots$) indicate the load time history.

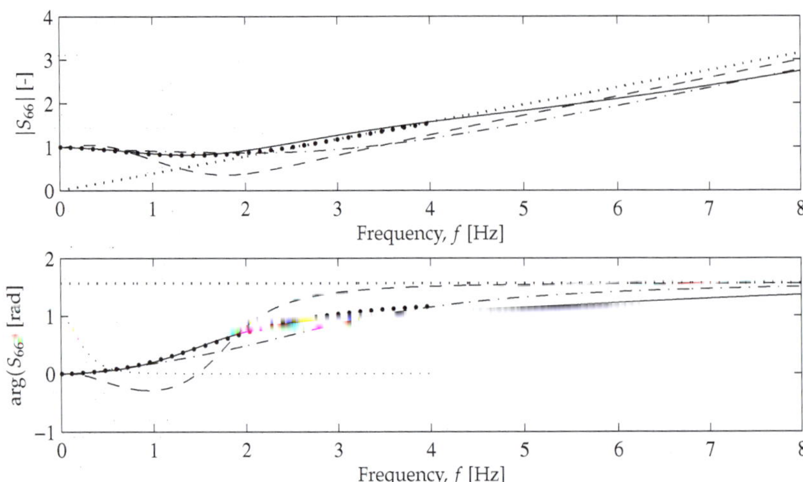

Fig. 16. Dynamic stiffness coefficient, S_{66}, obtained by the domain-transformation model (the large dots) and lumped-parameter models with $M = 2$ ($--$), $M = 4$ ($-\cdot-$), and $M = 6$ (——). The thin dotted line ($\cdots$) indicates the weight function w (not in radians), and the thick dotted line ($\cdots$) indicates the high-frequency solution, i.e. the singular part of S_{66}.

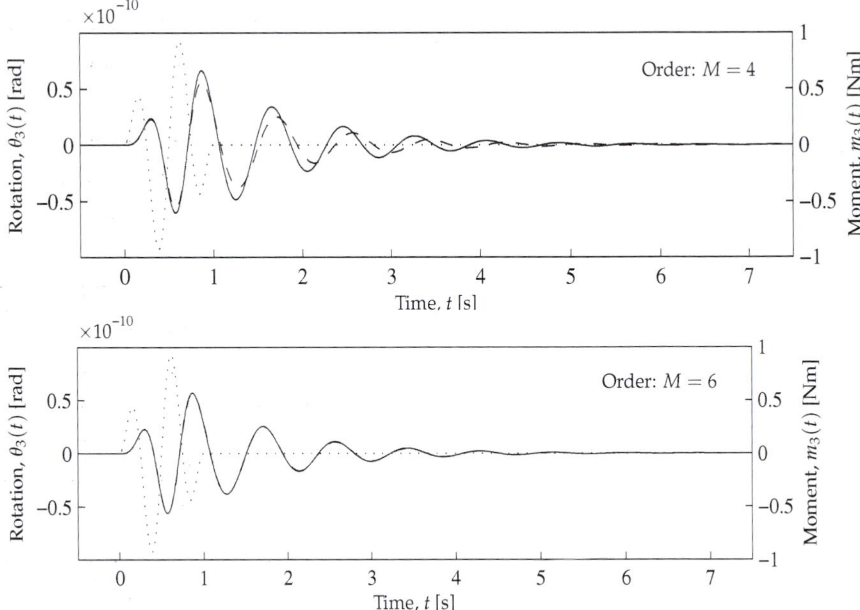

Fig. 17. Response $\theta_3(t)$ obtained by inverse Fourier transformation ($--$) and lumped-parameter model (——). The dots ($\cdots$) indicate the load time history.

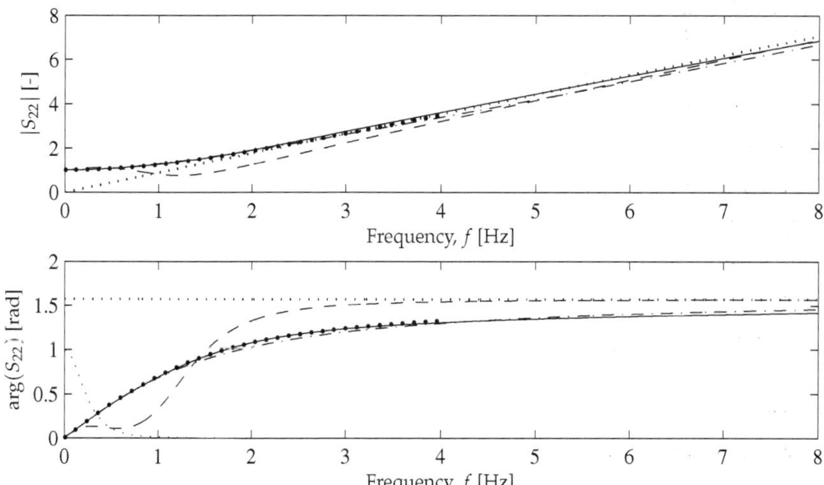

Fig. 18. Dynamic stiffness coefficient, S_{22}, obtained by the domain-transformation model (the large dots) and lumped-parameter models with $M = 2$ ($--$), $M = 4$ ($-\cdot-$), and $M = 6$ (——). The thin dotted line (   ) indicates the weight function w (not in radians), and the thick dotted line ($\cdots\cdots$) indicates the high-frequency solution, i.e. the singular part of S_{22}.

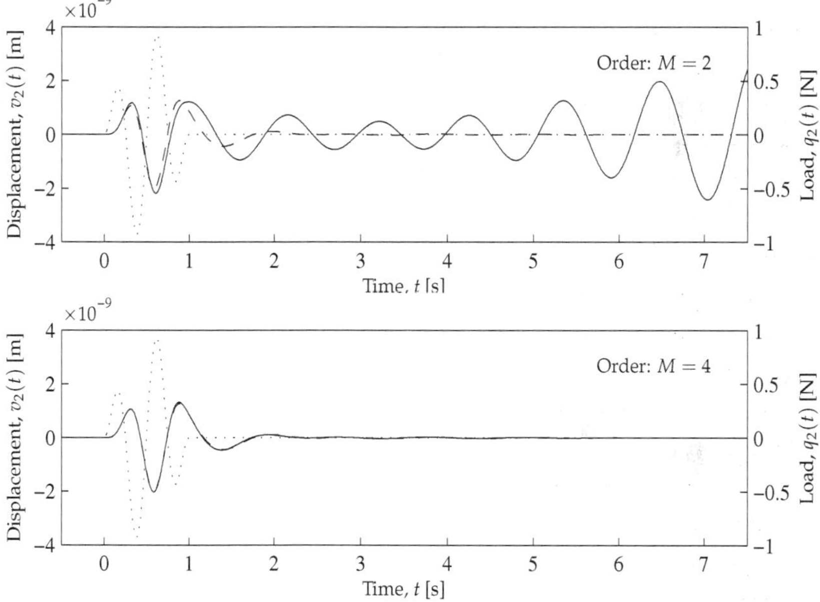

Fig. 19. Response $v_2(t)$ obtained by inverse Fourier transformation ($--$) and lumped-parameter model (——). The dots (   ) indicate the load time history.

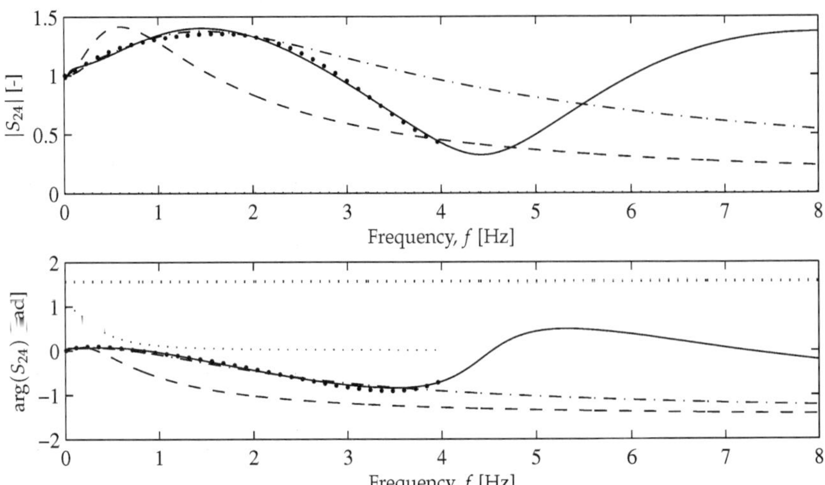

Fig. 20. Dynamic stiffness coefficient, S_{24}, obtained by the domain-transformation model (the large dots) and lumped-parameter models with $M = 2$ ($--$), $M = 4$ ($-\cdot-\cdot$), and $M = 6$ (——). The thin dotted line ($\cdots$) indicates the weight function w (not in radians), and the thick dotted line ($\cdots$) indicates the high-frequency solution, i.e. the singular part of S_{24}.

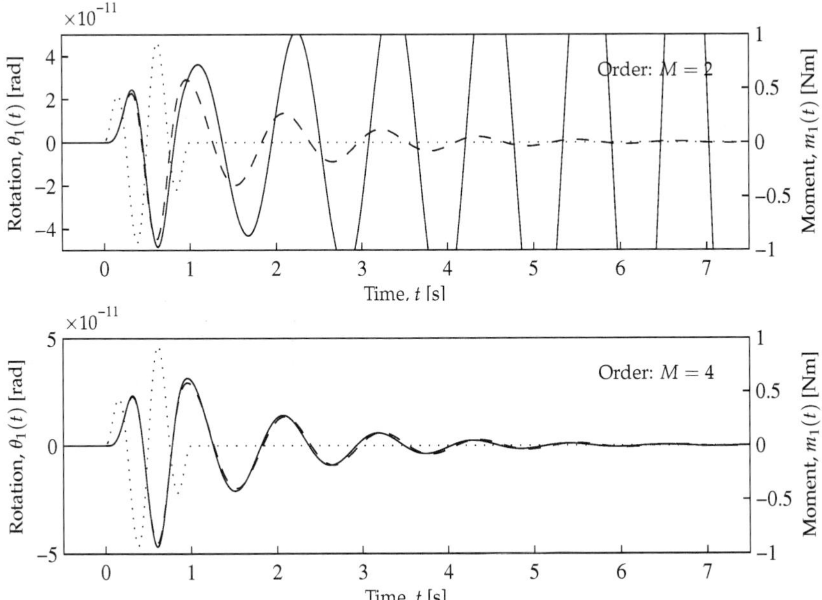

Fig. 21. Response $\theta_1(t)$ obtained by inverse Fourier transformation ($--$) and lumped-parameter model (——). The dots ($\cdots$) indicate the load time history.

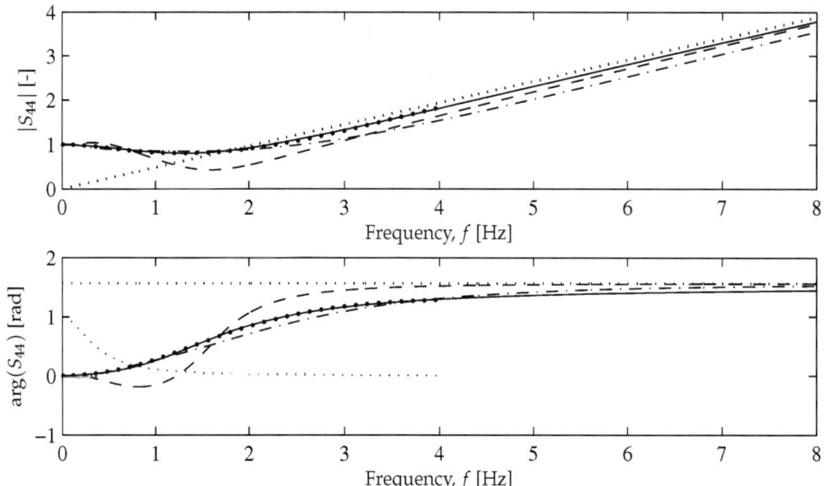

Fig. 22. Dynamic stiffness coefficient, S_{44}, obtained by the domain-transformation model (the large dots) and lumped-parameter models with $M = 2$ ($- -$), $M = 4$ ($- \cdot -$), and $M = 6$ ($——$). The thin dotted line ($\cdots$) indicates the weight function w (not in radians), and the thick dotted line ($\cdots$) indicates the high-frequency solution, i.e. the singular part of S_{44}.

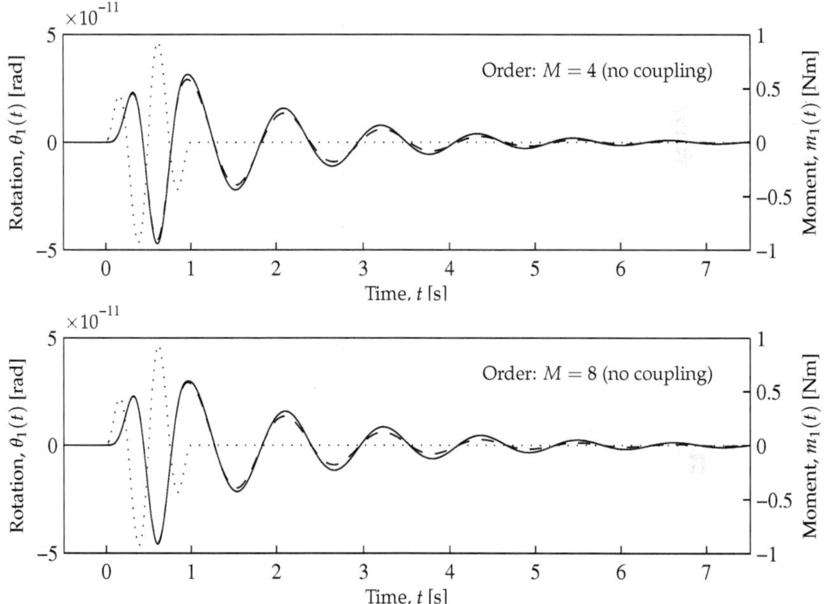

Fig. 23. Response $\theta_1(t)$ obtained by inverse Fourier transformation ($- -$) and lumped-parameter model ($——$). The dots ($\cdots$) indicate the load time history.

In conclusion, for the footing on the homogeneous soil it is found that an LPM with two internal degrees of freedom for the vertical and each sliding and rocking degree of freedom provides a model of great accuracy. This corresponds to fourth-order rational approximations for each of the response spectra obtained by the domain-transformation method. Little improvement is gained by including additional degrees of freedom. Furthermore, it is concluded that little accuracy is lost by neglecting the coupling between the sliding and rocking motion. However, a sixth-order model is necessary in order to get an accurate representation of the torsional impedance.

5.1.2 Example: A footing on a layered half-space

Next, a stratified ground is considered. The soil consists of two layers over homogeneous half-space. Material properties and layer depths are given in Table 3. This may correspond to sand over a layer of undrained clay resting on limestone or bedrock. The geometry and density of the footing are unchanged from the analysis of the homogeneous half-space.

The non-dimensional vertical and torsional impedance components, i.e. S_{33} and S_{66}, are presented in Figs. 24 and 26 as functions of the physical frequency, f. In addition to the domain-transformation method results, the LPM approximations are shown for $M = 2$, $M = 6$ and $M = 10$. Clearly, low-order lumped-parameter models are not able to describe the local tips and dips in the frequency response of a footing on a layered ground. However, the LPM with $M = 10$ provides a good approximation of the vertical and torsional impedances for frequencies $f < 2$ Hz. It is worthwhile to note that the lumped-parameter models of the footing on the layered ground are actually more accurate than the models of the footing on the homogeneous ground. This follows by comparison of Figs. 24 and 26 with Figs. 14 and 16. The time-domain solutions for an applied vertical force, $q_3(t)$, or torsional moment, $m_3(t)$, are plotted in Fig. 25 and Fig. 27, respectively. Evidently, the LPM with $M = 6$ provides an almost exact match to the solution obtained by inverse Fourier transformation—in particular in the case of vertical motion. However, in the case of torsional motion (see Fig. 27), the model with $M = 10$ is significantly better at describing the free vibration after the end of the excitation.

Next, the horizontal sliding and rocking are analysed. The non-dimensional impedance components S_{22}, $S_{24} = S_{42}$ and S_{44} are shown in Figs. 28, 30 and 32 as functions of the frequency, f. Again, the LPM approximations with $M = 2$, $M = 6$ and $M = 10$ are illustrated, and the low-order lumped-parameter models are found to be unable to describe the local variations in the frequency response. The LPM with $M = 10$ provides an acceptable approximation of the sliding, the coupling and the rocking impedances for frequencies $f < 2$ Hz, but generally the match is not as good as in the case of vertical and torsional motion.

The transient response to a horizontal force, $q_2(t)$, or rocking moment, $m_1(t)$, are shown in Figs. 29 and 31. Again, the LPM with $M = 6$ provides an almost exact match to the solution obtained by inverse Fourier transformation. However, the model with $M = 10$ is significantly better at describing the free vibration after the end of the excitation. This is the case for the sliding, $v_2(t)$, as well as the rotation, $\theta_1(t)$.

Layer no.	h (m)	E (MPa)	ν	ρ (kg/m^3)	η
Layer 1	8	10	0.25	2000	0.03
Layer 2	16	5	0.49	2200	0.02
Half-space	∞	100	0.25	2500	0.01

Table 3. Material properties and layer depths for layered half-space.

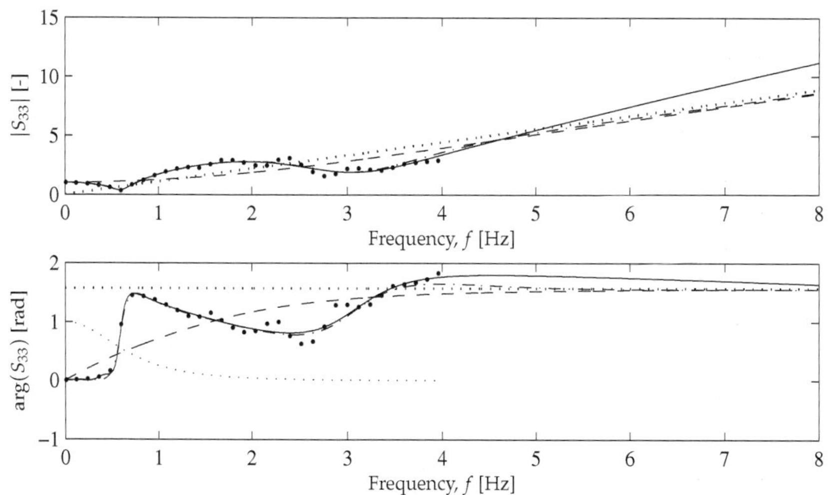

Fig. 24. Dynamic stiffness coefficient, S_{33}, obtained by the domain-transformation model (the large dots) and lumped-parameter models with $M = 2$ ($--$), $M = 6$ ($-\cdot-$), and $M = 10$ (——). The thin dotted line ($\cdots$) indicates the weight function w (not in radians), and the thick dotted line ($\cdots$) indicates the high-frequency solution, i.e. the singular part of S_{33}.

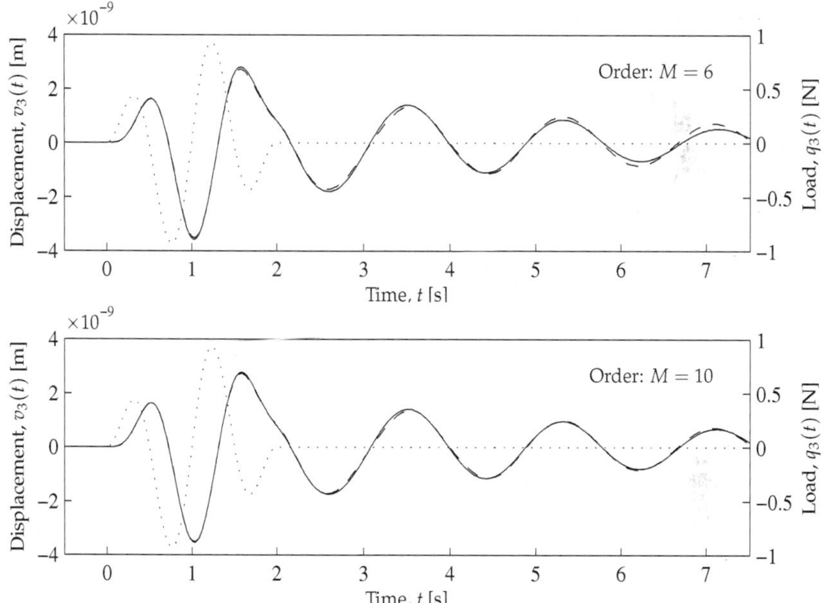

Fig. 25. Response $v_3(t)$ obtained by inverse Fourier transformation ($--$) and lumped-parameter model (——). The dots ($\cdots$) indicate the load time history.

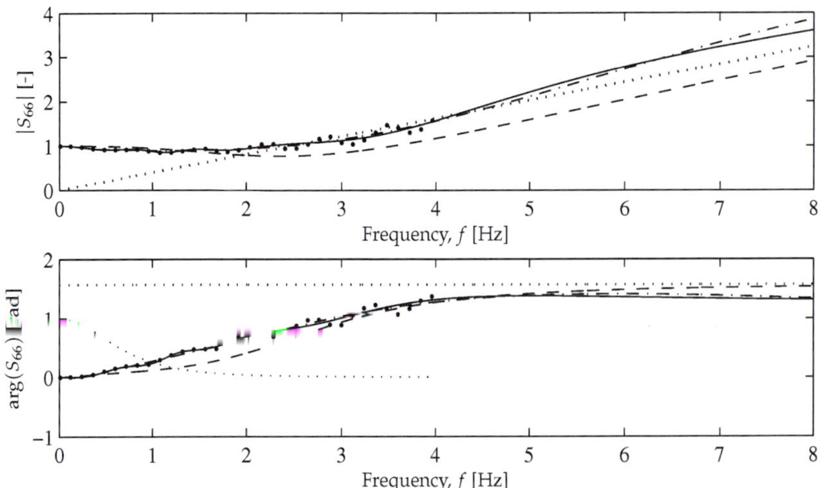

Fig. 26. Dynamic stiffness coefficient, S_{66}, obtained by the domain-transformation model (the large dots) and lumped-parameter models with $M = 2$ ($--$), $M = 6$ ($-\cdot-\cdot$), and $M = 10$ (——). The thin dotted line ($\cdots\cdots$) indicates the weight function w (not in radians), and the thick dotted line ($\cdots\cdots$) indicates the high-frequency solution, i.e. the singular part of S_{66}.

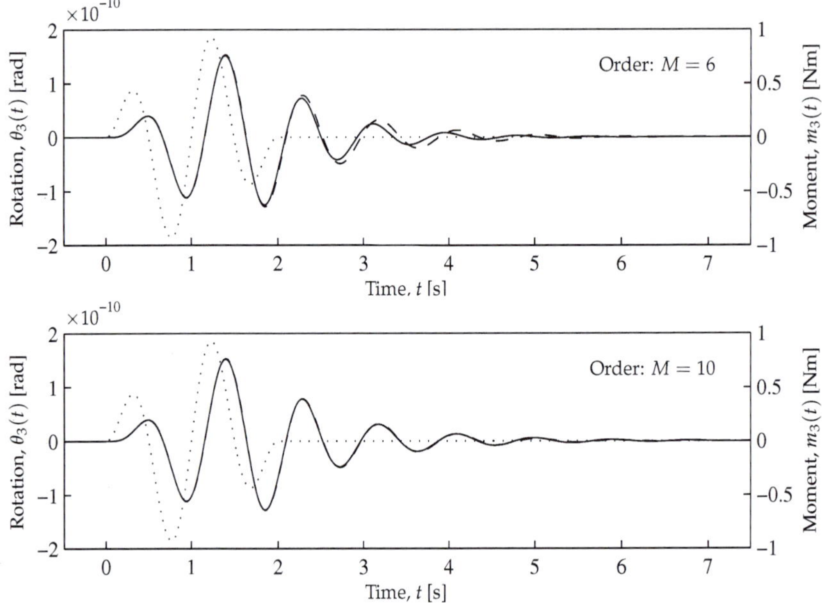

Fig. 27. Response $\theta_3(t)$ obtained by inverse Fourier transformation ($--$) and lumped-parameter model (——). The dots ($\cdots\cdots$) indicate the load time history.

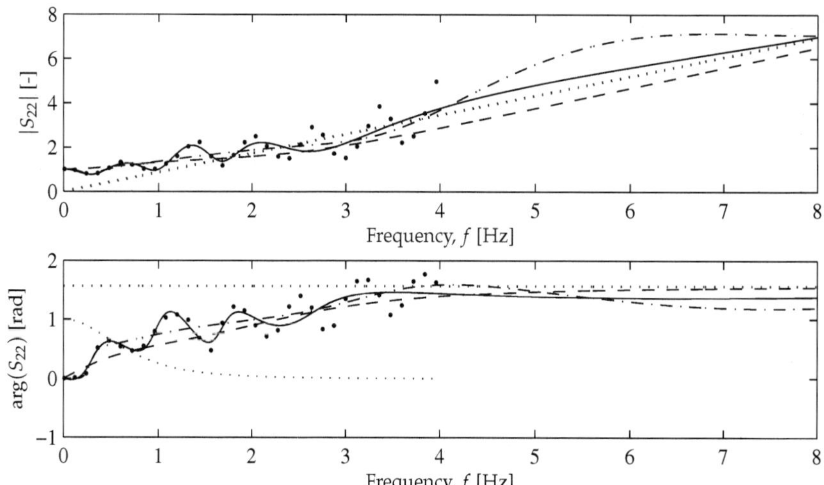

Fig. 28. Dynamic stiffness coefficient, S_{22}, obtained by the domain-transformation model (the large dots) and lumped-parameter models with $M = 2$ ($--$), $M = 6$ ($-\cdot-\cdot$), and $M = 10$ (——). The thin dotted line ($\cdots$) indicates the weight function w (not in radians), and the thick dotted line ($\cdots$) indicates the high-frequency solution, i.e. the singular part of S_{22}.

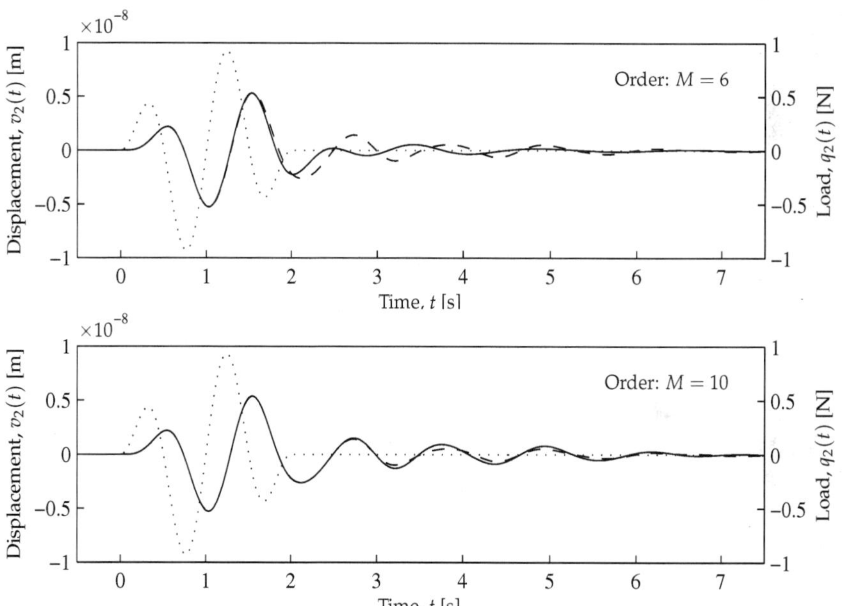

Fig. 29. Response $v_2(t)$ obtained by inverse Fourier transformation ($--$) and lumped-parameter model (——). The dots ($\cdots$) indicate the load time history.

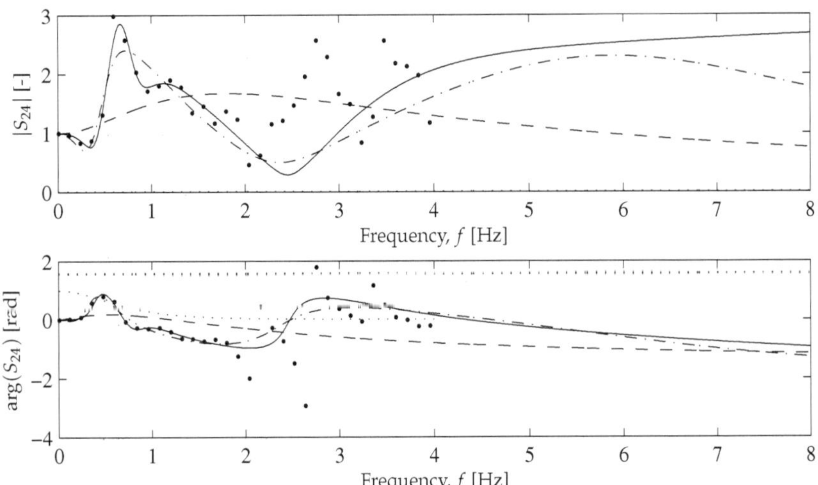

Fig. 30. Dynamic stiffness coefficient, S_{24}, obtained by the domain-transformation model (the large dots) and lumped-parameter models with $M = 2$ (– –), $M = 6$ (– · –), and $M = 10$ (——). The thin dotted line ( ) indicates the weight function w (not in radians), and the thick dotted line (······) indicates the high-frequency solution, i.e. the singular part of S_{24}.

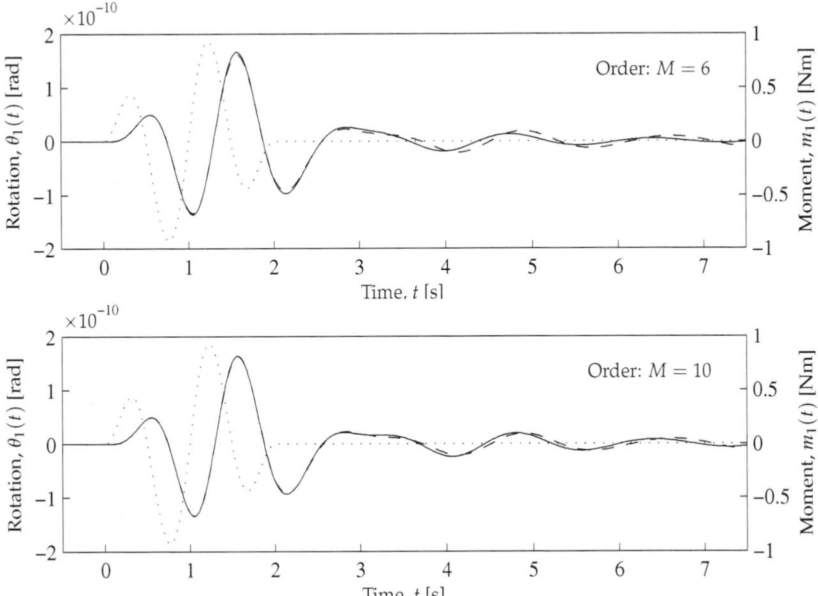

Fig. 31. Response $\theta_1(t)$ obtained by inverse Fourier transformation (– –) and lumped-parameter model (——). The dots ( ) indicate the load time history.

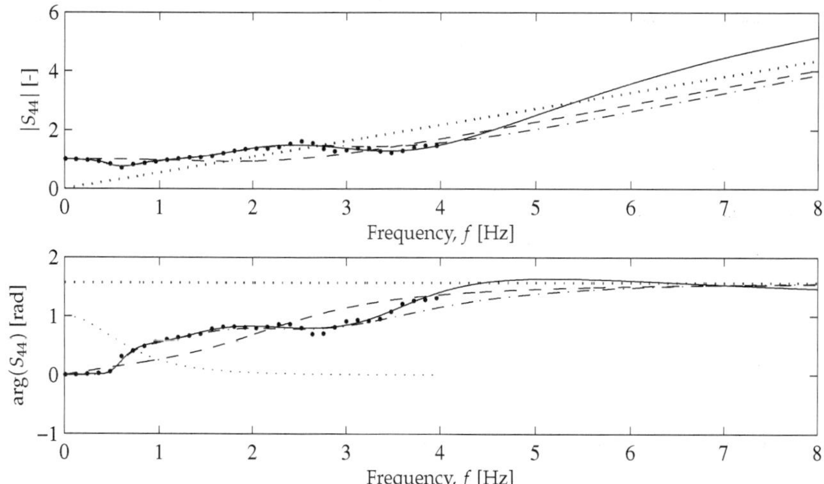

Fig. 32. Dynamic stiffness coefficient, S_{44}, obtained by the domain-transformation model (the large dots) and lumped-parameter models with $M = 2$ (− −), $M = 6$ (−·−), and $M = 10$ (——). The thin dotted line (······) indicates the weight function w (not in radians), and the thick dotted line (······) indicates the high-frequency solution, i.e. the singular part of S_{44}.

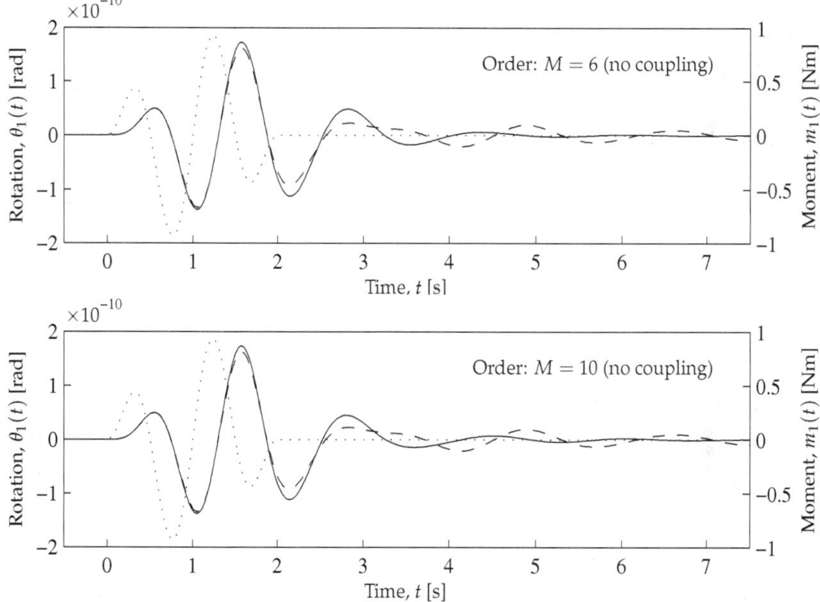

Fig. 33. Response $\theta_1(t)$ obtained by inverse Fourier transformation (− −) and lumped-parameter model (——). The dots (······) indicate the load time history.

Finally, in Fig. 33 the results are given for the alternative LPM, in which the coupling between sliding and rocking has been neglected. It is observed that the maximum response occurring during loading is predicted with almost the same accuracy as by the model in which the coupling is accounted for. However, the geometrical damping is badly described with regard to the decrease in magnitude and, in particular, the phase of the response during the free vibration. Hence, the response of the footing on the layered ground cannot be predicted with low order models, and an LPM with 3–5 internal degrees of freedom is necessary for each nonzero term in the impedance matrix, i.e. rational approximations of the order 6–10 are required. In particular, it is noted that the impedance term providing the coupling between sliding and rocking is not easily fitted by an LPM of low order, i.e. orders below six. The maximum response is well predicted without the coupling terms; however, if the coupling is not accounted for, the geometrical damping is poorly described. This may lead to erroneous conclusions regarding the fatigue lifespan of structures exposed to multiple transient dynamic loads, e.g. offshore wind turbines.

5.2 Example: A flexible foundation embedded in viscoelastic soil

In this example, a skirted circular foundation, also known as a bucket foundation, is analysed using the coupled finite-element method (FEM) and boundary-element method (BEM). The bucket has the radius $r_0 = 10$ m and the skirt length $h_0 = 12$ m, see Fig. 34. The lid has a thickness of $t_{lid} = 0.50$ m, whereas the thickness of the skirt is $t_{skirt} = 50$ mm. The bucket consists of steel, and the material properties are given in Table 4. The model of the lid is unrealistic and the real structure may be lighter and stiffer than the solid plate but with a complex geometry not easily modelled.

Material	E (MPa)	ν	ρ (kg/m^3)	η
Soil	20	0.25	2000	0.03
Steel (bucket)	200.000	0.30	7850	0.01

Table 4. Material properties of the bucket foundation and the subsoil.

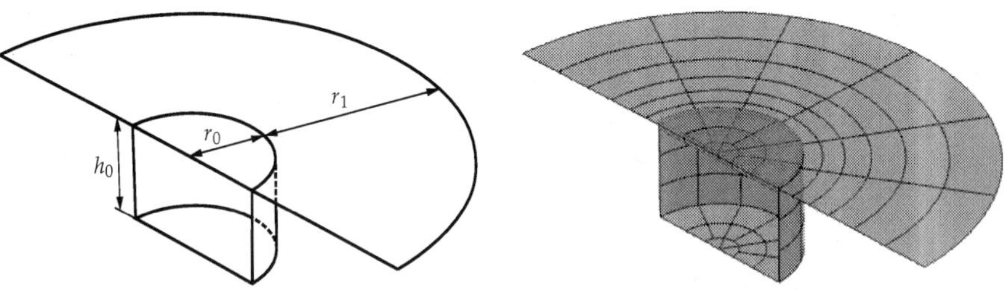

Fig. 34. Coupled finite-element/boundary-element model of skirted foundation: Geometry (left) and discretization (right). Only half the foundation is discretized, utilising the symmetry of the problem.

A thorough introduction to the FEM and BEM is beyond the scope of the present text. The reader is referred to the text books by Bathe (1996) and Petyt (1998) for a detailed explanation of the FEM, whereas information about the BEM can be found, for example, in the work by Brebbia (1982) and Domínguez (1993). The coupling of BEM and FEM schemes has been discussed by, among others, Mustoe (1980), von Estorff & Kausel (1989), Stamos & Beskos (1995), Jones et al. (1999), Elleithy et al. (2001) and Andersen (2002).

In the present example, a closed boundary-element (BE) domain is applied for the soil inside the bucket and an open BE domain is utilised for the remaining half-space. In either case, quadratic spatial interpolation is applied of the displacement and traction fields on the boundaries discretized by boundary elements, i.e. elements with nine nodes are applied. Finally, the lid and the skirt of the bucket foundation are discretized using shell finite elements. Again, quadrilaterals with nine nodes are used.

With ρ, c_P and c_S denoting the mass density, the P-wave velocity and the S-wave velocity of the soil, respectively, the high-frequency limit of the impedance components are given as

$$c_{33}^{\infty} = \rho c_P A_{lid} + \rho c_S A_{skirt}, \tag{109a}$$

$$c_{66}^{\infty} = 2\rho c_S \mathcal{I}_{lid} + \rho c_S A_{skirt} r_0^2, \tag{109b}$$

$$c_{22}^{\infty} = \rho c_S A_{lid} + \frac{1}{2} (\rho c_S + \rho c_P) A_{skirt}, \tag{109c}$$

$$c_{24}^{\infty} = c_{42}^{\infty} = -\frac{1}{2} (\rho c_S + \rho c_P) A_{skirt} \frac{h_0}{2} = -c_{15}^{\infty} = -c_{51}^{\infty}, \tag{109d}$$

$$c_{44}^{\infty} = c_{55}^{\infty} = \rho c_P \mathcal{I}_{lid} + \frac{1}{2} (\rho c_S + \rho c_P) \mathcal{I}_{skirt} + \frac{1}{2} \rho c_S A_{skirt} r_0^2, \tag{109e}$$

where $A_{lid} = \pi r_0^2$ and $A_{skirt} = 4\pi r_0 h_0$ are the areas of the lid and the skirt, respectively. The latter accounts for the inside as well as the outside of the skirt; hence, the factor 4 instead of the usual factor 2. Further, $\mathcal{I}_{lid}$ and $\mathcal{I}_{skirt}$ are the geometrical moments of inertia of the lid and the skirt around the centroid of the lid, defined as

$$\mathcal{I}_{lid} = \frac{\pi}{4} r_0^4, \qquad \mathcal{I}_{skirt} = \frac{4\pi}{3} r_0 h_0^3. \tag{110}$$

Again, the contributions from both sides of the skirt are included and it is noted that the torsional moment of inertia of the lid is simply $2\mathcal{I}_{lid}$.

In the following, an explanation is given of the terms in Eq. (109). Firstly, the vertical impedance is given as the sum of a contribution from the P-waves emanating from the bottom of the lid, represented by the first term in Eq. (109a), and a second contribution from the S-waves produced at the exterior and interior surfaces of the skirt when the foundation moves up and down as a rigid body. This may be an overestimation of the impedance for a flexible foundation and alternatively a high-frequency solution with no contributions from the skirt may be proposed.

Secondly, the torsional impedance provided by Eq. (109b) contains a contribution from the S-waves generated at the bottom of the lid, whereas the second term represents the S-waves stemming from the rotation of the skirt around the vertical axis. Thirdly, the horizontal impedance given by Eq. (109c) is composed of three terms. The first term represent S-waves initiated at the bottom of the lid, whereas the second term is due to the S- and P-waves arising at the skirt. Only S-waves are generated at the vertical lines at the "sides" of the foundation, and only P-waves are produced at the vertical lines on the "back" and "front"

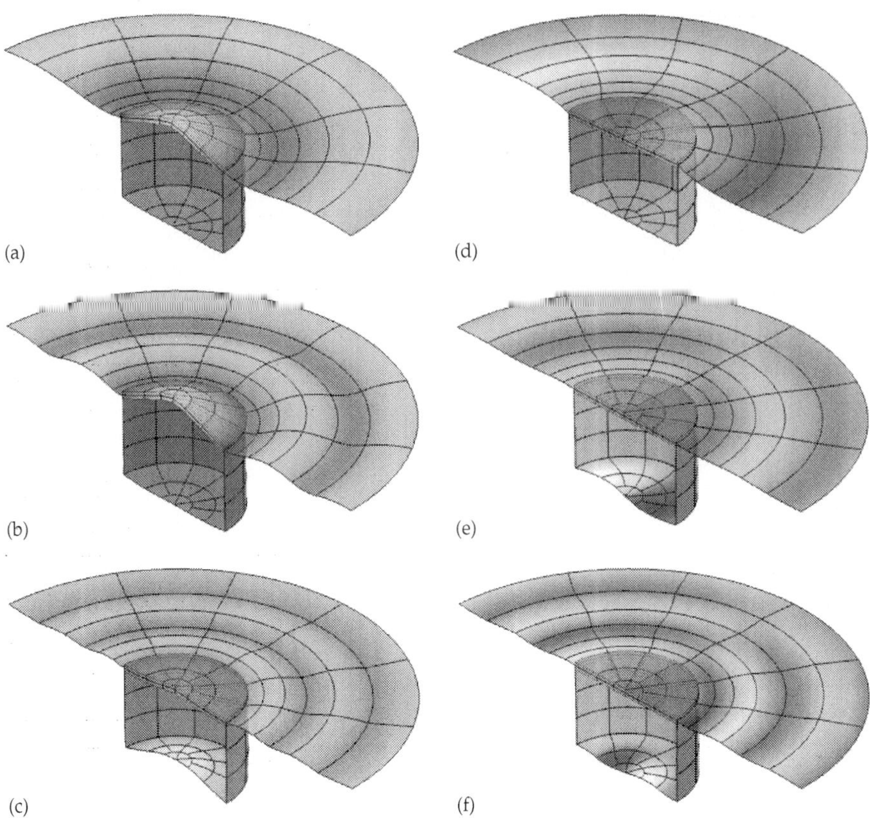

(a)   (d)

(b)   (e)

(c)   (f)

Fig. 35. Response in phase with the load for the skirted foundation. On the left, the results
are shown for a unit-magnitude harmonic vertical force acting at the centre of the foundation
at the frequencies (a) $f = 2\,\mathrm{Hz}$, (b) $f = 4\,\mathrm{Hz}$ and (c) $f = 6\,\mathrm{Hz}$. The displacements are scaled
by a factor of 10^9 and the light and dark shades of grey indicate positive and negative
vertical displacements. On the right, the results are shown for a unit-magnitude harmonic
torsional moment acting around the centre of the foundation at the frequencies (a) $f = 2\,\mathrm{Hz}$,
(b) $f = 4\,\mathrm{Hz}$ and (c) $f = 6\,\mathrm{Hz}$. The displacements are scaled by a factor of 10^{10} and the light
and dark shades of grey indicate positive and negative displacements in the direction
orthogonal to the plane of antisymmetry.

of the bucket. However, at all other vertical lines, a combination of P- and S-waves are
emitted. A formal mathematical proof of Eq. (109c) follows by integration along the perimeter
of the skirt. However, by physical reasoning one finds that half the area A_{skirt} emits P-waves,
concentrating around the "back" and "front" of the foundation, whereas the other half of
A_{skirt}, i.e. the "sides", emits S-waves. A similar reasoning lies behind the derivation of c_{24}^{∞},
and as indicated by Eq. (109d) there are no contributions from the lid to the sliding–rocking
coupling at high frequencies.

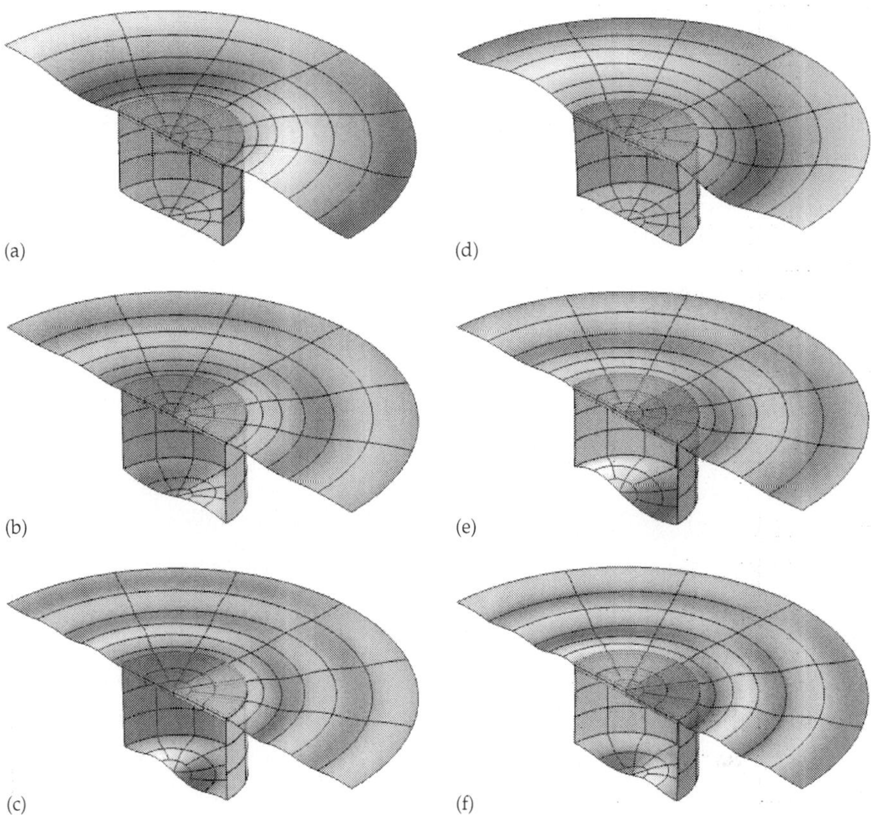

Fig. 36. Response in phase with the load for the skirted foundation. On the left, the results are shown for a unit-magnitude harmonic horizontal force acting at the centre of the foundation at the frequencies (a) $f = 2\,\mathrm{Hz}$, (b) $f = 4\,\mathrm{Hz}$ and (c) $f = 6\,\mathrm{Hz}$. The displacements are scaled by a factor of 10^9 and the light and dark shades of grey indicate positive and negative vertical displacements. On the right, the results are shown for a unit-magnitude harmonic rocking moment acting around the centre of the foundation at the frequencies (a) $f = 2\,\mathrm{Hz}$, (b) $f = 4\,\mathrm{Hz}$ and (c) $f = 6\,\mathrm{Hz}$. The displacements are scaled by a factor of 10^{10} and the light and dark shades of grey indicate positive and negative vertical displacements.

Finally, the rocking impedance provided by Eq. (109e) consists of three parts. The first one stems from P-waves originating from the bottom of the lid and the next term is a mixture of P- and S-waves generated at the skirts. However, only the S-waves polarised in the horizontal direction are included in the second term of Eq. (109e); but the rocking motion of the foundation also induces S-waves polarised in the vertical direction, in particular at the "back" and "front" of the bucket. Again, a strict proof follows by integration over the surface of the skirts, but a by physical reasoning it is found the half the area contributes to the generation of such S-waves.

The response of the bucket foundation is computed at 31 discrete frequencies from 0 to 6 Hz using the coupled boundary-element–finite-element model. The results for vertical and torsional excitation and three different frequencies are shown in Fig. 35, whereas the corresponding results for horizontal sliding and rocking are given in Fig. 36. It is noted that the light and dark shades of grey indicate vertical displacements upwards and downwards, respectively, in the plots for the vertical, horizontal and rocking-moment excitation. However, in the case of torsional excitation, no vertical displacements are generated. Hence, in this particular case the light/dark shades indicate horizontal displacements away from/towards the plane of symmetry.

Figures 37 to 40 show the frequency-domain solution obtained by the coupled finite-element–boundary-element schemes for the different impedance components, excluding the vertical impedance. The results of the corresponding lumped-parameter models of orders 2, 6 and 10 are plotted in the same figures. Clearly, the low-order models with $M = 2$, and having only a single internal degree-of-freedom, are inadequate. A similar conclusion can be made for the vertical impedance that is not shown here. However, the sixth-order lumped-parameter model provide a good fit to the FE–BE model in the low-frequency range and only an insignificant improvement is obtained by increasing the order of the LPMs to 10. It is of particular interest that the high-order LPM with $M = 10$ does not lead to a better fit at higher frequencies than the sixth-order model. Instead, wiggling occurs at low-frequencies with the tenth-order model, i.e. non-physical tips and dips arise in the results from the LPM between frequencies at which the target FE-BE solution has been computed. This behaviour should be avoided and thus the sixth-order model is preferred. In order to obtain a better fit without wiggling, a smaller frequency step should be applied in the FE-BE solution.

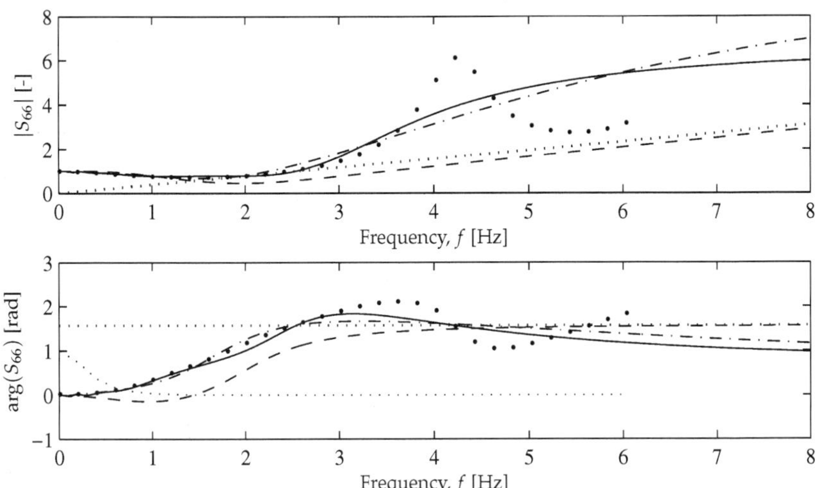

Fig. 37. Dynamic stiffness coefficient, S_{66}, obtained by finite-element–boundary-element (the large dots) and lumped-parameter models with $M = 2$ ($--$), $M = 6$ ($-\cdot-\cdot$), and $M = 10$ (——). The thin dotted line ($\cdots$) indicates the weight function w (not in radians), and the thick dotted line ($\cdots$) indicates the high-frequency solution, i.e. the singular part of S_{66}.

7

Determination of Rotor Imbalances

Jenny Niebsch
Radon Institute of Computational and Applied Mathematics, Austrian Academy of Sciences
Austria

1. Introduction

During operation, rotor imbalances in wind energy converters (WEC) induce a centrifugal force, which is harmonic with respect to the rotating frequency and has an absolute value proportional to the square of the frequency. Imbalance driven forces cause vibrations of the entire WEC. The amplitude of the vibration also depends on the rotating frequency. If it is close to the bending eigenfrequency of the WEC, the vibration amplitudes increase and might even be visible. With the growing size of new WEC, the structure has become more flexible. As a side effect of this higher flexibility it might be necessary to pass through the critical speed in order to reach the operating frequency, which leads to strong vibrations. However, even if the operating frequency is not close to the eigenfrequency, the load from the imbalance still affects the drive train and might cause damage or early fatigue on other components, e.g., in the gear unit. This is one possible reason why in most cases the expected problem-free lifetime of a WEC of 20 years is not achieved. Therefore, reducing vibrations by removing imbalances is getting more and more attention within the WEC community.

Present methods to detect imbalances are mainly based on the processing of measured vibration data. In practice, a Condition Monitoring System (CMS) records the development of the vibration amplitude of the so called 1p vibration, which vibrates at the operating frequency. It generates an alarm if a pre-defined threshold is exceeded. In (Caselitz & Giebhardt, 2005), more advanced signal processing methods were developed and a trend analysis to generate an alarm system was presented. Although signal analysis can detect the presence of imbalances, the task of identify its position and magnitude remains.

Another critical case arises when different types of imbalances interfere. The two main types of rotor imbalances are mass and aerodynamic imbalances. A mass imbalance occurs if the center of gravitation does not coincides with the center of the hub. This can be due to various factors, e.g., different mass distributions in the blades that can originate in production inaccuracies, or the inclusion of water in one or more blades. Mass imbalances mainly cause vibrations in radial direction, i.e., within the rotor plane, but also smaller torsional vibrations since the rotor has a certain distance from the tower center, acting as a lever for the centrifugal force. Aerodynamic imbalances reflect different aerodynamic behavior of the blades. As a consequence the wind attacks each blade with different force and moments. This also results in vibrations and displacements of the WEC, here mainly in axial and torsional direction, but also in contributions to radial vibrations. There are multiple causes for aerodynamic imbalances, e.g., errors in the pitch angles or profile changes of the blades. The major differences in the impact of mass and aerodynamic imbalances are the main directions of the induced vibrations and the fact that aerodynamic imbalance loads change with the wind velocity. Nevertheless, if the presence of aerodynamic imbalances is neglected in the modeling procedure, the determination of the mass imbalance can be faulty, and in the worst case, balancing with the determined weights can even increase the mass imbalance. As a consequence, the methods to determine mass imbalance need to ensure the absence of aerodynamic imbalances first.

In the field, the balancing process of a WEC is done as follows. An on-site expert team measures the vibrations in the radial, axial and torsion directions. Large axial and torsion vibrations indicate aerodynamic imbalances. The surfaces of the blades are investigated and optical methods are used to detect pitch angle deviation. The procedure to determine the mass imbalance is started after the cause of the aerodynamic imbalance is removed. In this procedure, the amplitude of the radial vibration is measured at a fixed operational speed, typically not too far away from the bending eigenfrequency. Afterwards a test mass (usually a mass belt) is placed at a distinguished blade and the measurements are repeated. From the reference and the original run, the mass imbalance and its position can be derived. Altogether, this is a time consuming and personnel-intensive procedure.

In (Ramlau & Niebsch, 2009) a procedure was presented that reconstructed a mass imbalance from vibration measurements without using test masses. The main idea in this approach is to replace the reference run by a mathematical model of the WEC. At this stage, only mass imbalances were considered. A simultaneous investigation of mass and aerodynamic imbalances was investigated by Borg and Kirchdorf, (Borg & Kirchhoff, 1998). The contribution of mass and aerodynamic imbalances to the 1p, 2p and 3p vibration was examined using a perturbation analysis in order to solve the differential equation that coupled the azimuth and yaw motion. Using the example of an NREL 15 kW turbine, the presence of 60 % mass imbalance and 40% aerodynamic imbalance explained by a 1 degree pitch angle deviation was observed. In (Nguyen, 2010) and (Niebsch et al., 2010) the model based determination of imbalances was expanded to the case of the presence of both mass imbalances and pitch angle deviation.

The main aim of this chapter is the presentation of a mathematical theory that allows the determination of mass and aerodynamical imbalances from vibrational measurements only. This task forms a typical inverse problem, i.e., we want to reconstruct the cause of a measured observation. In many cases, inverse problems are ill posed, which means that the solution of the problem does not depend continuously on the measured data, is not unique or does not exist at all. One consequence of ill-posedness is that small measurement errors might cause large deviations in the reconstruction. In order to stabilize the reconstruction, regularization methods have to be used, see Section 3.

Finding the solution of the inverse problem requires a good forward model, i.e., a model that computes the vibration of the WEC for a given imbalance distribution. This is realized by a structural model of the WEC, see Section 2. The determination of mass imbalances is briefly explained in Section 4. The mathematical description of loads from pitch angle deviations is considered in the same section as well. Section 5 presents the basic principle of the combined reconstruction of mass and aerodynamic imbalances.

2. Structural model of a wind turbine

2.1 The mathematical model

A structural dynamical model of an object or machine allows to predict the behavior of that object subjected to dynamic loads. There is a large variety of literature as well as software addressing this topic. Here, we followed the book (Gasch & Knothe, 1989), where the WEC tower is modeled as a flexible beam, the rotor and nacelle are treated as point masses. The computation of displacements from dynamic loads can be described by a partial differential equation (PDE) or an equivalent energy formulation. Usually, both formulations do not result in an analytical solution. Using Finite Element Methods (FEM), the energy formulation can be transformed into a system of ordinary differential equations (ODE). The object, in our case the wind turbine, has to be divided into elements, here beam elements, with nodes at each end of an element, see Figure 1. The displacement of an arbitrary point of the element is approximated by a combination of the displacements of the start and the end node. The ODE system connecting dynamical loads and object displacements has the form

$$M\mathbf{u}''(t) + S\mathbf{u}(t) = \mathbf{p}(t). \tag{1}$$

Here, t denotes the time. The displacements are combined in the vector $\mathbf{u}$, which contains the degrees of freedom (DOF) of each node in our FE model. The degrees of freedom in each node can be the displacement (u, v, w) in all three space directions as well as torsion around the x-axis and cross sections slopes in the (x, y)- and (x, z)-plane: $(u, v, w, \beta_x, \beta_y, \beta_z)$, cf. Figure 2. The physical properties of our object are represented by the mass matrix M and the stiffness matrix S. The load vector $\mathbf{p}$ contains the dynamic load in each node arising from forces and moments. For this calculation, damping is neglected. Otherwise the term $D\mathbf{u}'$ with damping matrix D adds to the left hand side of equation (1). Considering mass imbalances

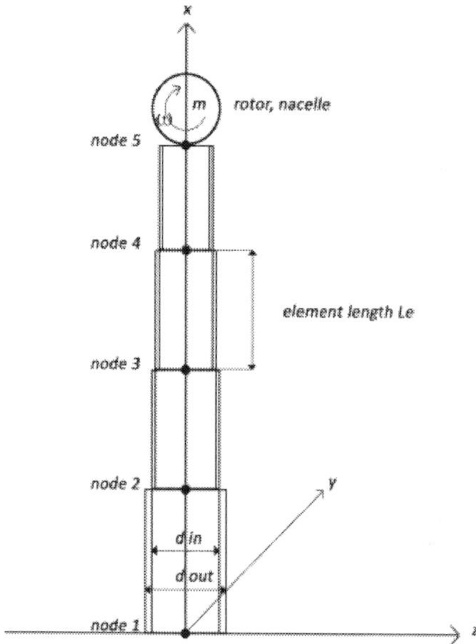

Fig. 1. Elements in a Finite Element model of a WEC

only, the forces and moments mainly act in radial direction, i.e., along the z-axis, and result in displacements and cross section slopes in that direction. Therefore, for each node we only consider the DOF (w, β_z). In order to construct the mass and the stiffness matrix each element

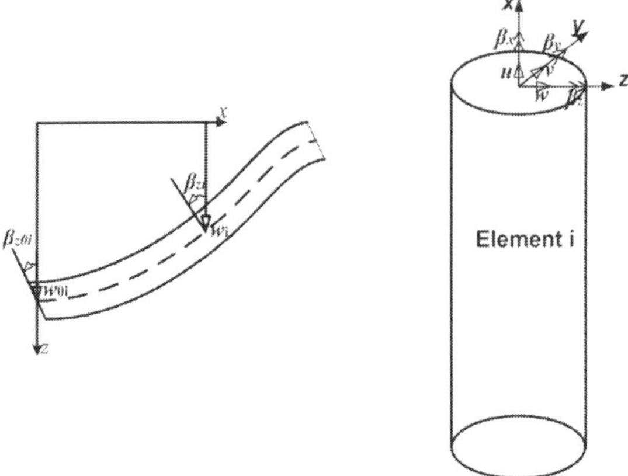

Fig. 2. Degrees of freedom in a Finite Element model of a WEC

is treated separately. The DOF of the bottom and the top node of the ith element are collected in the element DOF vector, cf. Figure 2,

$$\mathbf{u}_e^i = [w_{0i} \ \beta_{z0i} \ w_i \ \beta_{zi}]^T. \tag{2}$$

The derivation of the element mass and stiffness matrix M_e and S_e uses four shape functions scaled by the DOF of the bottom and top node to describe the DOF $(w_i(x), \beta_{zi}(x))$ of an arbitrary point x of the element. It is given in detail in (Gasch & Knothe, 1989). We only want to present the final formulas for the element matrices,

$$M_e = \frac{\mu L_e}{420} \begin{pmatrix} 156 & -22L_e & 54 & 13L_e \\ -22L_e & 4L_e^2 & -13L_e & -3L_e^2 \\ 54 & -13L_e & 156 & 22L_e \\ 13L_e & -3L_e^2 & 22L_e & 4L_e^2 \end{pmatrix}, \quad S_e = \frac{E \cdot I}{L_e^3} \begin{pmatrix} 12 & -6L_e & -12 & -6L_e \\ -6L_e & 4L_e^2 & 6L_e & 2L_e^2 \\ -12 & 6L_e & 12 & 6L_e \\ -6L_e & 2L_e^2 & 6L_e & 4L_e^2 \end{pmatrix}. \tag{3}$$

The length of the element is represented by L_e. E is Young's modulus, which is a material constant that can be found in a table . We assume our elements to be circular beam sections. The transverse moment of inertia I is given by $I = \pi/64 \cdot (d_{e,out}^4 - d_{e,in}^4)$ with outer and inner diameter of the beams section. μ is the translatorial mass per length $\mu = \varrho \cdot A$, where ϱ is the density of the material. $A = \pi/4 \cdot (d_{e,out}^2 - d_{e,in}^2)$ is the annulus area. To build the full system matrices S and M, the element matrices S_e and M_e are combined by superimposing the elements affecting the upper node of the ith element matrix with the ones belonging to the lower node of the $(i+1)$st element matrix, see Figure 3. The sum of rotor mass and nacelle mass m needs to be added to the last but one diagonal element of the full mass matrix. As mentioned above, the described model is restricted to radial displacements that are induced by radial forces, e.g., from mass imbalances. If we consider other types of load, e.g., aerodynamic, we have to deal with forces and moments in all three space directions. The derivation of the corresponding mass and stiffness matrix is a bit more comprehensive. In a general and abbreviated form it is given in (Gasch & Knothe, 1989). The application for a WEC is presented in Niebsch et al. (2010), and in a more detailed version in (Nguyen, 2010).

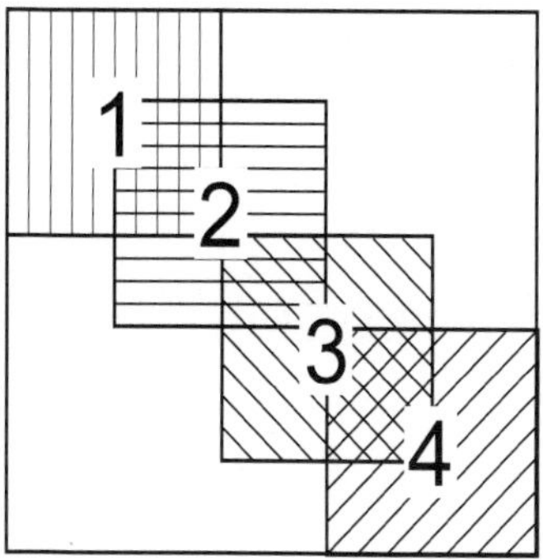

Fig. 3. System matrix and superimposed element matrices

2.2 Model optimization

Once M and S are determined, the solution of equation (1) for a given load $\mathbf{p}$ provides the displacement of each node in our model. We remark that the FEM is an approximative method. Additionally, the idealization of WEC as a flexible beam with a point mass as well as slight deviations in the geometric and physical parameters lead to model that approximates the reality but can not reproduce it exactly. Hence the system properties of our model, described by M and S, might differ slightly from the properties of the real WEC. In order to calibrate the model to the real WEC we have to chose one or more parameters that can be measured at the real WEC and then optimize our model according to those parameters. For our application the most important parameter of a WEC is the first (bending) eigenfrequency of the system. For each WEC type a range for the first eigenfrequency is given by the manufacturer, e.g., a VESTAS V80 of 100 m height has an eigenfrequency in the range $[0.21, \cdots , 0.255]$Hz. The actual eigenfrequency of a specific WEC of any type depends, e.g., on the grounding of the WEC and manufacturing tolerances in geometry and material. The eigenfrequency can be obtained from measurements during the performance of an emergency stop of the WEC. Thus our model, i.e., the matrices M and S, derived for a certain type of WEC from given geometrical and physical parameters as described above, can be optimized for specific WECs of that type with respect to the measured first eigenfrequency. The first eigenfrequency of the model can be computed using the assumption $\mathbf{u}(t) = \mathbf{u_0} \exp(\lambda t)$ and inserting it in the homogenous form of (1). Then we have to solve

$$\lambda^2 I \mathbf{u_0} = -M^{-1} S \mathbf{u_0} , \tag{4}$$

i.e., λ^2 are the eigenvalues of the matrix $-M^{-1}S$. For example, they can be obtained with the Matlab function *eig*. The eigenvalues are complex numbers. In the absence of damping, as in our case, the real part vanishes. The eigenfrequencies ω_{eig} are given by the imaginary part:

$$\omega_{eig} = \pm \Im \left(\sqrt{-eig(M^{-1}S)} \right) . \tag{5}$$

The rotational first eigenfrequency is then given by

$$\Omega_0 = \frac{\min\{\omega_{eig}\}}{2\pi} . \tag{6}$$

Usually there is no information of the foundation and grounding available whereas manufacturing tolerances in the geometry, i.e., the length and the inner and outer diameter of the beam elements are accessible in the modeling process. In fact, Ω_0 is a function of those parameters. We can chose the geometric parameters from realistic intervals of manufacturing tolerances in such a way that the new model eigenfrequency is very close to the measured one. Supposing Ω is the measured first eigenfrequency of the WEC, the optimal geometric parameters can be found by minimizing the functional

$$\min_{\mathbf{L},\mathbf{d}_{in},\mathbf{d}_{out}} |\Omega - \Omega_0(\mathbf{L}, \mathbf{d}_{in}, \mathbf{d}_{out})|, \tag{7}$$

where the vectors $\mathbf{L}, \mathbf{d}_{in}, \mathbf{d}_{out}$ contain the length, inner and outer diameter of each element.

3. Introduction to inverse problems

Within this Section, we would like to introduce some basic concepts from the theory of inverse and ill posed problems. We will focus in particular on regularization theory, which has been extensively developed over the last decades. As we will see, regularization is always needed when the solution of a problem does not depend continuously on the data, which causes in particular problems if the data originate from (noisy) measurements. For details, we refer to (Engl et al., 2000).

We assume that the connection of two terms f and g such as an imbalance and the displacements resulting from that imbalance, is described by an operator $\mathbf{A}$:

$$\mathbf{A}f = g. \tag{8}$$

The computation of g for given f is called the forward problem while the determination of f for given g is referred to as the inverse problem. In practical applications the exact data g are not known but a measured noisy version g^δ of that data. We assume that the noise level is bounded by an unknown number δ, i.e.,

$$\|g - g^\delta\| \leq \delta. \tag{9}$$

The computation of an imbalance from vibration/displacement data is an inverse problem. If the following three conditions are fulfilled, the Inverse Problem is called **well posed**:

1. For all data g there exists a solution f.

2. The solution f is unique.

3. The solution f depends continuously on the data g. ($\mathbf{A}^{-1}$ is continuous.)

The last condition ensures that small changes in the data g result in small changes in the solution f. A well posed inverse problem can be solved by applying the inverse operator to the data:

$$f = \mathbf{A}^{-1}g. \tag{10}$$

If one of the conditions is violated the inverse problem is called **ill posed**.

The violation of condition 1 can be fixed by the definition of a generalized solution. We compute our solution as the least-squares solution taking f as the element that minimizes the distance of $\mathbf{A}f$ to the data g:

$$f^\dagger = arg \min_f \|\mathbf{A}f - g\|^2. \tag{11}$$

The operator that maps the data g to the least-squares solution $f^\dagger$ is denoted by $\mathbf{A}^\dagger$ and called generalized inverse of $\mathbf{A}$. The violation of condition 2 can be rectified by distinguishing one solution from the set of all solutions. It can be the solution with the smallest norm or the one that best fits prior known properties of the desired solution.

In condition 3 we have to deal with the discontinuous inverse or generalized inverse operator. Small errors in the data can result in huge errors in the solution. To avoid this behavior, the discontinuous inverse is approximated pointwise by a family of continuous operators. To be more precise, we have to find a family of operators $\mathbf{T}_\alpha$, with a regularization parameter $\alpha = \alpha(\delta, g^\delta)$, that fulfills the conditions

$$\alpha(\delta) \xrightarrow{\delta \to 0} 0, \quad \lim_{\delta \to 0} \mathbf{T}_\alpha g^\delta = \mathbf{A}^\dagger g. \tag{12}$$

This implies that for very small data error δ the parameter α becomes small and the corresponding continuous $\mathbf{T}_\alpha$ is a good approximation to $\mathbf{A}^\dagger$. The right choice of α is difficult because the error we get by computing $f_\alpha^\delta = \mathbf{T}_\alpha g^\delta$ as an approximate solution of $f^\dagger = \mathbf{A}^\dagger g$ has two parts that behave very differently:

$$\|\mathbf{T}_\alpha g^\delta - \mathbf{A}^\dagger g\| \leq \underbrace{\|\mathbf{T}_\alpha g^\delta - \mathbf{T}_\alpha g\|}_{propagated\ data\ error} + \underbrace{\|\mathbf{T}_\alpha g - \mathbf{A}^\dagger g\|}_{approximation\ error} . \tag{13}$$

The approximation error decreases with α while the propagated data error increases with decreasing α, cf. Figure 4. This is due to the fact that for small α the operator $\mathbf{T}_\alpha$ is closer to $\mathbf{A}^\dagger$ and thus "less continuous" than for bigger α. The total error has a minimum away from $\alpha = 0$. To find the parameter α with minimal error $\|\mathbf{T}_\alpha g^\delta - \mathbf{A}^\dagger g\|$, a parameter choice rule is necessary. The operator family defined in (12) combined with a parameter choice rule is called **regularization method**.

A widely used example for a regularization method is Tikhonov's regularization where the operator $\mathbf{T}_\alpha$ is given by

$$\mathbf{T}_\alpha = (\mathbf{A}^*\mathbf{A} + \alpha\mathbf{I})^{-1}\mathbf{A}^*, \tag{14}$$

where $\mathbf{I}$ is the identity and $\mathbf{A}^*$ denotes the adjoint operator of $\mathbf{A}$. In case $\mathbf{A}$ is a matrix, $\mathbf{A}^*$ is the transpose of $\mathbf{A}$. Alternatively, $f_\alpha^\delta = \mathbf{T}_\alpha g^\delta$ can be characterized as the unique minimizer of the Tikhonov functional

$$J_\alpha(f) = \|\mathbf{A}f - g^\delta\|^2 + \alpha\|f\|^2. \tag{15}$$

The characterization of f_α^δ via the Tikhonov functional is in particular important as it allows a straightforward generalization for nonlinear operators. The linear operator can simply be replaced by a nonlinear operator. We mention this because the consideration of aerodynamic imbalances leads to a nonlinear operator $\mathbf{A}$. The determination of the regularization parameter α depends on properties of the operator and the choice of the regularization method, (Engl et al., 2000). In principle, there are a-priori parameter choice rules, where α can be determined from prior information, and a-posteriori rules. A well known a posteriori parameter choice rule is Morozov's discrepancy principle where α is chosen s.t.

$$\delta \leq \|g^\delta - \mathbf{A}f_\alpha^\delta\|^2 \leq c\delta \tag{16}$$

holds (Morozov, 1984). The application of the discrepancy principle requires the computation of the approximate solution f_α^δ for a chosen α first. Afterwards (16) is checked and α has to be

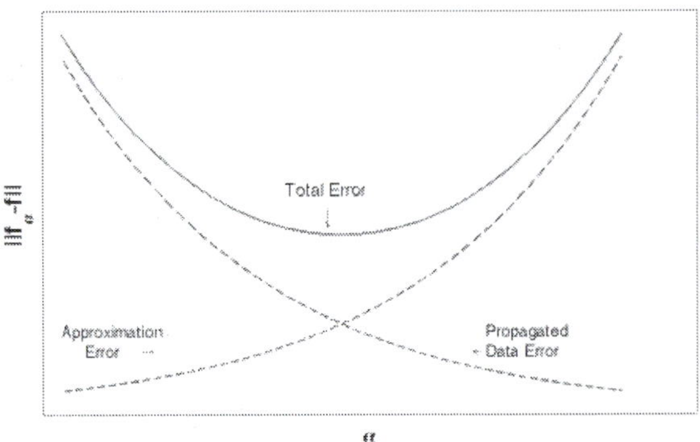

Fig. 4. Regularization error

changed if the condition does not hold. All a-posteriori parameter choice rules depend on the data error level δ and the data g^δ. Very popular are heuristic parameter choice rules, where the regularization parameter is independent of the noise level δ. Examples are the L-curve method (Hansen P., 1992) or the quasi-optimality rule (Kindermann, 2008). Please note that heuristic parameter choice rules do not lead to convergent regularization methods, although they perform well in many applications.

4. Imbalance determination

The determination of imbalances from measurements of the induced vibrations (or displacements) is an inverse problem as explained above.

4.1 Mass imbalance

First, we restrict ourself to the determination of mass imbalances and assume that aerodynamic imbalances are insignificant. In the structural model section we mentioned that in this case we only need a model that considers DOF in radial or z-direction. The knowledge of the mass and stiffness matrix provides us with a connection of the loads from imbalances $\mathbf{p}$ and the resulting displacements $\mathbf{u}$ in the nodes of our model via equation (1).

A mass imbalance can be described by a mass m that is located at a distance r from the rotor center and has an angle φ to a certain zero mark of the rotor, usually blade A, cf. Figure 5. If the rotor revolves with revolutionary frequency Ω, the mass imbalance induces a centrifugal force of absolute value $\omega^2 mr$, with the angular velocity $\omega = 2\pi\Omega$. The force or load vector is given by:

$$p(t) = \omega^2 mre^{i(\omega t+\varphi)} =: p_0\omega^2 e^{i\omega t}, \tag{17}$$

where $p_0 = mre^{i\varphi}$ defines the mass imbalance in absolute value and phase location. Harmonic loads of the form (17) cause harmonic vibration $\mathbf{u} = \mathbf{u}_0 e^{i\omega t}$ of the same frequency ω. Inserting $\mathbf{u}$, its second derivative and $\mathbf{p} = \mathbf{p}_0 e^{i\omega t}$ into equation (1), time dependency cancels out and we get an explicit solution for the vibration amplitudes $\mathbf{u}_0$:

$$\mathbf{u}_0 = (-M + \omega^{-2}S)^{-1}\mathbf{p}_0. \tag{18}$$

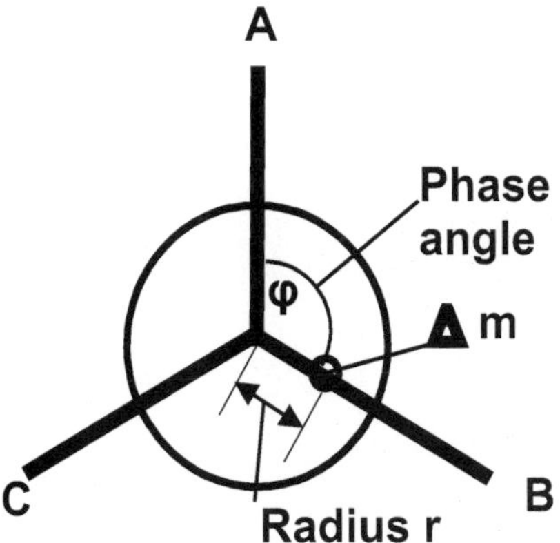

Fig. 5. Mass imbalance

The matrix $(-M + \omega^{-2}S)^{-1}$ would define our forward operator in (8) if we would assume that the vibration amplitudes could be measured in every node of the model. Usually this is not possible, measurements are taken in the nacelle which is represented by the last model node, cf. Figure 1. Additionally, the rotor and its load are located at that node, too. Thus the load vector $\mathbf{p}_0$ would have only one entry, p_0 from (17), at the last but one position that corresponds to the displacement DOF w of the last node. Hence in (8) now $f = p_0, g = u_0$ the displacement of the last node, and $\mathbf{A}$ is just the element in the last but one row and last but one column of $(-M + \omega^{-2}S)^{-1}$. Denoting the number of DOF by N we have

$$\mathbf{A}p_0 = u_0, \quad \mathbf{A} = (-M + \omega^{-2}S)^{-1}_{(N-1,N-1)}. \tag{19}$$

We remark that u_0 is the complex amplitude containing the absolute value and the phase angle $u_0 = u_a e^{i\phi}$.

The measured values for u_a and ϕ are denoted by u_a^δ and ϕ^δ. Since $\mathbf{A}$ is a complex number we deal with the simplest well posed inverse problem possible. It is solved by

$$p_0^\delta = \frac{1}{\mathbf{A}}u_0^\delta. \tag{20}$$

4.2 Aerodynamic imbalance from pitch angle deviation

The main cause for aerodynamic imbalances is a deviation between the pitch angles of the blades, e.g., from assembling inaccuracies. Depending on the wind conditions, even a small deviation of one of the pitch angles can cause large forces and moments to be transferred onto the rotor. This results in displacements in direction of the rotor axis (the y-axis) as well as torsion around the tower axis (x-axis). But there are also forces in radial direction that add to the forces from mass imbalances and are not negligible. Hence, neglecting aerodynamic imbalances could result in an inaccurate determination of mass imbalances. In the worst case, the computed balancing mass and position could increase the mass imbalance. The mass imbalance estimation described in the former section can only be applied if aerodynamic imbalances are small enough. Currently, the WEC is checked for axial and torsional vibrations. If large corresponding amplitudes indicate an aerodynamic imbalance, the surfaces of the blades are checked and photographic measurements are carried out to find a possible pitch angle deviation. After its correction the mass imbalance can be determined with the usual method.

The simultaneous reconstruction of mass and aerodynamic imbalances was considered in (Nguyen, 2010) and (Niebsch et al., 2010). The principle is the same as in the reconstruction of mass imbalances but now the structural model of the WEC is extended to DOF in radial and axial direction as well as torsion around the tower axis. Additionally, we have to describe the loads from aerodynamic imbalances mathematically. This was done using the Blade Element Momentum (BEM) theory, which is commonly used for simulations of WECs, see, e.g., (Hansen, 2008; Ingram, 2005). The result of the BEM theory are the tangential and normal (or thrust) force distributed over the blades that are divided into elements. The distributed forces are summed up to an equivalent normal force F_i with a distance l_i from the rotor center as well as an equivalent tangential force T_i, cf. Figure 6.

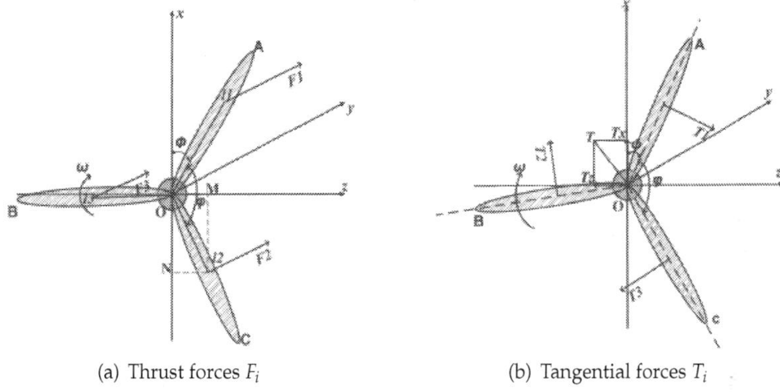

(a) Thrust forces F_i (b) Tangential forces T_i

Fig. 6. Normal (thrust) and tangential forces on the rotor blades

The forces depend on the pitch angle of the blade, the airfoil data, the angle of attack of the wind, and the relative wind velocity, as well as a lift and drag coefficient table. For details we refer to (Niebsch et al., 2010).

The force to the rotor in the axial (y-) direction is calculated by:

$$F_y = F_1 + F_2 + F_3. \tag{21}$$

The moments induced by this forces are given by

$$M_x^1 = F_1 l_1 \sin(\omega t + \phi) + F_2 l_2 \sin(\omega t + \phi + \varphi) + F_3 l_3 \sin(\omega t + \phi + 2\varphi),$$
$$M_z^1 = F_1 l_1 \cos(\omega t + \phi) + F_2 l_2 \cos(\omega t + \phi + \varphi) + F_3 l_3 \cos(\omega t + \phi + 2\varphi), \tag{22}$$

where M_x^1 and M_z^1 denote the moments around the x- and the z-axis on the rotor and $\varphi = \frac{2\pi}{3} (\equiv 120°)$ is the angle between the rotor blades. Note that if all blades have the same pitch angle, we have $F_1 = F_2 = F_3$ and $l_1 = l_2 = l_3$. This means that the moments M_x^1 and M_z^1 vanish. The projection of the total tangential force $T = T_1 + T_2 + T_3$ onto the z-axis and the x-axis is given by

$$T_z = T_1 \cos(\omega t + \phi) + T_2 \cos(\omega t + \phi + \varphi) + T_3 \cos(\omega t + \phi + 2\varphi), \tag{23}$$
$$T_x = T_1 \sin(\omega t + \phi) + T_2 \sin(\omega t + \phi + \varphi) + T_3 \sin(\omega t + \phi + 2\varphi).$$

Since we have a small distance D between rotor plane and the tower center, T_z and T_x also produce moments around the x- and the z-axes:

$$M_x^2 = T_z \cdot D, \quad M_z^2 = T_x \cdot D. \tag{24}$$

With the formulas (21) - (24) we can describe the load vector $\mathbf{p}(t)$ in (1), which has only entries at the last node. We recall that the node has the DOF $(v, w, \beta_x, \beta_y, \beta_z)$, hence

$$\mathbf{p} = (0, \cdots, 0, F_y, F_z, M_x, M_y, M_z)^T, \tag{25}$$

with $F_z = T_z$, $M_x = M_x^1 + M_x^2$, and $M_z = M_z^1 + M_z^2$. We remark that M_y is converted into the rotational movement and finally into electrical energy. M_y should not add any contribution to the load vector.

5. Combination of mass and aerodynamic imbalances

The simultaneous consideration of mass and aerodynamic imbalances is also based on equation (1). As mentioned in the last section, for aerodynamic imbalances a model is required that includes DOF in the radial and axial directions as well as torsion around the tower axis. The combined presence of both imbalance types also requires a combination of the associated load vectors. We recall that the centrifugal force from a point mass imbalance is given by $\omega^2 mr$, the location of the eccentric mass is given by the radius r and the angle ϕ_m measured from a zero mark (blade A). The projections of the force onto the z- and x-axis are

$$F_z^2 = \omega^2 mr \cos(\omega t + \phi + \phi_m), \tag{26}$$
$$F_x = \omega^2 mr \sin(\omega t + \phi + \phi_m).$$

Here, ϕ is the angle between blade A and the x-axis. Because the rotational plane has a distance D to the tower, the forces F_z^2 and F_x also produce moments around the x- and the z-axes:

$$M_x^3 = F_z^2 \cdot D, \tag{27}$$
$$M_z^3 = F_x \cdot D.$$

The force in x-direction is of no consequence since we assume the tower to be rigid. But the moments M_x^3, M_z^3 and the force F_z^2 have to be added to the moments and forces from pitch angle deviation as described in (25). The forces and moments of the combined load vector add to

$$\begin{aligned} F_y &= F_1 + F_2 + F_3 \\ F_z &= T_z + F_z^2 \\ M_x &= M_x^1 + M_x^2 + M_x^3 \\ M_z &= M_z^1 + M_z^2 + M_z^3. \end{aligned} \tag{28}$$

We observe that the forces and moments in (28) are either constant (F_y) or harmonic. Therefore, equation (1) with a load vector of the form (25) and entries (28) can be solved explicitly. Details on the solution are given in (Niebsch et al., 2010).

Starting from the pitch angles of the three blades $(\theta_1, \theta_2, \theta_3)$, and from the characteristics of a mass imbalance (mr, ϕ_m) and assuming given values for angular speed $\omega = 2\pi\Omega$, wind speed, and airfoil data, we have all the tools to determine the corresponding imbalance load $\mathbf{p}$ using the BEM method for the pitch angle deviation and by projecting (17) onto the x- and the z-axis. Solving (1) produces the resulting displacements $\mathbf{u}$. The restriction of the vector $\mathbf{u}$ onto the DOF that can be measures are denoted by $\mathbf{g} = \mathbf{u}_{|sensor}$. We combine all these operations into the forward operator $\mathbf{A}$:

$$\mathbf{A}(\theta_1, \theta_2, \theta_3, mr, \phi_m) = \mathbf{g}. \tag{29}$$

We remark that the BEM uses nonlinear optimization routines to compute parameter values in the equations for the normal and tangential force. Therefore, the final operator $\mathbf{A}$ is nonlinear. The vector $(\theta_1, \theta_2, \theta_3, mr, \phi_m)$ plays the role of f in (8). Usually, the radial and axial vibration, and the torsion around the tower axis are measurable using three acceleration sensors. Since the acceleration sensors do not measure the initial offset arising from the constant force F_y we have to rely on radial and torsion measurements only. For a known or estimated noise level δ of the measurements we can compute the solution $(\theta_1, \theta_2, \theta_3, mr, \phi_m)_\alpha^\delta$ as the minimizer of the Tikhonov functional (15). Since $\mathbf{A}$ is a nonlinear operator, minimization methods have to be employed to find the minimizing element, like e.g., the MATLAb implemented routines like *fminsearch* or gradient based methods. The regularization parameter α can be chosen iteratively using Morozov's Discrepancy Principle (16). First results on the simultaneous reconstruction of $(\theta_1, \theta_2, \theta_3, mr, \phi_m)$ from noisy data $\mathbf{g}^\delta$ were obtained in (Niebsch et al., 2010) with data errors of about 10%. Several experiments showed that the simultaneous reconstruction is successful provided we have a fairly good initial value for the mass imbalance. This can be obtained in a first step by reconstructing the mass imbalance neglecting pitch angle deviations with the method described in Section 4. The result is not the true mass imbalance but a sufficiently accurate initial estimate for the simultaneous reconstruction carried out as a second step. To present an example, a pitch angle deviation of 3 degree of the blade B as well as a mass imbalance of 350 kgm located at blade B. The data $\mathbf{g}$ were calculated by the forward computation of $\mathbf{A}(0°, 3°, 0°, 350\,kgm, 120°)$ and contaminated with 10% noise. The two step reconstruction from the noisy data resulted in $(-0.25°, 2.8°, 0.43°, 342\,kgm, 121°)$. The correction of the pitch angles and the setting of balancing weights according to that reconstruction lead to a significant reduction of the vibration, cf. Figure 7.

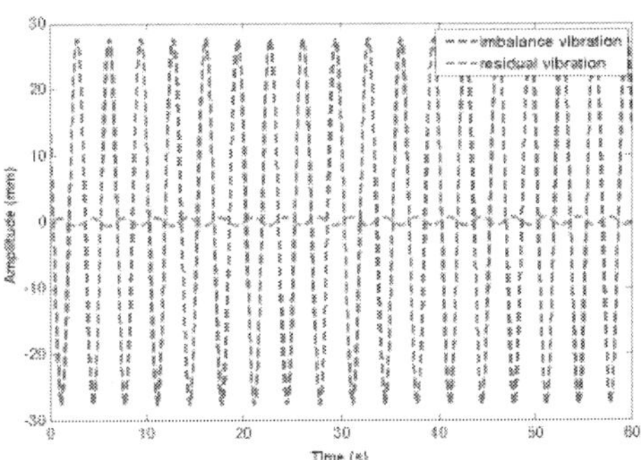

Fig. 7. Vibrations in z-direction before and after balancing

6. Conclusion

We presented a method to determine imbalances from vibration measurements based on a relatively simple model of the WEC under consideration. In contrast to detection methods based on signal processing, the imbalance can be localized and quantified. Moreover, mass imbalances and pitch angle deviations that cause aerodynamic imbalances can be discovered simultaneously. The model of the WEC is used to describe the connection of an imbalance load and the caused vibrations or displacements mathematically. The reverse direction of recovering an imbalance from given noisy vibration measurements is an inverse (ill-posed) problem and has to be treated accordingly. For this reason, we presented a short introduction to the main ideas and issues of the inverse problem theory.

In addition to the model parameters, the presented method requires the mathematical description of the loads from the different types of imbalances. Whereas for mass imbalances this is quite simple, the forces from pitch angle deviation are computed via the BEM method which uses idealizations that do not cover all effects that might arise during the operation of the WEC. Another drawback of the determination of pitch angle deviations is the fact that the BEM method requires the airfoil data of the WEC's blades. For most of the newer wind turbines this data are a well kept secret of the manufacturers. The restriction of the imbalance estimation to mass imbalances is much easier to implement into an existing condition monitoring system and does not require "sensitive" data.

We propose two main questions for future research. First, the assumption of a constant revolution frequency is not realistic. Therefore we have to consider equation (1) for a variable (time dependent) angular velocity ω, which allows the use of vibration data measured at variable speed. Presently, only data collected with constant or almost constant operational speed can be used for the imbalance determination. The second question is how to avoid the sensitive information on airfoil data that is necessary to reconstruct pitch angle deviations.

7. References

Borg, J. P. & Kirchhoff, R. H. (1998). Mass and Aerodynamic Imbalance of a Horizontal Axis Wind Turbine. *ASME Journal of Solar Energy Engineering* Vol. 120, 66-74.

Ciang, C. C., Lee, J., Bang, H. (2008). Structural health monitoring for a wind turbine system: a review of damage detection methods. *Meas.Sci.Technol.* , Vol. 19,122001(20pp).

Caselitz, P. & Giebhardt, J. (2005). Rotor Condition Monitoring for Improved Operational Safety of Offshore Wind Energy Converters. *ASME Journal of Solar Energy Engineering* Vol. 127, 253-261.

Engl, H. W..; Hanke, M. & Neubauer, A. (2000). *Regularization of Inverse Problems*, Kluwer, Dortrecht.

Gasch, R. & Knothe, K. (1989). *Strukturdynamik 2*, Springer, Berlin.

Hau, E. (2006). *Wind Turbines: Fundamentals, Technologies, Application, Economics*; Springer, Berlin, Heidelberg.

Hansen, M. (2008). *Aerodynamics of Wind Turbines*; Earthscan, London.

Hansen, P. C. (1992). Analysis of discrete ill-posed problems by means of the L-curve. *SIAM Rev*; Vol. 34 561-580.

Ingram, G. (2005). Wind Turbine Blade Analysis using the Blade Element Momentum Method. *Note on the BEM method*, Durham University, Durham.

Kindermann, S. & Neubauer, A. (2008) On the convergence of the quasi-optimality criterion for (iterated) Tikhonov regularization. *Inverse Problems and Imaging* Vol. 2, Nr. 2, 291-299.

Morozov, V. A. (1984). *Methods for Solving Incorrectly posed Problems*, Springer, New York, Berlin Heidelberg.

Niebsch, J.; Ramlau, R. & Nguyen, T. T. (2010). Mass and aerodynamic Imbalance Estimates of Wind Turbines. *Engergies*, Vol. 3, p 696-710, ISSN 1996-1073.

Nguyen, T. T. (2010). Mass and aerodynamic Imbalance Estimates of Wind Turbines. *Diploma thesis at the Johannes Kepler University Linz, Austria.*

Ramlau, R. (2002). Morozov's Discrepancy Principle for Tikhonov regularization of nonlinear operators. *Numerical Functional Analysis and Optimization*, Vol. 23 No.1&2, 147-172.

Ramlau, R. & Niebsch, J. (2009). Imbalance Estimation Without Test Masses for Wind Turbines. *ASME Journal of Solar Energy Engineering* Vol. 131, No. 1, 011010-1- 011010-7.

Scherzer, O. (1993). Morozov's Discrepancy Principle for Tikhonov regularization of nonlinear operators. *Computing*, Vol. 51 , 45-60.

8

Wind Turbine Gearbox Technologies

Adam M. Ragheb[1] and Magdi Ragheb[2]
[1]*Department of Aerospace Engineering*
[2]*Department of Nuclear, Plasma and Radiological Engineering,*
University of Illinois at Urbana-Champaign, 216 Talbot Laboratory
USA

1. Introduction

The reliability issues associated with transmission or gearbox-equipped wind turbines and the existing solutions of using direct-drive (gearless) and torque splitting transmissions in wind turbines designs, are discussed. Accordingly, a range of applicability of the different design gearbox design options as a function of the rated power of a wind turbine is identified. As the rated power increases, it appears that the torque splitting and gearless design options become the favored options, compared with the conventional, Continuously Variable Transmission (CVT), and Magnetic Bearing transmissions which would continue being as viable options for the lower power rated wind turbines range.

The history of gearbox problems and their relevant statistics are reviewed, as well as the equations relating the gearing ratios, the number of generator poles, and the high speed and low speed shafts rotational speeds.

Aside from direct-drive systems, the topics of torque splitting, magnetic bearings and their gas and wind turbine applications, and Continuously Variable Transmissions (CVTs), are discussed.

Operational experience reveals that the gearboxes of modern electrical utility wind turbines at the MegaWatt (MW) level of rated power are their weakest-link-in-the-chain component. Small wind turbines at the kW level of rated power do not need the use of gearboxes since their rotors rotate at a speed that is significantly larger than the utility level turbines and can be directly coupled to their electrical generators.

Wind gusts and turbulence lead to misalignment of the drive train and a gradual failure of the gear components. This failure interval creates a significant increase in the capital and operating costs and downtime of a turbine, while greatly reducing its profitability and reliability. Existing gearboxes are a spinoff from marine technology used in shipbuilding and locomotive technology. The gearboxes are massive components as shown in Fig. 1.

The typical design lifetime of a utility wind turbine is 20 years, but the gearboxes, which convert the rotor blades rotational speed of between 5 and 22 revolutions per minute (rpm) to the generator-required rotational speed of around 1,000 to 1,600 rpm, are observed to commonly fail within an operational period of 5 years, and require replacement. That 20 year lifetime goal is itself a reduction from the earlier 30 year lifetime design goal (Ragheb & Ragheb, 2010).

2. Gearbox issues background

The insurance companies have displayed scrutiny in insuring wind power generation. The insurers joined the rapidly-growing market in the 1990s before the durability and long term maintenance requirements of wind turbines were fully identified. To meet the demand, a number of units were placed into service with limited operational testing of prototypes.

During the period of quick introduction rate, failures during wind turbines operation were common. These included rotor blades shedding fragments, short circuits, cracked foundations, and gearbox failure. Before a set of internationally recognized wind turbine gearbox design standards was created, a significant underestimation of the operational loads and inherent gearbox design deficiencies resulted in unreliable wind turbine gearboxes.

The lack of full accounting of the critical design loads, the non-linearity or unpredictability of the transfer of loads between the drive train and its mounting fixture, and the mismatched reliability of individual gearbox components are all factors that were identified

Fig. 1. Top view of a Liberty Quantum Drive 2.5 MW rated power wind turbine gearbox (Source: Clipper Windpower).

by the National Renewable Energy Laboratory (NREL) as contributing to the reduced operating life of gearboxes (Musial et al., 2007).

In 2006, the German Allianz reportedly received 1,000 wind turbine damage claims. An operator had to expect damage to his facility at a 4-5 years interval, excluding malfunctions and uninsured breakdowns.

As a result of these earlier failures, insurers adopted provisions that require the inclusion by the operator of maintenance requirements into their insurance contracts. One of the common maintenance requirements is to replace the gearbox every 5 years over the 20-year design lifetime of the wind turbine. This is a costly task, since the replacement of a gearbox accounts for about 10 percent of the construction and installation cost of the wind turbine, and will negatively affect the estimated income from a wind turbine (Kaiser & Fröhlingsdorf, 2007). Figure 1 depicts the size of the Quantum Drive gearbox of a Liberty 2.5 MW wind turbine (Clipper Windpower, 2010)

The failure of wind turbine gearboxes may be traced to the random gusting nature of the wind. Even the smallest gust of wind will create an uneven loading on the rotor blades, which will generate a torque on the rotor shaft that will unevenly load the bearings and misalign the teeth of the gears. This misalignment of the gears results in uneven wear on the teeth, which in turn will facilitate further misalignment, which will cause more uneven wear, and so on in a positive feedback way.

The machine chassis will move, which will misalign the gearbox with the generator shaft and may eventually cause a failure in the high speed rear gearing portion of the gearbox. Further compounding the problem of uneven rotor blade loading is the gust slicing effect, which refers to multiple blades repeatedly traveling through a localized gust (Burton et al., 2004). If a gust of wind were to require 12 seconds to travel through the swept area of a wind turbine rotor operating at 15 rpm, each of the three blades would be subject to the gust three times, resulting in the gearbox being subjected to a total of nine uneven loadings in a rapid succession.

The majority of gearboxes at the 1.5 MW rated power range of wind turbines use a one- or two-stage planetary gearing system, sometimes referred to as an epicyclic gearing system. In this arrangement, multiple outer gears, planets, revolve around a single center gear, the sun. In order to achieve a change in the rpm, an outer ring or annulus is required.

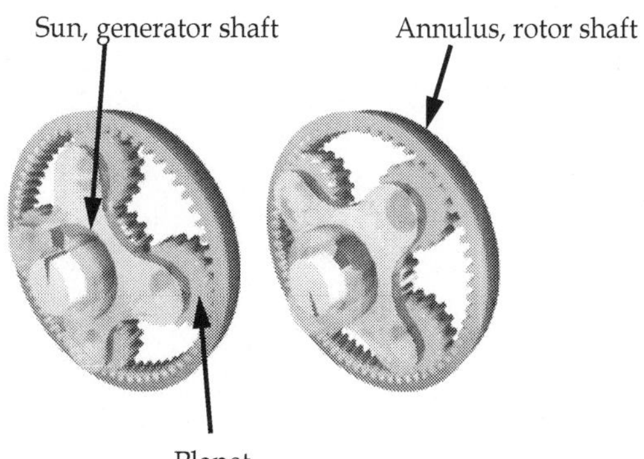

Fig. 2. Planetary gearing system.

As it would relate to a wind turbine, the annulus in Fig. 2 would be connected to the rotor hub, while the sun gear would be connected to the generator. In practice however, modern gearboxes are much more complicated than that of Fig. 2, and Fig. 3 depicts two different General Electric (GE) wind turbine gearboxes.

Fig. 3. GE 1P 2.3 one-stage planetary and two-stage parallel shaft (top) and 2P 2.9 two-stage planetary and one-stage parallel shaft (bottom) wind turbine gearboxes (Image: GE).

Planetary gearing systems exhibit higher power densities than parallel axis gears, and are able to offer a multitude of gearing options, and a large change in rpm within a small volume. The disadvantages of planetary gearing systems include the need for highly-complex designs, the general inaccessibility of vital components, and high loads on the shaft bearings. It is the last of these three that has proven the most troublesome in wind turbine applications.

In order to calculate the reduction potential of a planetary gear system, the first step is to determine the number of teeth, N, that each of the three component gears has. These values will be referred to as:

$$N_{sun}, N_{annulus}, and\, N_{planet}$$

as they relate to the number of teeth on the sun, annulus, and planet gears, respectively. Using the relationship that the number of teeth is directly proportional to the diameter of a gear, the three values should satisfy Eqn. 1, which shows that the sun and annulus gears will fit within the annulus.

$$N_{sun} + 2N_{planet} = N_{annulus} \tag{1}$$

With Eqn. 1 satisfied, the equation of motion for the three gears is,

$$\left(2 + \frac{N_{sun}}{N_{planet}}\right)\omega_{annulus} + \frac{N_{sun}}{N_{planet}}\omega_{sun} - 2\left(1 + \frac{N_{sun}}{N_{planet}}\right)\omega_{planet} = 0 \tag{2}$$

where: ω_{sun}, $\omega_{annulus}$, and ω_{planet} are the angular velocities of the respective gears.

Since the angular velocity and is directly proportional to the revolutions per minute (rpm), Eqn. 2 may be modified to Eqn. 3 below.

$$\left(2 + \frac{N_{sun}}{N_{planet}}\right)rpm_{annulus} + \frac{N_{sun}}{N_{planet}}rpm_{sun} - 2\left(1 + \frac{N_{sun}}{N_{planet}}\right)rpm_{planet} = 0 \tag{3}$$

Known values may be substituted into Eqn. 3 in order to determine the relative rpm values of the sun and annulus gears, noting the two equalities of Eqns. 4 and 5 below (Ragheb & Ragheb, 2010).

$$rpm_{sun} = \left(-\frac{N_{sun}}{N_{planet}}\right)rpm_{planet} \tag{4}$$

$$rpm_{planet} = \left(-\frac{N_{planet}}{N_{annulus}}\right)rpm_{annulus} \tag{5}$$

Historically, the gearbox has been the weakest link in a modern, utility scale wind turbine. Following the current trend of larger wind turbines for offshore applications with their larger rotor diameters and heavier rotor blades, gearboxes are being subject to significantly increased loads.

Minor improvements in the gearbox lubrication and oil filtration system have increased the reliability of wind turbines, but to significantly improve the gearbox reliability, the design must be changed from the current planetary gear design. This improved reliability is especially important for offshore applications, as the wind turbines are generally much larger and the cost of maintenance is much greater.

3. Gearless / direct-drive wind turbines

The Enercon Company of Germany and ScanWind of Norway have served as the pioneers of the Gearless, or Direct-Drive, wind turbine generator. By increasing the number of magnetic pole pairs in a generator from the 4 or 6 of conventional generators to 100 or more, the need for a gearbox may be eliminated.
In order to produce the 60 or 50 Hz electrical power for the United States or Europe, a 4-pole generator would have to operate at 1,800 or 1,500 rpm respectively. Increasing the number of poles to 6 would decrease the generator rpm to 1,200 and 1,000, respectively. The relationship between the generator rpm, the number of poles n, and the frequency f is given by:

$$f = n\frac{rpm}{120}[Hz] \tag{6}$$

A four-pole, 1,800 rpm generator has a frequency:

$$f = 4\frac{1,800}{120} = 60[Hz]$$

The generator's rpm can be expressed as:

$$rpm_{generator} = 120\frac{f(Hz)}{n} \tag{6}'$$

Increasing the number of poles n in a generator to 160 to produce 60 Hz electricity allows it to rotate at a smaller rpm value of:

$$rpm_{generator} = 120\frac{60}{160} = 45$$

The gearing ratio, G, is defined in Eqn. 7.

$$G = \frac{rpm_{generator}}{rpm_{rotor\ blade}} \tag{7}$$

Using the example of producing electricity at a frequency of 60 Hz with a 4-pole generator and a rotor blade operating at 15 rpm, Eqn. 7 shows that a gearing ratio G of :

$$G = \frac{1,800}{15} = 120$$

between the rotor blade shaft and the generator shaft would be required.
If however, a 160 pole generator is used, the gearing ratio drops to:

$$G = \frac{45}{15} = 3$$

with all other values held constant.

Finally, a 400-pole generator operating on a rotor blade at 18 rpm would yield a gearing ratio of:

$$rpm_{generator} = 120\frac{60}{400} = 18$$

$$G = \frac{18}{18} = 1$$

A gearing ratio of unity implies that a gearbox would not be needed.

The first entrant with a direct-drive wind turbine is widely cited as being Enercon GmbH of Aurich, Germany. They suggest that their annular generator, in addition to precluding the need for a gearbox, contains a smaller number of moving parts, further contributing the increased reliability and reduction in frictional losses. Because the operational speeds of the generator are much lower, the generator is subjected to little, if any, wear, allowing to achieve a longer operational life and to handle larger loads.

Unique to Enercon, is their manual winding of the copper wire in the stator portion of the annular generator, justified by their use of a continuous wire strand in each generator, which reduces resistive losses. Enercon is very proud of their closed varnish-insulated wires, rated to Temperature Tolerance Class F (155° F), suggesting that breaks in the insulation, especially at joint locations, may have been a significant problem in early generator construction. In order to maintain high levels of quality control, Enercon manufactures its annular generators in the company's own production facilities.

Beginning with their first direct-drive wind turbine in 1993, Enercon has dominated the direct-drive wind turbine market, and in 2007 was fourth in terms of worldwide wind turbine market share, capturing 14 percent of the market behind Vestas, GE, and Gamesa. While in the past a number of competitors in the direct-drive market have filed for bankruptcy or been bought and sold repeatedly, as was the case for Lagerwey being sold to Zephyros which was sold to Harakosan, and then ended up in the hands of STX heavy Industries of Korea, the marketplace appears to have settled with the entrance of industry giants GE and Siemens. Figures 4 and 5 show annular generators under assembly at the Enercon's company manufacturing facilities.

One requirement for a direct-drive wind turbine is to not have direct coupling with the electrical grid. This is due to the fact that wind turbine rotor blades operate within an rpm range, and with a direct-coupled generator, the output voltage and frequency vary slightly. A DC link and inverter convert the produced energy to parameters suitable for transmission to the electrical grid. Prior to the development of these active electronic systems, wind turbines used capacitors and static Volt-Ampère-Reactive (VAR) systems that were far from optimal.

Another consideration of direct-drive wind turbines is their increased manufacturing and material costs. When the German company Siemens embarked on a two-year testing program for its 3.6 MW direct-drive turbine, Henrik Stiesdal commented that direct-drive wind turbines may become competitive with their geared counterparts near the upper end of turbine sizes (at that time in the 4-6 MW range), and with the test rigs they determined at what level direct-drive could be made competitive (Ragheb & Ragheb, 2010). Three years later, these comments turned out to be almost prophetic, as the majority of the wind turbine designs with rated powers around 1.5 MW still utilize a planetary gear system, while the

Fig. 4. Multipole annular generators under assembly (Photo: Enercon).

Fig. 5. Rotor and stator of Enercon E-70 wind turbine (Photo: Enercon).

more recent and larger designs, many of which are marketed for offshore applications, are designed around direct-drive generators. It appears that this changeover point between geared and direct-drive wind turbines lies in the 1.5-3 MW rated power range.

According to the United States Department of Energy, direct-drive generators require large diameters, which necessitates the use of large amounts of rare earth elements magnets, and consequently are expensive and require a larger and heavier drivetrain. In addition to this, a small air gap on the millimeter scale is required to be maintained between the rotor and stator to yield sufficiently high flux densities. As a consequence, the tight manufacturing tolerances required and the detailed design to handle the complex loads encountered add another set of challenges that may set an upper limit on the size of such generators (Department of Energy, 2010).

While it was the first, Enercon is not the only company marketing direct-drive wind turbines. Japan Steel Works (JSW) has licensed Enercon's technology, and competes against the likes of Vensys, Leitwind, MTorres, and ScanWind, now acquired by General Electric (GE). Table 1 presents a summary of some of the utility scale direct-drive wind turbines designs currently available.

In late 2009, GE acquired ScanWind for approximately 15 million Euros in what appeared to be a technology-driven move to reenter the offshore wind turbine sector. GE's previous foray into offshore wind turbines was the Arklow Bank Wind Park (Ireland) project, in which seven 3.6 MW technology demonstrator wind turbines were installed. In contrast to the ScanWind direct-drive turbines, these wind turbines utilized a three step planetary gear system.

The direct-drive approach to the gearbox problem appears to be taking hold quite well on the largest capacity wind turbines. Due to its lower amount of moving parts, it seems ideally suited for large offshore applications.

Direct-drive may not however solve the existing gearbox problems for all wind turbines over the rated power range between 1.5 and 10 MW, as it brings with it weight increases of around 25 percent and a cost increase of around 30 percent.

A further development of the direct-drive solution couples a direct-drive concept with superconducting materials, with potential benefits being reduced mass and volume, and consequently smaller transportation costs and lower loadings on the tower. Cost advantages of superconducting direct-drive generators will most likely not exist for turbines below a rated power of 5 MW, but with turbines of 10 MW already being constructed for offshore use, this drivetrain concept may soon become a reality (Department of Energy, 2010).

4. Torque splitting

A different attempt at solving the gearbox problem on 2+ MW sized wind turbines was undertaken by Clipper Windpower of Carpineria, California. Under a partnership with the United States Department of Energy (DOE) and the National Renewable Energy Laboratory (NREL), Clipper developed their 2.5 MW Liberty Wind Turbine, the largest manufactured in the United States. These turbines, manufactured in 2006, were put into service in early 2007 as part of the Steel Winds Project, a superfund location along Lake Erie in Lackawana, New York, and the company received a DOE Outstanding Research and Development Partnership Award in 2007. After only a few months of service, problems were observed in their distributed gearing-style gearboxes. A subsequent analysis traced the fault to improper drivetrain timing caused by incorrect gear tolerances on parts arriving from suppliers (Robb, 2008).

Company	Country	Turbine Rated Power (MW)	Rotor blade Diameter (m)
Clipper [Windpower] Marine deployment date: 2012	UK	10	144
Sway	Norway	10	145
Enercon E-126	Germany	7.5	127
Nordex N150/6000	Germany	6.0	150
Xingtan Electric Manufacturing Corporation	China	5.0	-
GE 4.1-113	United States	4.1	113
GE 4.0-110	United States	4.0	110
ScanWind 3.5 (evolved into GE 4.X series)	Norway	3.5	90
Enercon E-101	Germany	3.0	101
Leitwind LTW101	Italy	3.0	101
Siemens SWT 3.0-101	Germany	3.0	101
Enercon E-82	Germany	2.0 - 3.0	83
Guangxi Yinhe Avantis Wind Power (in testing)	China	2.5	-
Vensys 2.5 MW	Germany	2.5	90 - 100
Enercon E-70	Germany	2.3	71
Leitwind LTW70	Italy	1.7 - 2.0	70.1
Leitwind LTW80	Italy	1.5 - 1.8	80.3
MTorres TWT 1.65/70	Spain	1.65	70
MTorres TWT 1.65/77	Spain	1.65	77
MTorres TWT 1.65/82	Spain	1.65	82
Leitwind LTW77	Italy	1.5	76.6
Leitwind LTW86	Italy	1.5	86.3
Vensys 1.5 MW	Germany	1.5	70 - 82

Table 1. Direct-drive wind turbine rated power (MW) and rotor blade diameters.

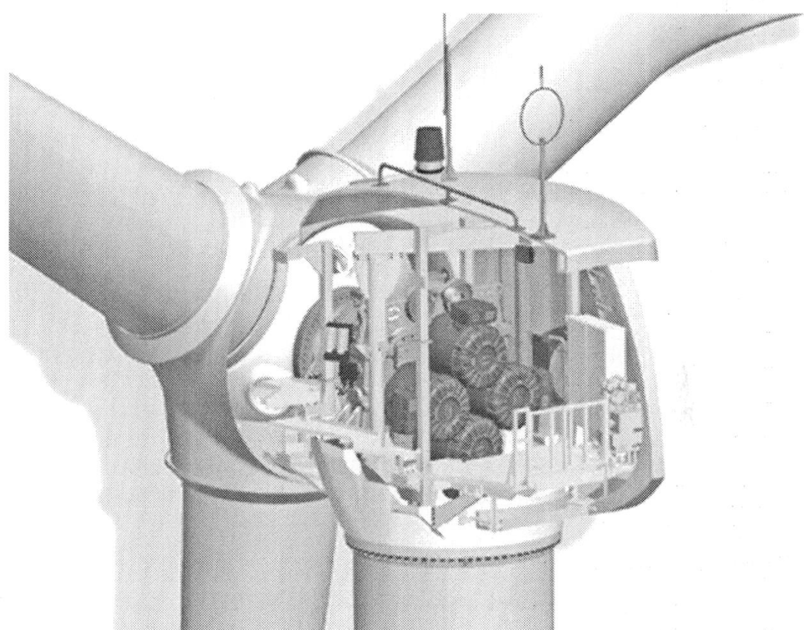

Fig. 6. Torque splitting between four electrical generators on the 2.5 MW Clipper Liberty (Image: Clipper Windpower).

Using its patented Quantum Drive Distributed Generation Powertrain, the 2.5 MW Liberty wind turbine uses a multiple-path gearbox design to split the torque from its 89– 99 meter rotor blades evenly between four generators that are operated in parallel. In contrast to a planetary gearing system, Clipper utilizes external double helical gears in order to allow for wide faces with their lower deflection sensitivities, smaller diameters, and reduced manufacturing costs due to lower required tolerances. The gear set for each of the generators is designed in "cartridge" form so as to allow for replacement without requiring the removal of the gearbox. Additionally, if a fault were to develop in one of the generators or cartridged gear sets, the production capacity of the wind turbine is reduced by only 25 percent until the problem can be corrected (Mikhail & Hahlbeck, 2006).

After selling 370 turbines in 2006, and 825 in 2007, the company appeared to have recovered from their early quality control problems. Clipper Wind was acquired in December 2010 by United Technologies Corporation. On March 24, 2011, Clipper Wind dedicated the first large-scale wind farm on the island of Oahu, which consists of 12 2.5 MW wind turbines coupled to a 15 MW batter storage system to smooth power output fluctuations. This project was developed by the Boston-based First Wind, one of Clipper Windpower's long standing customers. As of early 2011, a total of 375 Clipper Windpower turbines are featured in 17 projects across the US, with a cumulative rated power of 938 MW.

Torque splitting appears to be a cheaper alternative to the direct-drive solution, although it appears that the upper viable limit of torque splitting may lie below that of direct-drive machines.

In addition to Clipper Windpower, CWind of Ontario, Canada is introducing a 2 MW, 8-generator wind turbine design. They were testing a 65 kW wind turbine, and have announced plans to develop a 7.5 MW turbine. Their design concept may be a hybrid between torque splitting and a Continuously Variable Transmission (CVT), as they allude to a "friction drive system" to absorb sudden wind spikes. A frictional contact drive is one of the many types of CVTs. Finally, it should be noted that as shown in Table 1, the subsidiary of Clipper Windpower, Clipper Marine, has opted for a direct-drive system on its 10 MW turbine. This may provide clues as to the maximum economical size for a wind turbine built around a torque splitting concept.

5. Magnetic bearings

A very promising potential solution to the shaft misalignment problem may come from the aerospace and centrifuge uranium enrichment industries in the form of magnetic bearings or Active Magnetic Bearings (AMBs).

Recent research by NASA, MTU and others point to research in the area of high temperature magnetic bearings for use in gas turbine engines to propel aircraft. What appears to be the next large leap in terms of powering commercial transport aircraft is the Geared Turbofan (GTF) engine, which is slated to power the Mitsubishi MRJ, Bombarider C-Series, and A320neo, and may serve as the platform on which AMBs may be used in aerospace applications. An AMB system consists of a magnetic shaft, a controller, multiple electromagnetic coils attached to a stator shaft location as shown in Fig. 7. In the event of a failure of the control system, AMBs typically have a passive backup bearing system, which defaults to a rolling element bearing for the "limp home" operational mode sensors (Clark et al., 2004).

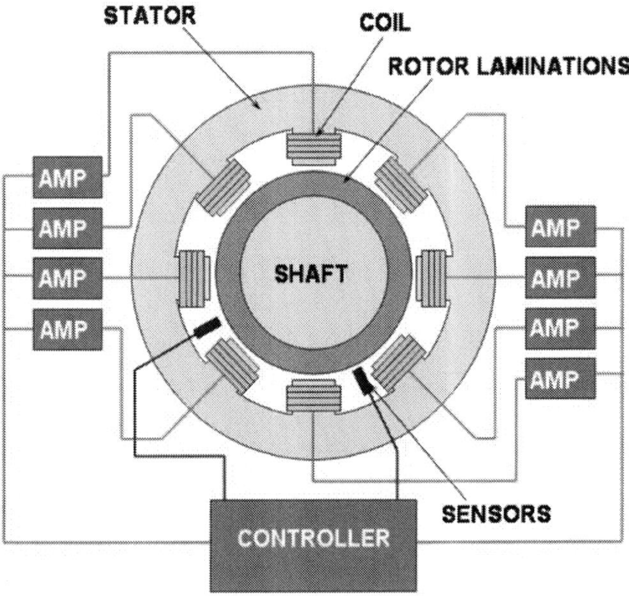

Fig. 7. Schematic of an Active Magnetic Bearing (Clark et al., 2004).

The GTF engine is by no means a new concept, as engine maker Pratt and Whitney understood the theoretical justification behind the concept in the early 1980s. The level of technology and materials development necessary to meet the stringent safety, reliability, and ruggedness requirements of modern gas turbine engines has been achieved lately. The Pratt and Whitney company suggests that through thousands of hours of development, advances in bearing, gear system, and lubrication design have been made and incorporated into their new family of GTFs, with initial reports suggesting promising heat and efficiency data.

SAE International reports that Pratt and Whitney uses a self-centering bearing technology that has all but eliminated the problems of gear misalignment and stress in the gearbox of the PW8000 GTF. It seems to be more likely that this has been achieved through their patented squirrel-cage bearing (Kostka, 2010), but based on the high temperature tolerance of AMBs, a magnetic bearing in a gas turbine engine does not appear to be too far off.

The use of magnetic bearings for gas turbine engines has been studied in depth, and papers on the topic point out a number of their potential benefits, as well as their shortcomings. Benefits of magnetic bearings include durability and damage tolerance (Clark et al., 2004), much smaller frictional losses (Schweitzer, 2002), and increased reliability at a reduced weight. Magnetic bearings also offer the potential to eliminate lubricating oil systems and avoid bearing wear, and have already demonstrated their successful application in machine spindles, mid-sized turbomachinery, and large centrifugal compressors (Becker, 2010). Eliminating the oil system in a wind tunnel gearbox provides a very large potential benefit, as numerous wind turbine fires have been attributed to the oil in an overheated gearbox catching fire. Figure 8 is a photograph of one of many wind turbines whose overheated gearboxes caused the lubricating oil to catch fire.

Fig. 8. A utility scale wind turbine on fire (Photo: flickr).

Rolling element bearings, currently used in wind turbines, are hindered by their relatively short lifetime when subjected to high loads. Both foil and magnetic bearings offer longer lifetimes, with magnetic bearings outperforming foil bearings when used in large rotating machinery under high loads and a relatively low speed (Clark, 2004). Large, heavily loaded, and relatively slow rotating provides a nearly perfect description of a modern utility scale wind turbine generator.

A common criticism of magnetic bearings is the high power requirement to generate ample current to generate a magnetic field great enough to yield an ample magnetic force to handle the large loads. This criticism is simply outdated, as recent advances in permanent magnets allow similarly strong magnetic fields to be generated by said magnets instead of via a current. It is these same permanent magnet advances that have allowed the construction of the aforementioned direct-drive generators.

Magnetic bearings appear well-poised to mitigate some of the current gearbox problems, but their application to wind turbines lies well behind the current state of development of direct-drive and torque splitting solutions. This solution has the potential to aid in the solution of gearbox problems on the lower end of utility scale wind turbines, as it may be adaptable to existing gearbox designs with minimal design changes required. As the technology matures, magnetic bearings have the potential to allow conventional gearbox designs to approach turbine rated powers of as much as 4 MW, if specific design constraints call for the use of a conventional gearbox.

6. Continuously Variable Transmissions, CVTs

Another option for solving the gearbox problem is the use of a Continuously Variable Transmission (CVT). This gearing design has only recently reached mass production in passenger vehicles, although it has been in use for a long time on farm machinery, drill presses, snowmobiles, and garden tractors. Transmissions of the CVT type are capable of varying continuously through an infinite number of gearing ratios in contrast to the discrete varying between a set number of specific gear ratios of a standard gearbox.

It is this gearing flexibility that allows the output shaft, connected to the generator in wind turbine applications, to maintain a constant rate of rotation for varying input angular velocities. The variability of wind speed and the corresponding variation in the rotor rpm combined with the fixed phase and frequency requirements for electricity to be transmitted to the electrical grid make it seem that CVTs in concert with a proportional Position, Integral, Derivative (PID) controller have the potential to significantly increase the efficiency and cost-effectiveness of wind turbines.

One disadvantage of CVTs is that their ability to handle torques is limited by the strength of the transmission medium and the friction between said medium and the source pulley. Through the use of state of the art lubricants, the chain-drive type of CVT has been able to adequately serve any amount of torque experienced on buses, heavy trucks, and earth-moving equipment. In fact, the Gear Chain Industrial B.V. Company of Japan appears to have initiated work on a wind application for chain-driven CVTs.

In addition to being able to handle minor shaft misalignments without being damaged, CVTs offer two additional potential benefits to wind turbines. As reported by Mangliardi and Mantriota (1996, 1994), a CVT-equipped wind turbine is able to operate at a more ideal tip speed ratio in a variable speed wind environment by following the large fluctuations in the wind speed. When simulated in a steady wind stream, a power increase with the addition of a CVT was observed for wind speeds above 11 m/s, and at 17 m/s, the CVT-equipped turbine power was double that of a conventional configuration, while exhibiting only a 20 percent increase in torque. These results suggest that the typical cut-out wind speed of 25 m/s, set to limit the shaft stress and other stresses, may possibly be reevaluated, to reflect the lower shaft stresses and higher rotor efficiencies at higher wind speeds (Mangliardi & Mantriota, 1994). The dynamic results were even more promising, as a CVT-equipped turbine subjected to a turbulent wind condition demonstrated increased efficiencies of on average 10 percent relative to the steady wind stream CVT example. Additionally, the CVT-equipped turbine simulation produced higher quality electrical energy, as the inertia of the rotor helped to significantly reduce the surges that are ever-present in constant-speed wind turbines subjected to rapid changes in wind speed (Mangliardi & Mantriota, 1996). Mangliardi and Mantriota go on to determine the extraction efficiency of a CVT-equipped and a CVT-less wind turbine as a function of wind speed, and this is presented below in Fig. 9.

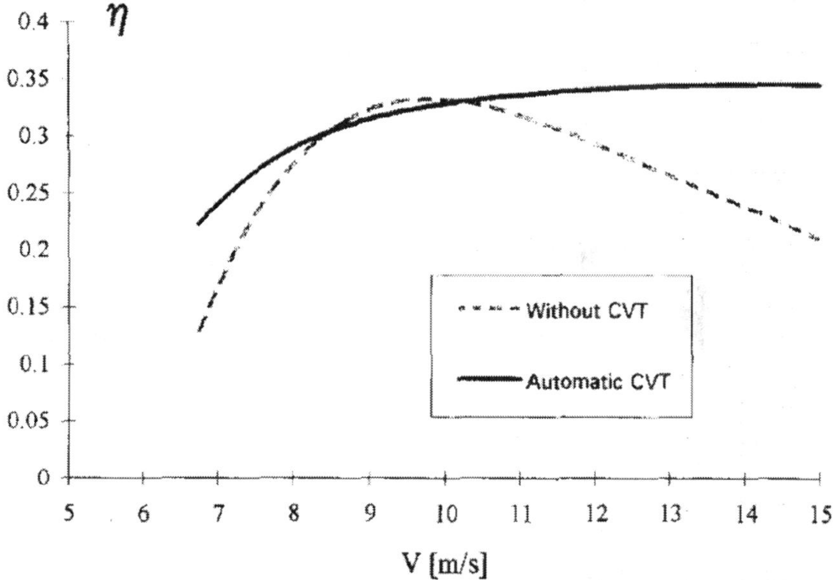

Fig. 9. Extraction efficiency η of standard and CVT-equipped wind turbines as a function of wind speed in a turbulent wind field (Mangliardi and Mantriota, 1996).

As observable in Fig. 9, a CVT-equipped wind turbine is more efficient than a conventional wind turbine at extracting the energy of the wind over all but a narrow range of wind speeds. The wind speed range where the CVT-equipped turbine is at a disadvantage is centered on the design point of the conventional wind turbine, where both turbines exhibit similar aerodynamic efficiencies, but the CVT-equipped turbine is hampered by energy losses in its gearing system. It should be noted that this is a rather narrow range, and the value by which the CVT wind turbine trails the conventional wind turbine is much smaller when compared to its benefits over the rest of the range of wind speeds.

As one moves from an ideal constant and uniform wind field to a turbulent wind field, the potential benefits of a CVT-equipped wind turbine increase. The ratio of efficiency of a CVT wind turbine to a conventional one, R_η, increases (Mangliardi and Mantriota, 1996).

Potential challenges to turbines equipped with a CVT center mainly on the lack of knowledge about the scalability of such designs. Questions such as what is the upper limit to the amount of torque that may be transmitted through a belt drive have yet to be answered. The potential benefits exist, but it appears that more research and turbine test platforms are needed before the range of applicability of CVTs on wind turbines is known (Department of Energy, 2010) and their commercial benefits quantified. Hydrostatic drives are one type of CVT that has been studied for wind turbine applications, but it appears, at least initially, that this may replace one problem, gearbox oil filtration, with another, increased maintenance and hydraulic fluid cleanliness requirements.

7. Discussion

According to Fig. 10, gearbox failures account to 5 percent of wind turbine failures. However, they are costly compared with the other failures when they occur.

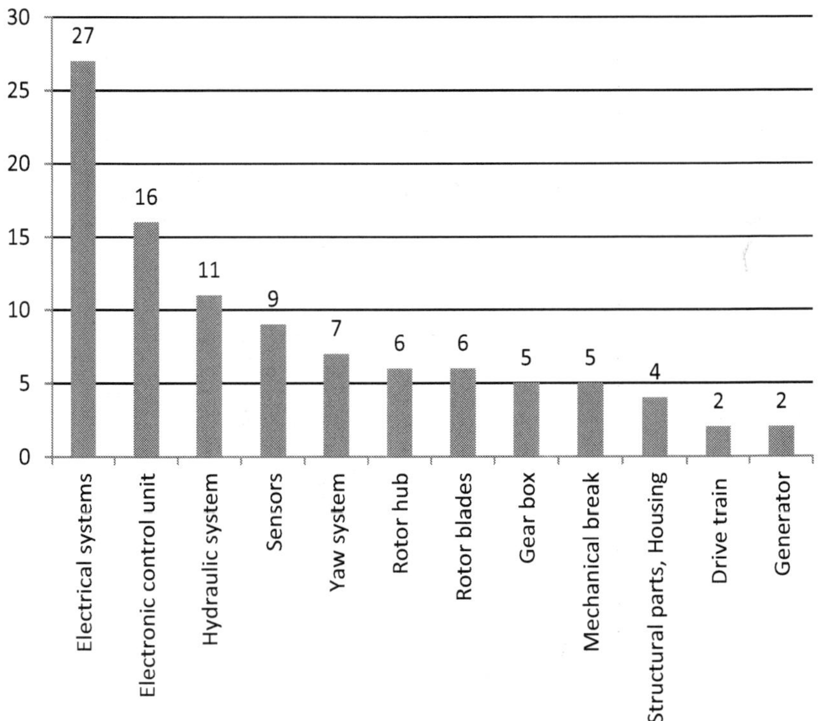

Fig. 10. Percentage of needed repairs and maintenance on utility scale wind turbines. Data: AWEA.

While wind turbines are designed for a lifetime of around 20 years, existing gearboxes have exhibited failures after about 5 years of operation. The costs associated with securing a crane large enough to replace the gearbox and the long downtimes associated with such a repair affect the operational profitability of wind turbines. A simple gearbox replacement on a 1.5 MW wind turbine may cost the operator over $250,000 (Rensselar, 2010). The replacement of a gearbox accounts for about 10 percent of the construction and installation cost of the wind turbine, and will negatively affect the estimated income from a wind turbine (Kaiser & Fröhlingsdorf, 2007).

Additionally, fires may be started by the oil in an overheated gearbox. The gusty nature of the wind is what degrades the gearbox, and this is unavoidable.

Figure 11 summarizes the estimates of the economic rated power ranges of applicability for each of the considered wind turbine gearbox solutions.

The direct-drive approach to the current wind turbine gearbox reliability problem seems to be taking a strong hold in the 3 MW and larger market segment, although torque splitting is also being used in this range.

For the 1.5 to 3 MW range however, multiple viable options exist or show potential, including torque splitting, magnetic bearings, and Continuously Variable Transmissions (CVTs). These options may gain traction over direct-drive solutions due to the approximately 30 percent cost premium of a direct-drive system, and the larger sizes and capital costs associated with such a system.

If the magnetic bearing route is to be used, the answer may lie with gas turbine manufacturers, as their design criteria already call for bearings that are highly reliable, damage tolerant, and capable of handling large loads. CVTs appear to also offer aerodynamic efficiency benefits

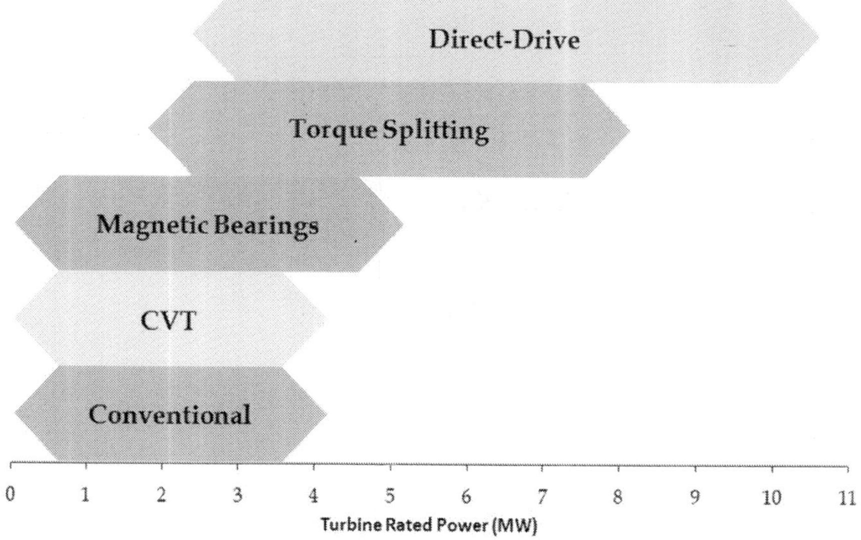

Fig. 11. Identified rated power applicability ranges of existing and possible wind turbine gearbox options. CVT: Continuously Variable Transmission.

to wind turbines, but they may be limited by the amount of torque that may be transmitted by chain, belt, or hydrostatic means. For this reason, magnetic bearings appear to provide a potential solution to a slightly wider range of turbine rated powers than CVTs would.

8. References

Becker, K.H. (2010) Magnetic Bearings for Smart Aero Engines (MAGFLY). *Proceedings of the 13th International Symposium on Transport Phenomena and Dynamics of Rotating Machinery (ISROMAC-13)*, G4RD-CT-2001-00625, Honolulu, Hawaii, April 2010.

Burton, T., Sharpe, D, Jenkins, N, Bossany, E. (2004). *Wind Energy Handbook (3rd Ed.)*. John Wiley & Sons Ltd., ISBN: 0-471-48997-2, West Sussex, England.

Clark, D.J. Jansen, M.J., Montague, G.T. (2004). An Overview of Magnetic Bearing Technology for Gas Turbine Engines. *National Aeronautics and Space Administration*, NASA/TM-2004-213177.

Department of Energy (2010). Advanced Wind Turbine Drivetrain Concepts: Workshop Report. *Key Findings from the Advanced Drivetrain Workshop*, Broomfield, Colorado, June 2010.

Enercon (2010). Enercon Wind Energy Converters: Technology & Service. Available from: <http://www.enercon.de/p/downloads/EN_Eng_TandS_0710.pdf>

Kaiser, S., Fröhlingsdorf, M. (August 20, 2007). The Dangers of Wind Power, In: *Spiegel Online*, May 2010, Available from:
<http://www.spiegel.de/international/germany/0,1518,500902,00.html>

Kostka, R.A., Kenawy, N. Compact Bearing Support. United States Patent Number 7,857,519. Issued December 28, 2010.

Musial, W. Butterfield, S., McNiff, B. (2007). Improving Wind Turbine Gearbox Reliability, *Proceedings of the 2007 European Wind Energy Conference*, NREL: CP-500-41548, Milan, Italy, May 2007.

Mangliardi, L, Mantriota, G. (1994). Automatically Regulated C.V.T. in Wind Power Systems. *Renewable Energy*, Vol. 4, No. 3, (1994), pp. 299-310, 0960-1481(93)E0004-B.

Mangliardi, L., Mantriota, G. (1996). Dynamic Behaviour of Wind Power Systems Equipped with Automatically Regulated Continuously Variable Transmission. *Renewable Energy*, Vol. 7, No. 2, (1996), pp. 185-203, 0960-1481(95)00125-5.

Mikhail, A.S., Hahlbeck, E.C. Distributed Power Train (DGD) With Multiple Power Paths. United States Patent Number 7,069,802. Issued July 4, 2006.

Ragheb A., Ragheb, M. (2010). Wind Turbine Gearbox Technologies, *Proceedings of the 1st International Nuclear and Renewable Energy Conference (INREC'10)*, ISBN: 978-1-4244-5213-2, Amman, Jordan, March 2010.

Rensselar, J. (2010). The Elephant in the Wind Turbine. *Tribology & Lubrication Technology*, June 2010, pp.2-12.

Robb, D. "The Return of the Clipper Liberty Wind Turbine." Power: Business and Technology for the Global Generation Industry. (December 1, 2008)

Schweitzer, G. (2002). Active Magnetic Bearings – Chances and Limitations. *Proceedings of the 6th International Conference on Rotor Dynamics*, Sydney Australia, September 2002.

9

Monitoring and Damage Detection in Structural Parts of Wind Turbines

Andreas Friedmann, Dirk Mayer, Michael Koch and Thomas Siebel
Fraunhofer Institute for Structural Durability and System Reliability LBF
Germany

1. Introduction

Structural Health Monitoring (SHM) is known as the process of in-service damage detection for aerospace, civil and mechanical engineering objects and is a key element of strategies for condition based maintenance and damage prognosis. It has been proven as especially well suited for the monitoring of large infrastructure objects like buildings, bridges or wind turbines. Recently, more attention has been drawn to the transfer of SHM methods to practical applications, including issues of system integration.

In the field of wind turbines and within this field, especially for turbines erected off-shore, monitoring systems could help to reduce maintenance costs. Off-shore turbines have a limited access, particularly in times of strong winds with high production rates. Therefore, it is desirable to be able to plan maintenance not only on a periodic schedule including visual inspections but depending on the health state of the turbine's components which are monitored automatically.

While the monitoring of rotating parts and power train components of wind turbines (known as Condition Monitoring) is common practice, the methods described in this paper are of use for monitoring the integrity of structural parts. Due to several reasons, such a monitoring is not common practice. Most of the systems proposed in the literature rely only on one damage detection method, which might not be the best choice for all possible damage.

Within structural parts, the monitoring tasks cover the detection of cracks, monitoring of fatigue and exceptional loads, and the detection of global damage. For each of these tasks, at least one special monitoring method is available and described within this work: Acousto Ultrasonics, Load Monitoring, and vibration analysis, respectively.

Farrar & Doebling (1997) describe four consecutive levels of monitoring proposed by Rytter (1993). Starting with „Level 1: Determination that damage is present in the structure", the complexity of the monitoring task increases by adding the need for localising the damage (level two) and the „quantification of the severity of the damaged" for level three. Level four is reached when a „prediction of the remaining service life of the structure" is possible.

By using the monitoring systems described above, in our opinion only level 1 or in special cases level 2 can be attained. For most customers, the expected results do not justify the efforts that have to be made to install such a monitoring system.

In general, our work aims at developing a monitoring system that is able to perform monitoring up to level 4. Therefore, we think it is necessary to combine different methods. Even though the different monitoring approaches described in this paper differ in the type of sensors used or whether they are active or passive, local or global methods, their common feature is that they can be implemented on smart sensor networks.

The development of miniaturized signal processing platforms offer interesting possibilities of realizing a monitoring system which includes a high number of sensors widely distributed over the large mechanical structure. This approach should considerably reduce the efforts of cabling, even when using wire-connected sensor nodes, see Fig. 1. However, the use of communication channels, especially wireless, raises challenges such as limited bandwidth for the transmission of data, synchronization and reliable data transfer. Thus, it is desirable to use the nodes of the sensor network not only for data acquisition and transmission, but also for the local preprocessing of the data in order to compress the amount of transmitted data. For instance, basic calculations like spectral estimation of the acquired data sequences can be implemented. The microcontrollers usually applied in wireless sensor platforms are mostly not capable of performing extensive calculations. Therefore, the algorithms for local processing should involve a low computational effort.

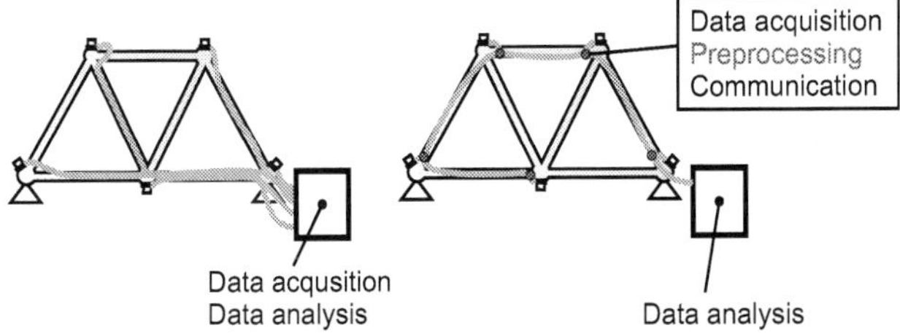

Fig. 1. SHM system with centralized acquisition and processing unit versus system with smart sensors.

2. Load Monitoring

2.1 Basic principles

The concept Load Monitoring is of major interests for technical applications in two ways:

- The reconstruction of the forces to which a structure is subjected (development phase)
- The determination of the residual life time of a structure (operational phase).

A knowledge of the forces resulting from ambient excitation such as wind or waves enables the structural elements of wind turbines like towers, rotor blades or foundations to be improved during the design phase. External forces must be reconstructed by using indirect measuring techniques since they can not be measured directly. Reconstruction measurement techniques are based on the transformation of force related measured quantities like acceleration, velocity, deflection or strain. In general, this transformation is conducted via the solution of the inverse problem:

$$y(t) = \int_0^t H(t - \tau)F(\tau)\,d\tau \tag{1}$$

where the system properties $H(t)$ and the responses $y(t)$ are known and the input forces $F(t)$ are unknown (Fritzen et al., 2008).

Thus the inverse identification problem consists of finding the system inputs from the dynamic responses, boundary conditions and a system model. The different methods for identifying structural loads can be categorized into deterministic methods, stochastic methods, and methods based on artificial intelligence. A review of methods for force reconstruction is given by Uhl (2007).

Since monitoring wind turbines typically concerns the operational phase, the reconstruction of forces is of secondary interest. In turn, more stress must be focussed on determining the residual life time of a structure.

Determining the residual life time is based on the evaluation of cyclic loads. Structures errected in the field are typically subjected to cyclic loads resulting from ambient excitation such as wind or wave loads in the case of off-shore wind turbines or traffic loads for bridges. A cyclic sinusoidal load is characterized by three specifications. Two specifications define the load level, e.g. maximum stress S_{max} and minimum stress S_{min} or mean stress S_m and stress amplitude S_a. The third specification is the number of load cycles N. Depending on the stress ratio $R = S_{min}/S_{max}$, the cyclic load can be divided into the following cases: pulsating compression load, alternating load, pulsating tensile load or static load, and into intermediate tensile/compressive alternating load (Haibach, 2006).

A measure for the capacity of a structural component to withstand cyclic loads with constant amplitude and constant mean value is given by its S/N-curve, also reffered to as Wöhler curve, Fig. 2. Basically, the S/N-curve reveals the number of load cycles a component can withstand under continuous or frequently repeated pulsating loads (DIN 50 100, 1978).

Three levels of endurance characterize the structural durability: low-cycle, high-cycle or finite-life fatigue strength and the ultra-high cycle fatigue strength. The endurance strength corresponds to the maximum stress amplitude S_a and a given mean stress S_m which a structure can withstand when applied arbitrarily often (DIN 50 100, 1978) or more frequent than a technically reasonable, relatively large number of cycles. However, the existence of an endurance strength is contentious issue, since it has been demonstrated that component failures are also caused in the high-cycle regime (Sonsino, 2005). The transition to finite-life fatigue strength is characterized by a steep increase of the fatigue strength. The knee-point, which seperates the long-life and the finite-life fatigue strength corresponds to a cycle number of about $N_D = 10^6 - 10^7$. Both, long-life and finite-life fatigue strength are dominated by elastic strains. In contrast to this, low-cycle fatigue strength is dominated by plastic strains. The transition from finite-life to low-cycle fatigue strength is in the area of the yield stress (Radaj, 2003).

S/N-curves are derived from cyclic loading tests. The tests are carried out on unnotched or notched specimens or on component-like specimens. Load profiles applied to the specimen are either axial, bending or torsional. To derive one curve, the mean stress S_m or the minimum stress S_{min} is left constant for all specimens, only the stress amplitude S_a or maximum stress S_{max} is modified. The numbers of load cycles a specimen withstands up to a specific failure criterion, e.g. rupture or a certain stiffness reduction, is plotted horizontaly against the corresponding stress amplitude values (DIN 50 100, 1978; Radaj, 2003).

The previous considerations concern the durability of structures subjected to constant amplitude loading. However, most structures under operational conditions experience loading environments with variable mean loads and load amplitudes. This differentiation is important since fatigue response may be very sensitve to the specifics of the loading type (Heuler & Klätschke, 2005).

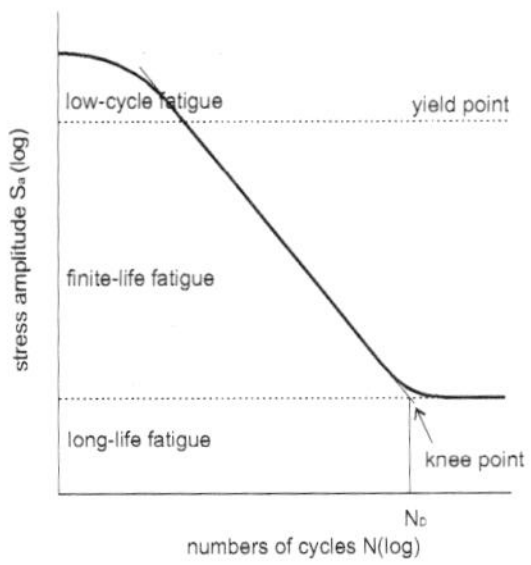

Fig. 2. S/N-curve.

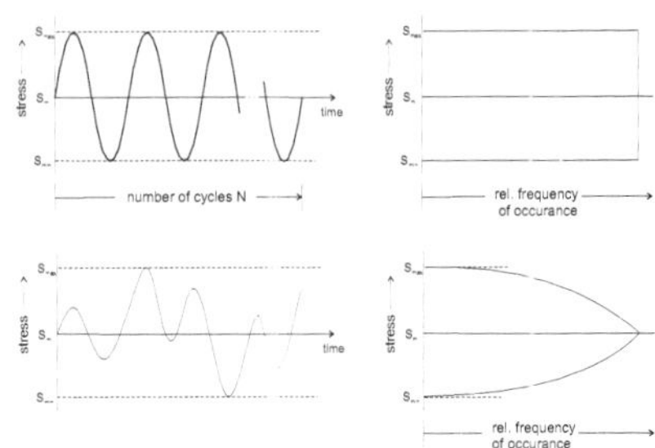

Fig. 3. Load spectra derived by level crossing counting from constant and variable amplitude load-time histories, after Haibach (1971).

The fatigue life curve, or: Gassner curve, is determined in similar tests but using defined sequences of variable amplitude loads (Sonsino, 2004). Depending on the composition of the load spectrum, the fatigue life curve deviates from the S/N-curve. The relation between both curves is represented by a spectrum shape factor (Heuler & Klätschke, 2005).

The content of a variable load time history, e.g. the relative frequency of occurance of each amplitude, can be illustrated in a load spectrum. Different cycle counting methods exist to derive load spectra. An overview of the so called one-parameter counting methods is given in DIN 45 667 (1969) and Westermann-Friedrich & Zenner (1988). Level crossing and range-pair counting are historically well established one-parameter counting methods (Sonsino, 2004).

As an example Fig. 3 shows load spectra derived from constant amplitude and variable amplitude load-time-history. The mean stress S_m and the maximum stress amplitude S_{max} is common to the resulting load spectra. However, the constant amplitude load spectrum reveals a high cycle number for the maximum amplitude, while the variable amplitude load spectrum reveals high cycle numbers only for smaller amplitudes. Fig. 4 shows examples of load spectra. In general, the fuller a load spectrum is, i.e. the more relatively large amplitudes it contains, the less load cycles a structure withstands without damage, Fig. 4 (Haibach, 2006). From today's perspective, most of the one-parameter counting methods can be considered as special cases of the two-parameter rainflow counting method (Haibach, 2006). The rainflow

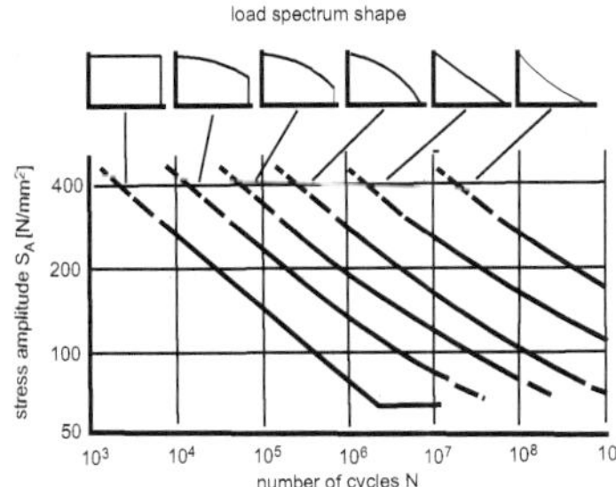

Fig. 4. Effect of different load spectra on the fatigue life curve, after Haibach (2006).

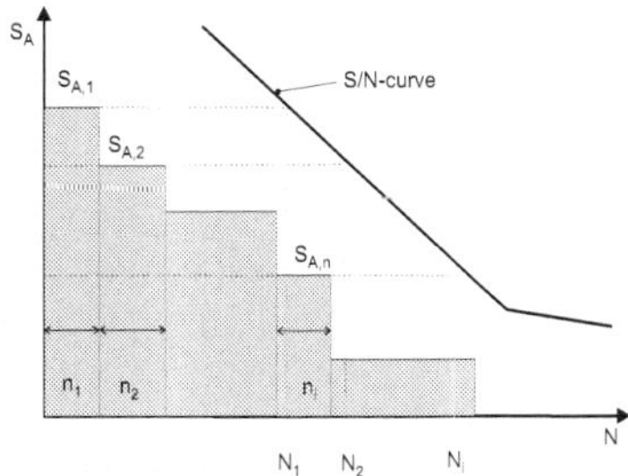

Fig. 5. Load spectrum and S/N-curve.

counting method is the most recent and possibly the most widely accepted procedure for load cycle counting (Boller & Buderath, 2006). Each loading cycle can be defined as a closed hysteresis loop along the stress-strain path. The maximum and the mininum values or the amplitude and the mean values of the closed hysteresis loops are charted as elements into the rainflow matrix. Rainflow matrices allow distribution of the hysteresis loops to be determined. A detailed description of the method is given by Haibach (2006), Radaj (2003) and Westermann-Friedrich & Zenner (1988).

In order to provide representative load data, standardized load-spectra and load-time histories (SLH) have been previously be developed. SLH are currently available for various fields of application such as in the aircraft industry, automotive applications, steel mill drives, as well as for wind turbines and off-shore structures. In particular, the SLH WISPER/WISPERX (Have, 1992) and WashI (Schütz et al., 1989) exist for wind turbines and off-shore structures, respectively (Heuler & Klätschke, 2005).

The residual life time of a component is estimated by means of a damage accumulation hypothesis. For this reason, a load spectrum for the specific component, a description of the stress concentration for notches in the component and an appropriate fatigue life curve are required (Boller & Buderath, 2006). Within numerous damage accumulation hypotheses the linear hypothesis of Palmgren and Miner plays an important role in practical applications. Different modifications of the hypothesis exist. Basically, the idea of the hypothesis is to determine the residual life time of a component via the sum of the load cycles which the component experiences in relation to its corresponding fatigue life curve, Fig. 5. A partial damage D_j is calculated for a certain load amplitude S_j with

$$D_j = \frac{\Delta N_j}{N_{Fj}}, \tag{2}$$

where ΔN_j is the number of load cycles corresponding to S_j and N_{Fj} is the number of load cycles with the same load amplitude corresponding to the fatigue failure curve, up to failure. The total damage D resulting from the load cycles of different load amplitudes S_j with $j = 1, 2, ..., n$ is yielded by:

$$D = \sum_{j=1}^{n} D_j. \tag{3}$$

According to the hypothesis the component fails when its total damage attaines unity, $D = 1$. However, several studies demonstrate that in particular cases, the real total damage may deviate significantly from unity. Prerequestives for the right choice of the S/N-curve are that the material, the surface conditions, the specimen geometry and loading conditions are appropriate (Radaj, 2003).

2.2 System description

To measure the operational loading, strain gauges are applied to the structure's hot spots determined during the development phase. The strain gauges can be connected to smart sensor nodes placed adjacent to the spots (see Fig. 6). Running a rainflow counting algorithm on the smart sensor nodes yields an analysis of the real operational loads the structure is subjected to. Due to the data reduction using the rainflow counting, the amount of data to be transmitted to a central unit is reduced to a minimum. Furthermore, the data are only transmitted when an update of the residual life time is requested by the operator (e.g. every 10 minutes).

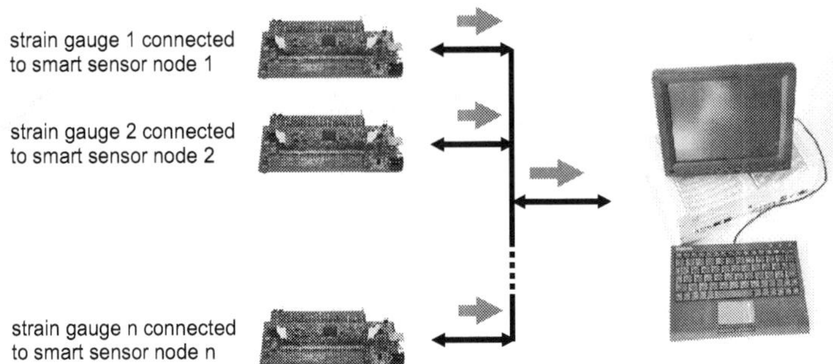

Fig. 6. Load Monitoring system based on decentralized preprocessing with smart sensor nodes.

Using the load spectra as input, damage accumulation is calculated on the central unit (see Fig. 6 for the connection scheme of smart sensor nodes and the central unit). Next, the residual life time can be determined by comparing those damage accumulation with the endurable loading. Furthermore, with this Load Monitoring concept, exceptional loads can be determined and used to trigger analyses of the structure's integrity using other monitoring methods.

2.3 Application: Model of a wind turbine

To test the performance of the decentralized Load Monitoring in the field, a structure is chosen which is exposed to actual environmental excitations by wind loads. A small model of a wind turbine (weight approx. 0.5 kg) is mounted on top of an aluminum beam, which serves as a model for the tower (see Fig. 7). Although quite simple and small, the wind turbine model possesses a gearbox with several stages which may serve as a potential noise source during operation. For a test, the beam is instrumented with four strain gauges which are wired to a Wheatstone bridge and mounted close to the bottom of the beam as shown in Fig. 7.

The left side of Fig. 8 shows a rainflow matrix calculated under wind excitation and the associated damage accumulation. The implementation has been conducted to the effect, that the smart sensor node determines the turning points from the strain signal before calculating the rainflow matrix by using the rainflow counting algorithm. Following this, the rainflow matrix was periodically (in this case every 10 minutes) sent to a central unit. By means of this, the communication effort and the real time requirements between the two participants in network have been reduced. Based on the data-update, the central unit (a desktop PC in this case) estimates the current damage accumulation. The result of this is a damage accumulation function growing over time steps of 10 minutes, as shown on the right side of Fig. 8. The function rises slowly, caused by a small number of load cycles or a minor strain signal amplitude. In contrast to this, the steep rising at about $t = 200$ min, is due to a large number of load cycles or a large strain amplitude, as the case may be.

The Palmgren-Miner rule states that failure occurs when the value of the accumulate damage is unity. In this test, with only a wind loading measured over a period of 220 minutes, the calculated damage accumulation is remote from this critical value.

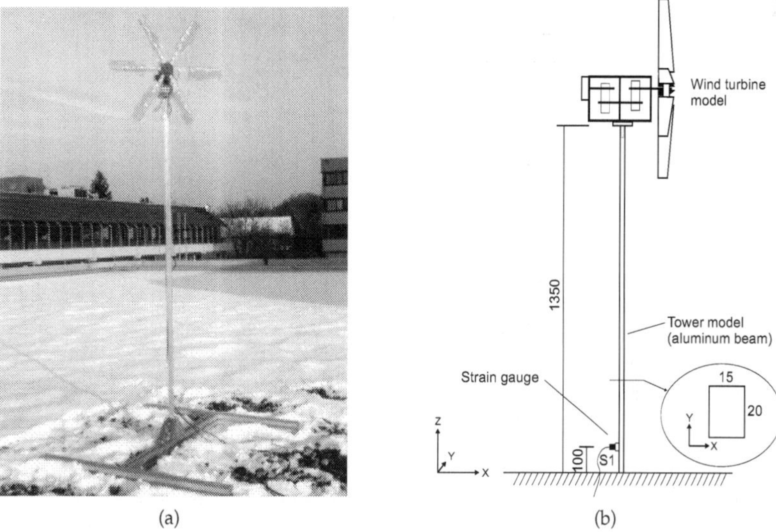

Fig. 7. Experimental set-up. (a) A model of a wind turbine was exposed to actual environmental excitations by wind loads on the top of a building. (b) Strain gauges were mounted nearly at the bottom of the beam.

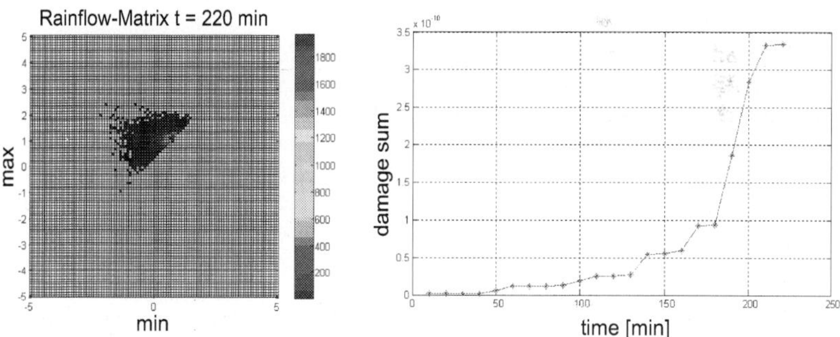

Fig. 8. Left: A matrix calculated under wind excitation. Right: The damage accumulation growing over time.

3. Vibration analysis

3.1 Basic principles

The process of monitoring the health state of a structure involves the observation of a system over time by the means of dynamic response measurements, the extraction of damage-sensitive features from these measurements, and the statistical analysis of these features. Features are damage sensitive properties of a structure which allow difference between the undamaged and the damaged structure to be distinguished (Sohn et al., 2004).

Some vibration-based damage sensitive properties are described in the following.

Resonant frequencies. Monitoring methods based on resonant frequencies can be categorized into the forward and the inverse problem. The forward problem consists of determining frequency shifts due to known damage cases. Damage cases are typically simulated using numerical models. Damages can then be identified by comparing the simulated to the experimentally measured frequencies.

The inverse problem consists in determining damage parameters from shifts in resonant frequencies. However, the major drawback of monitoring methods based on frequency shifts is the low sensitivity of resonant frequencies to damage (Montalvão et al., 2006).

Mode shapes. Using mode shapes as a feature for damage detection is advantageous over using methods based on resonant frequencies since mode shapes contain local information about a structure which makes them sensitive to local damage. This is a major advantage over resonant frequencies which are global parameters and thus can only detect global damage. Furthermore, mode shapes are less sensitive to environmental effects such as temperature. 'Traditional mode shape change methods' are based on damage identification from mode shape changes which are deteremined by comparing them to finite element models or preliminary experimental tests. On the other hand, 'modern signal processing methods' can be applied alone to actual mode shape data from the damaged structure. Here, the damage location can be revealed by detecting the local disconinuity of a mode shape. A promissing feature of enhancing the damage sensitivity of mode shape data is the use of mode shape curvatures. A curvature is the second derivative of a mode shape. The modal strain energy-based methods can be considered as a special case of the curvature-based method. The method uses changes of fractional modal strain energies, which are directly related to mode shape curvatures. Within the modal strain energy-based methods particular attention is paid to the damage index method (DIM) (Fan & Qiao, 2010).

Frequency domain response functions. The use of non-modal frequency domain response functions for monitoring can be advantageous over modal-based methods. In contrast to response functions modal parameters are indirectly measured. Thus they can be falsified by measurement errors and modal extraction faults. Further, the completeness of modal data can not be guaranteed in most practical applications because a large number of sensors is required (Lee & Shin, 2001). One approach to detecting the damage locations on complex structures using directly measured frequency-domain response functions is based on the Transmissibility Function (TF) (Siebel & Mayer, 2011). The TF for one element of a structure is built from the ratio of the response functions of two adjacent points. TFs can be calculated from system outputs without knowing the exact input forces. Hence, in order to apply the method, information about the source of excitation is not neccessarily required (Chesné & Deraemaeker, 2010).

In the technical literature, the topic of feature extraction has received much attention. Reviews and case studies have been written by Sohn et al. (2004), Carden & Fanning (2004), Montalvão et al. (2006), Humar et al. (2006), Fan & Qiao (2010) and, with special emphasis on the monitoring of wind turbines, by Ciang et al. (2008) and Hameed et al. (2007).

The large size of wind turbines and the difficult accessibility complicates maintenance and repair work. Thus, in order to guarantee safety and to improve availability, the implementation of an autonomous monitoring system that regularly delivers data about the structural health is of practical interest. Furthermore, easy handling and installation, i.e. low cabling effort, are required for practicability.

The implementation of vibration-based monitoring methods requires frequency or modal data. Modal data of real structures can be extracted by methods of system identification. The demands for autonomy of the monitoring system and for a regular data transfer makes system identification a challenging task. That is, because vibration must be measured under operating conditions and no defined force is applied. On off-shore wind turbines, vibration is induced into the structure by ambient excitation, e.g. by wind or wave loads. System identification methods handling these conditions are referred to as output-only system identification methods, in-operation or as Operational Modal Analysis (OMA).

Important output-only system identification methods for the extraction of modal data (i.e. natural frequency, damping ratio and mode shapes) are outlined below. The first four categories of methods presented are based on time-domain responses, whereas the methods of the last category use frequency-domain responses.

NExT. Traditional time domain multi-input multi-output (MIMO) Experimental Modal Analysis (EMA) uses impulse response functions (IRF) to extract modal parameters. The Natural Excitation Technique (NExT) adopts the EMA methodology by employing Correlation Functions (COR) instead of IRF. COR can be obtained by different techniques such as the Random Decrement (RD) technique, inverse Fourier Transform of the Auto Spectral Density, or by direct estimates from the random response of a structure subjected to broadband natural excitation. Both COR and IRF are time domain response functions that can be expressed as sum of exponentially-decayed sinusoidals. The modal parameters of each decaying sinusoidal are identical to those of the corresponding structural mode (Zhang, 2004).

ARMA. The dynamic properties of a system in the time domain can be described by an Auto-Regressive Moving Average (ARMA) model. In the case of a multivariate system, the model is called Auto-Regressive Moving-Average Vector (ARMAV) (Andersen, 1997).

The parametric ARMAV model describes the relation of the system's responses to stationary zero-mean Gaussian white noise input and to AR (auto-regressive) and MA (moving-average) matrices. The AR component describes the system dynamics on the basis of the response history and the MA component regards the effect of external noise and the white noise excitation. The ARMAV model can be expressed in the state-space from which natural frequencies, damping ratios and mode shapes can be extracted (Bodeux & Colinval, 2001).

A number of algorithms for the Prediction-Error Method (PEM) have been proposed in order to identify modal parameters from ARMAV models. However, the PEM-ARMAV type OMA procedures have the drawback of being computationally intensive and of requiring an initial 'guess' for the parameters which are to be identified (Zhang et al., 2005).

Covariance-driven SSI. The covariance-driven Stochastic Subspace Identification (SSI) methods allows state-space models to be estimated from measured vibration data. For system realization, measured impulse response or covariance data are used to define a Hankel matrix. The Hankel matrix can be factorized into the observability matrix and the controllability matrix. The system matrices of the state-space model can then be extracted from the observability and the controllability matrices. Various SSI methods have been developed for the system realization, e.g. Principal Component (PC) method,

Canonical Variant Analysis (CVA) and Un-weighted Principal Component (UPC) method. The methods differ in the way the observability matrix is estimated, and also in how it is used for finding the system matrices. The last step of the stochastic realization-based OMA is the calculation of the modal parameters from the system matrices. These methods are also referred to as stochastic realization-based procedures (Viberg, 1995; Zhang et al., 2005).

Data-driven SSI. The advantage of the data-driven over the covariance-driven SSI is that it makes direct use of stochastic response data without an estimation of covariance as the first step. Furthermore, it is not restricted to white noise excitation, as is the covariance-driven SSI, but it can also be employed for colored noise. The data-driven SSI predicts the future system response from the past output data. Making use of state prediction leads to a Kalman filter for a linear time-invariant system. It can be expressed by a so-called innovation state-space equation model, where the state vector is substituted by its prediction and where the two inputs (i.e. process noise and measurement noise) are converted to an input process – the innovations. After computing the projection of the row space of the future outputs on the row space of the past outputs and estimating a Kalman filter state, the modal parameters are calculated as before with UPC, PC or CVA (Andersen & Brincker, 2000; Zhang et al., 2005).

Frequency Domain Decomposition. The classical frequency-domain approach for OMA is the Peak Picking (PP) technique. Modal frequencies can be directly obtained from the peaks of the Auto Spectral Density (ASD) plot, the mode shapes can be extracted from the column of the ASD matrix which corresponds to the same frequency. The PP method gives reasonable estimates of the modes, it is fast and simple to use. However, PP can be inaccurate when applied to complex structures, especially in the case of closely spaced modes. The Frequency Domain Decomposition (FDD) is an extension of PP which aims to overcome this disadvantage. The FDD technique estimates modes from spectral density matrices by applying a Singular Value Decomposition. This corresponds to a single-degree-of-freedom identification of the system for each singular value. The modal frequencies can then be obtained from the singular values and the singular vectors are an estimation of the corresponding mode shapes. The Enhanced Frequency Domain Decomposition (EFDD) is a further development of the FDD. In addition to modal frequencies and shapes, the EFDD can estimate modal damping. For this, singular value data is transferred to time-domain by an inverse Fourier transformation. Modal damping can then be estimated from the free decay (Herlufsen et al., 2005; Zhang et al., 2005).

3.2 System description

The system presented here is designed to autonomously acquire the development of the modal properties of the system under test. In this, it just delivers data useable for Structural Health Monitoring (SHM), but it does not autonomously carry out analyses of them. The system accounts for the requirements mentioned above in the following way: To satisfy the demand for autonomy and to be able to use ambient instead of artificial excitation, the measured data is analysed using the algorithms of OMA. To lower the cabling effort, a network of smart sensors is designed in which the sensor nodes are capable of preprocessing the data using the RD method. Due to the reduction of the amount of data to be transmitted, the sensor nodes can be connected in a bus structure with the central processing unit.

The microcontrollers usually applied in (wireless) sensor platforms are typically not capable of performing extensive computations. For instance, only basic calculations like spectral estimation of the acquired data sequences can be implemented (Lynch et al., 2006). In this paper, the RD method is evaluated with respect to distributed signal processing on sensor nodes. It is a simple, yet effective method for estimating correlation functions (Brincker et al., 1991; Cole, 1973), and was originally devised for detecting damage in aerospace structures subjected to random loading. It can also successfully be applied as a component of a structural parameter identification, e.g. in OMA (Asmussen, 1998; Rodrigues & Brincker, 2005). Having estimated averaged correlation functions by means of the smart sensors, these functions are transferred to a central processing unit. There, the algorithms of Frequency Domain Decomposition (FDD) are applied to estimate the eigenfrequencies and mode shapes of the system under test.

3.2.1 Description of the Random Decrement technique

As mentioned above, most of the parts of a wind turbine being of interest to SHM cannot be artificially excited for structural analyses, because they might be too large (e.g. the whole tower) or because it is impractical to apply a vibration exciter during its operation. Thus only the output signals, i.e. the vibrations excited by operational loads can be used in order to estimate the system's behavior. Extracting this information can be done using the RD method, which is a simple technique that averages time data series $x(t_n)$ measured on the system under random input loads when a given trigger condition is fulfilled (see Equation 4 as an example of a level crossing trigger at trigger level a). The result of this averaging process from $n = 1$ to N is called an RD signature $D_{XX}(\tau)$.

$$D_{XX}(\tau) = \frac{1}{N} \sum_{n=1}^{N} x(t_n + \tau) \bigg| x(t_n) = a \qquad (4)$$

The method can be explained descriptively in the following way: At each time instant, the response of the system is composed of three parts: The response to an initial displacement, the response to an initial velocity and the response to the random input loads during the time period between the initial state and the time instant of interest (Rodrigues & Brincker, 2005). By averaging many of those time series, the random part will disappear, while the result can be interpreted as the system's response to the inital condition defined by the trigger, thus, containing information about the system's behavior. A depictive example can be found in Friedmann et al. (2010).

Asmussen (1998) proved that a connection between RD signatures D_{XX} and correlation functions R_{XX} can be established. For a level crossing trigger conditon, the factors used to derive correlation functions from RD signatures are the trigger level a itself and the variance σ_x^2 of the signal triggered (see Equation 5).

$$R_{XX}(\tau) = D_{XX}(\tau) \frac{\sigma_x^2}{a} \qquad (5)$$

These factors can be derived from the measurements without incurring high computational or memory costs so that the estimation of correlation functions can be implemented in a decentralized network of smart sensors.

The concept may be extended from autocorrelation functions as described above to the estimation of cross-correlation functions between two system outputs. This is simply achieved by averaging time blocks from one system output (here y) while the averaging process is triggered by another output (here x). If a simple level crossing trigger is assumed, the mathematical expression of the RD technique as established by Asmussen (1998) is:

$$D_{YX}(\tau) = \frac{1}{N} \sum_{n=1}^{N} y(t_n + \tau) \bigg| x(t_n) = a. \tag{6}$$

The conversion of a cross RD signature to a cross-correlation function is expressed by Equation 7.

$$R_{YX}(\tau) = D_{YX}(\tau)\frac{\sigma_x^2}{a} \tag{7}$$

3.2.2 Description of Operational Modal Analysis based on the Random Decrement method

The idea behind the described data acquisition system is to estimate RD signatures by the smart sensors and to transfer them to a central unit. On this central unit, the correlation functions are calculated and the modal analysis is performed. In the application described here, this central unit is a common desktop computer where the matrices $\mathbf{R}$ of the correlation functions are evaluated by a Matlab routine. Having calculated all correlation functions of the matrix $\mathbf{R}$, an intermediate step is needed before starting with the modal decomposition. Because the algorithms of frequency domain based OMA need a matrix $\mathbf{G}(f)$ of spectral densities as an input, a single block Discrete Fourier Transform (DFT) has to be applied to the correlation functions (McConnell, 1995). It should be mentioned that within this DFT, no use is made of time windowing. For the subsequent OMA, the FDD algorithm is used. This algorithm, first described by Brincker et al. (2000), is based on a Singular Value Decomposition of the matrix $\mathbf{G}(f)$. For every frequency f, this process leads to two fully populated matrices $\mathbf{U}(f)$ and a diagonal matrix $\mathbf{S}(f)$ holding the spectra of the so-called singular values $S_{ii}(f)$ in decreasing order (see Equation 8).

$$\mathbf{G}(f) = \mathbf{U}(f)\mathbf{S}(f)\mathbf{U}^{H}(f) \tag{8}$$

The peak values of the first singular values are then interpreted as indicators for the systems' eigenfrequencies. Furthermore, using the FDD algorithm, it is possible to estimate the mode shapes for the found frequencies. The eigenvectors describing the mode shapes corresponding to the eigenfrequencies determined by the spectra of the singular values can be found in the corresponding columns of the matrix $\mathbf{U}(f)$.

Performing an OMA in the usual way, picking the peaks from the spectra of the singular values $S_{ii}(f)$, requires users' input. However, because the modal decomposition has to be automated for the use in SHM, the need for such users' input must be eliminated. Therefore, an algorithm is needed that is able to pick the spectral peaks in a similar way to an educated user. Much work has been done in this area and some solutions are implemented in commercial software, e.g. by Peeters et al. (2006), Andersen et al. (2007), and Zimmerman et al. (2008).

In the implementation used here, an algorithm is employed for the peak picking that only operates using the numerical data of the given spectra. In doing so, the numerical data of the given spectra is analyzed automatically for potential eigenfrequencies. The procedure regards three parameters, a lower noise threshold, an upper signal bound and a value for the sensitivity in the frequency dimension. The sensitivity enables a residual random part of peaks in the spectra to be eliminated which are due to the finite number of averages used to calculate the RD signatures $R_{YX}(\tau)$.

3.3 Applications

3.3.1 Model of a wind turbine

To test the performance of the vibration analysis approach in the field, the model of a wind turbine described in Section 2.3 is used. The cross-section and length of the beam serving as the tower are appropriately chosen, such that the resonant frequencies of the assembled system are in a range similar to those of a full scale structure. Since the cross-section of the beam is rectangular, the bending eigenmodes in the x and y directions should possess different eigenfrequencies in order to alleviate the structural analysis. For the tests, the beam is instrumented with two triaxial, laboratory accelerometers with a sensitivity of 0.1 V/g and a measuring range of ± 50 g. One sensor is mounted close to the top of the beam (sensor 1) and the other mid-way along the length (sensor 2). In this first application, only the x-axis is used for data acquisition using the network described.

As a reference for the measurements under operating conditions, an EMA and an OMA are conducted. The results of those analyses (first bending modes around 4.5 Hz, second bending modes around 30.0 Hz) are listed and compared to other results in Table 1. For the OMA, the commercial software ARTeMIS has been used. It has to be mentioned that for the first and the second bending modes the frequency in the x and y directions fit each other even though the aluminum beam has no quadratic cross-section. This can be explained with the asymmetric fixture at the lower end of the beam.

Both sensor signals are processed on one hardware platform but by means of a decentralized implementation. This is considered as a first step for functional prototyping of the algorithms and the system layout in general. The estimation of the RD signatures follows Equation 4 and Equation 6. The estimation of the signal's variance σ_x^2 is done using an autoregressive power estimator (Kuo & Morgan, 1996). Its implementation requires only one storage bin and its use is possible due to the fact that signals without zero mean (as is the case with the used accelerations), the signals' power or mean square value equals the variance (Bendat & Piersol, 2000). A sampling rate of 200 Hz was chosen, which is high enough to acquire vibrations related to the first few bending modes. The RD signatures' length was set to 1000 elements; the signatures shown are averaged 8192 times.

The estimated RD signatures D_{YX} are shown in Fig. 9. A comparison of the cross-correlation functions R_{11} and R_{22} is shown in Fig. 10 to check that the assumption of reciprocity holds like it should be for every mechanical system.

The matrix of spectral densities $\mathbf{G}(f)$ derived from the correlation functions is shown in Fig. 11 and the peaks selected by the peak picking algorithm are marked by circles in Fig. 12. The results are the same as an educated user would have guesstimated. Only the peak at 50 Hz found within the experimental set-up would not have been chosen by a user because it clearly originates from a power line pickup. The eigenfrequencies found for the experimental set-up are 4.2 Hz and 33.4 Hz, respectively. Those results deviate max. $\pm$ 6% from the frequencies calculated or measured directly (see Table 1).

mode	measured by EMA	estimated by FDD in ARTeMIS	estimated with the network described here
1.	4.3 Hz / 4.4 Hz	4.2 Hz / 4.4 Hz	4.2 Hz / y not estimated
2.	31.6 Hz / 29.7 Hz	31.5 Hz / 29.0 Hz	33.4 Hz / y not estimated

Table 1. Comparison of eigenfrequencies (x-axis / y-axis) measured for the wind turbine using different approaches.

Fundamental and Advanced Topics in Wind Power

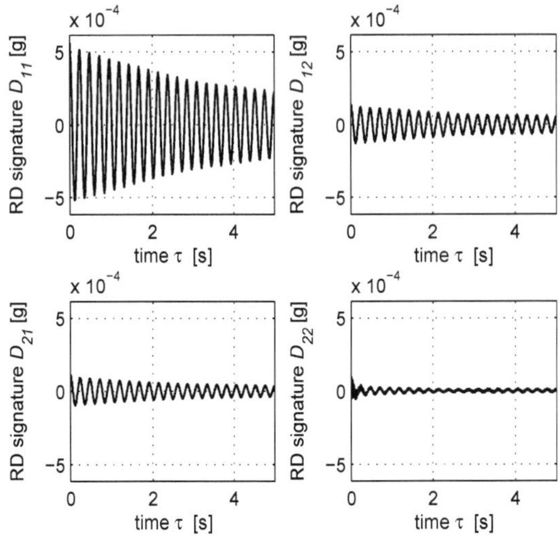

Fig. 9. Matrix **D** of the RD signatures measured on the model of a wind turbine.

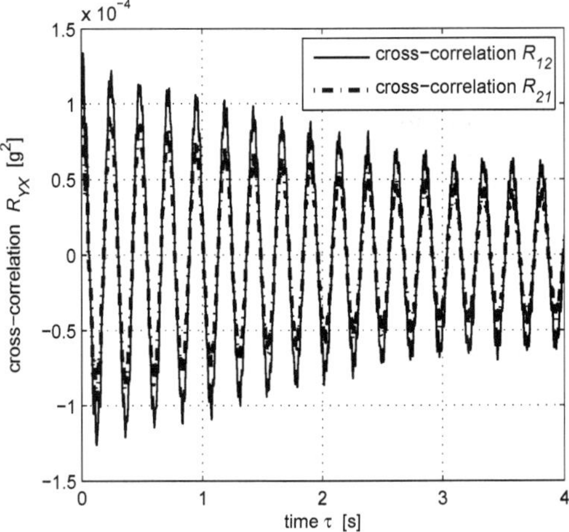

Fig. 10. Comparison of cross-correlation functions R_{12} and R_{21} of the wind turbine.

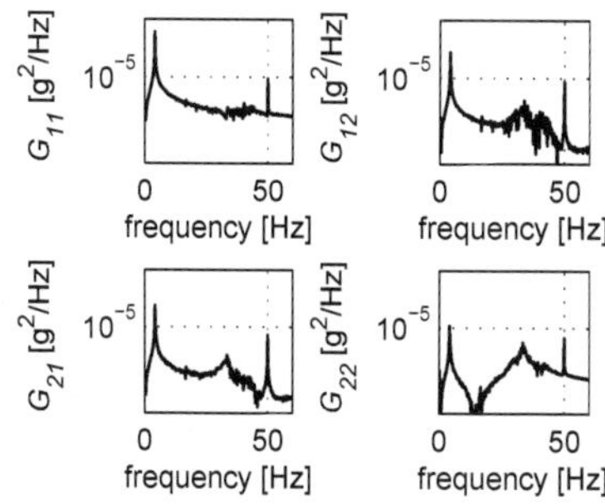

Fig. 11. Matrix of spectral densities $\mathbf{G}(f)$.

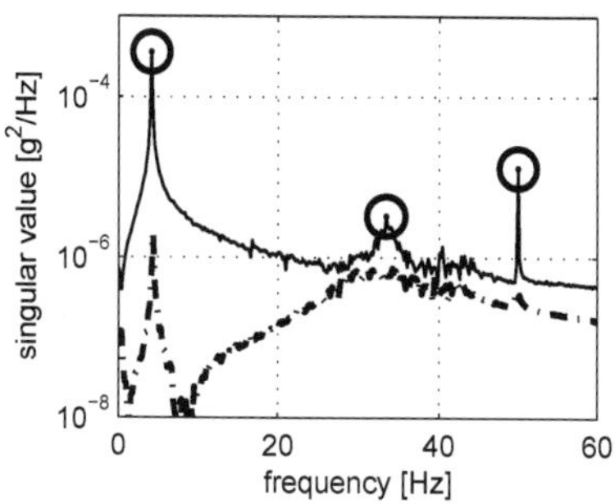

Fig. 12. Spectra of the first (-) and second (- -) singular values $S_{11}(f)$ and $S_{22}(f)$ of the wind turbine.

3.3.2 Pedestrian bridge

As the first real application for the proposed system, the monitoring of a pedestrian bridge is described. This bridge shown in Fig. 13 connects two buildings and has a length of 20 m.
To gain fundamental insight into the behavior of this bridge, an EMA has been performed. From this analysis, the main parameters of the monitoring system have been analytically derived (see Table 2). The steps needed to calculate this parameters are described below.

Fig. 13. North view of the bridge subjected to monitoring.

Parameter	Symbol	Value
Frequencies of modes of interest	$[f_1, ..., f_{HMoI}]$	$[8.1, 9.8, 15.1]$ Hz
Frequency of highest mode of interest	f_{HMoI}	15.1 Hz
Structural damping loss factor of modes of interest	$[\eta_1, ..., \eta_{HMoI}]$	$[0.63, 0.43, 1.23]$ %
Desired decay of the sigantures	$\frac{X_{end}}{X}$	1%
Sampling frequency	f_s	128 Hz
Block length of RD signatures (in samples)	n	2048
Block length of RD signatures (in seconds)	T	16 s

Table 2. Setup of the monitoring system derived from the EMA.

The first step in defining the setup of the monitoring system from the EMA is to select the range of modes that should be monitored. This selection should be based on the experience of a monitoring engineer as well as on knowledge about the hot spots of the system being monitored. These are given by the user or manufacturer. Having chosen the modes of interest, the sampling frequency and block length can be calculated using the output coming from the EMA as well as some values defined by the user (see Equations 9 and 10).

$$f_s = [2...8] \cdot f_{HMoI} \tag{9}$$

$$T \geq \max \left\{ \left. \frac{\ln\left(\frac{X_{end}}{X}\right)}{-\frac{\eta_f}{2} \frac{2\pi f}{\sqrt{1-\left(\frac{\eta_f}{2}\right)^2}}} \right| f \in [f_1, ..., f_{HMoI}] \right\}$$

$$T = a\frac{1}{f_s} \left.\right| a \in 2^{\mathbb{N}} \tag{10}$$

The quality of the determined mode shapes strongly depends on the accuracy of the estimated phase relationships acquired using the cross RD signatures (see Section 3.2.1). Therefore the acquired cross RD signatures should include good quality information about all the modes of interest. This might be archieved by choosing the triggering sensors or reference degrees of freedom (refDOF), in a way that each mode of interest is excitable at least at one refDOF. A similar requirement is described by the controllability criterion of a controlled system. Aiming at optimal controllability, the input matrix of a state-space-model can be used (Lunze, 2010). The state-space-model can be derived from the EMA of the bridge's structure (Bartel et al., 2010; Buff et al., 2010; Herold, 2003).

For monitoring the pedestrian bridge under operating conditions, the signal processing algorithm presented in section 3.2.1 is implemented on an embedded PC by automatic code generation from Simulink. Generally, the resources of the used embedded PC are limited. Based on the PC capacity and the setup parameters, the maximum number of sensors that can be handled by the hardware can be calculated. To guarantee real-time capability, the worst-case sensor-task execution time must be less than the sensor signal's sampling time. From this it follows that for the bridge application, the hardware can handle a maximum number of 14 sensors (2 reference and 12 response sensors). In contrast, an EMA is usually performed with a huge number of DOFs. Hence, out of these EMA candidate sensor positions, the best possible set of 14 sensor positions has to be chosen. The Effective Independence is a sensor placement method based on the linear independence of candidate sensor sets. In an iterative process, a large starting set is reduced to a given number of 14 sensors (Buff et al., 2010; Kammer, 1991).

The second step includes the initialisation of the operational monitoring system on the pedestrian bridge. Prior to starting the data acquisition, the system needs information about the characteristics of the bridge's natural excitation to determine sensor parameters like sensitivity, resolution and optimum trigger levels for both reference nodes. Therefore the acceleration data of the sensors have been recorded and analysed (see Fig. 14). Two characteristic types of ambient excitations are effective on the bridge: passing pedestrians and wind excitation.

According to Asmussen (1998), the optimum trigger level is $a = \sqrt{2}\sigma_x$, in the case of the level crossing trigger condition. In the real application it is inappropriate to calculate the standard

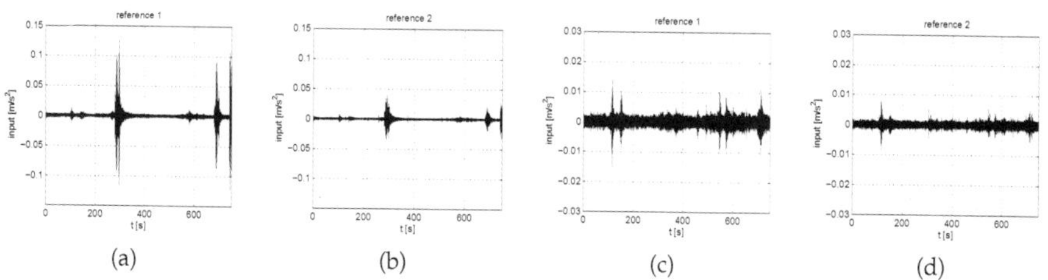

Fig. 14. At the reference nodes there are two characteristic types of natural excitations on the bridge. (a) pedestrian excitation reference 1; (b) pedestrian excitation reference 2; (c) wind excitation reference 1; (d) wind excitation reference 2.

deviation σ_x from the total sensor signal period, because of the poor signal-to-noise ratio (SNR) due to the measuring hardware over a wide range. For this reason, an experimental trigger level a_e has been calculated from only those sections with a sufficient SNR. For the bridge application, this will be the case if a pedestrian passes over the bridge (see Fig. 15).

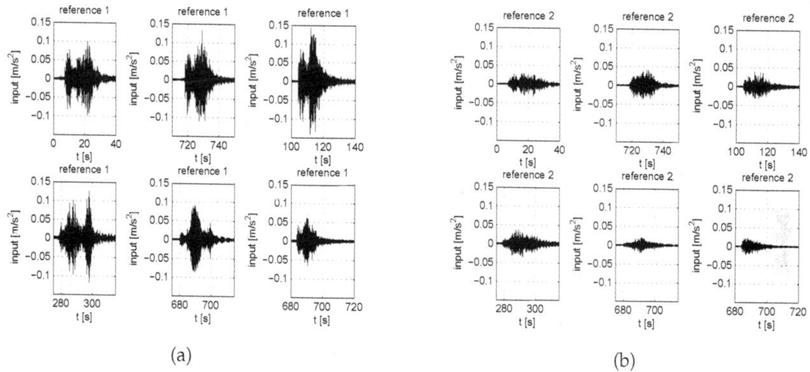

Fig. 15. If a pedestrian passes over the bridge, the signal-to-noise ratio of the reference sensor signals is sufficient to calculate the experimental trigger levels. (a) signal sections of reference 1 to calculate $a_{e,1}$; (b) signal sections of reference 2 to calculate $a_{e,2}$.

Consequently, the experimental trigger levels can be calculated as follows:

$$a_{e,i} = \frac{\sqrt{2}}{N} \sum_{k=1}^{N} \sigma_k \tag{11}$$

Having finished the preprocessing and setup up of the smart sensor network, the first measurements were conducted using the 14 DOFs as described. Those measurements yielded consecutive sets of 28 RD signatures. Using the two trigger levels and the two variances acquired together with those signatures, consecutive sets of 28 correlation functions have been calculated and transformed into spectral density matrices. For two correlation functions a reciprocity check could be made. This yielded a good agreement in amplitude and position of the zero crossings. Decomposing these matrices into singular value spectra and using the peak picking algorithm, the time-dependent development of several eigenfrequencies can be derived. Furthermore, clear mode shapes could be derived from the measured data. Fig. 10 shows a torsional mode of the bridge's deck. The mode shape and the frequency showed excellent agreement to the shapes and frequencies derived from the EMA.

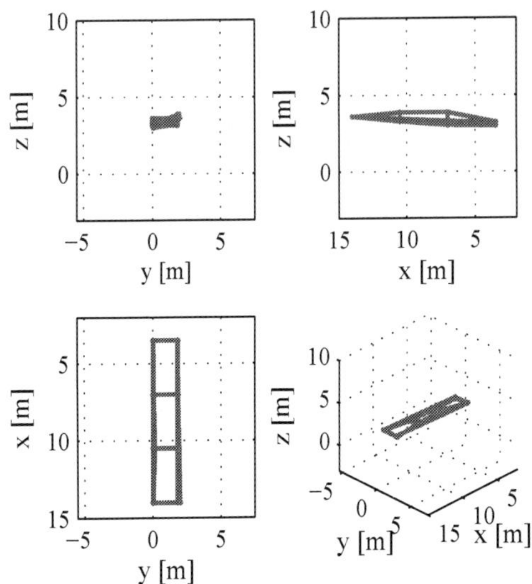

Fig. 16. Mode shape of a torsional mode of the bridge's deck.

4. Acousto Ultrasonics

4.1 Basic principles

Damage detection methods based on ultrasonics rely on the propagation and reflection of elastic waves within a material. The purpose is to identify local damage and flaws via wave field disturbances (Giurgiutiu & Cuc, 2005).

For ultrasound non-destructive evaluation (NDE) a transmitter transfers ultrasound waves into a material and a receiver picks up the waves when they have passed through the material. Transmitter and receiver can be combined into a single component (Lading et al., 2002).

Embedding piezoelectric transducers into a structure allows methods for active Structural Health Monitoring (SHM) to be implemented. Piezoelectric transducers are inexpensive, lightweight and unobtrusive and, due to these properties, appropriate for embedding into a structure with minimal weight penalty and at affordable costs. Compared to conventional NDE ultrasound transducers, active piezoelectric transducers interact with a structure in a similar way. One piezoelectric transducer can act as both transmitter and receiver of ultrasonic waves (Giurgiutiu & Cuc, 2005).

Among the different types of elastic waves that can propagate in materials, the class of the guided waves plays an important role for SHM. The advantage of guided waves is, that they can travel large distances in structures; even in curved structures, with only little energy loss. Guided waves are being used for the detection of cracks, inclusions, and disbondings in metallic and composite structures. Lamb waves, which are a type of guided wave, are appropriate for thin plates and shell structures. Due to their variable mode structure, multi-mode character, sensitivity to different types of flaws, their ability to propagate over long distances, their capability of following curvatures and accessing concealed components, the guided Lamb waves offer an improved inspection potential over other ultrasonic methods. Another type of guided waves are Rayleigh waves, which travel close to the surface with very little penetration into the depth of a solid, and therefore are more useful for detecting surface defects (Giurgiutiu, 2008).

Lamb waves are guided by the two parallel free boundary surfaces of a material. They propagate perpendicularly to the plate thickness. The boundary surfaces only impose low damping to the Lamb waves, thus they can travel long distances. Lamb waves are interesting from the monitoring point of view since damage such as boundary, material or geometric discontinuities which generate wave reflections can be detected (Silva et al., 2010).

The Lamb waves can be classified into symmetrical and anti-symmetrical wave shapes, Fig. 17, both satisfying the wave equation and boundary conditions for this problem. Each wave shape can propagate independently of the other (Kessler et al., 2002).

Piezoelectric transducers interact with a structure in a similar way to that of conventional ultrasonic transducers. Hence, methods used in traditional ultrasonic monitoring, like pitch-catch or pulse-echo, can be adopted to the application of piezoelectric transducers. The pitch-catch method can be used to detect damage which is located between a transmitter and a receiver transducer. Lamb waves travelling through the damaged region undergo changes in amplitude, phase, dispersion or time of transit. The changes can be identified by comparing with a 'pristine' situation. The pulse-echo method is based on a single transducer attached to a structure. The transducer functions as both transmitter and receiver of waves. The transducer sends out a Lamb wave which is reflected by a crack. The reflected wave (the echo) is then captured by the same transducer. In contrast to conventional pulse-echo testing where testing is traditionally applied across-the-thickness, the guided wave pulse-echo is more appropriate to cover a wide sensing area. Further ultrasonic approaches for damage detection, which are not detailed, are time reversal methods or migration techniques (Giurgiutiu, 2008).

Another technique for acousto-ultrasonic testing is referred to as electro-mechanical impedance method (EMI). Pioneering work on the use of EMI has been carried out by Liang et al. (1994). Only a single piezo transducer is required for its implementation. The technique is based on the direct relation between the electrical impedance (or admittance), which can be measured at the piezo transducers by commercially available impedance analysers, and the

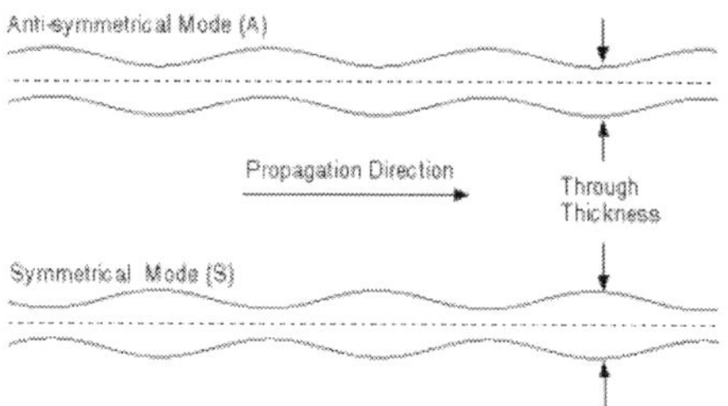

Fig. 17. Graphical representation of anti-symmetrical and symmetrical wave shapes (Kessler et al., 2002).

mechanical impedance of the host structure. The real part of the complex electrical impedance plotted over a sufficiently high frequency range, typically of the order of kHz, is called the 'signature' of the structure (Ballah & Soh, 2003).

Since the imaginary part is more sensitive to temperature variations than the real part, the real part is mainly being used for monitoring applications. In order to achieve high sensitivity to minor changes in the structural integrity the electrical impedance is measured at high frequency ranges of 30-400 kHz. For the choice of the frequency range for a given structure, which is commonly determined by trial-and-error, it must be taken into account that at high frequencies, the sensing area is limited adjacent to the piezoeletric transducer. Estimates of the sensing areas of single PZTs vary anywhere from 0.4 m for composite structures to 2 m for simple metal beams (Park et al., 2003). Furthermore, the modification of a signature increases as the distance between the sensor and the crack decreases (Zagrai & Giurgiutiu, 2002). A comprehensive overview of the topic of SHM with piezoelectric wafer active sensors is given in Giurgiutiu (2008).

A method for (passive) SHM, which can be implemented with similar piezoelectric transducer arrangements, is the acoustic emission (AE) method. Sources of acoustic emission within a structure can be cracking, deformation, debonding, delamination, impacts and others. Stress waves of broad spectral content are produced by localized transient changes in stored elastic energy (Ciang et al., 2008). Piezoelectric tranducers can then be used to convert the resulting accelerations into an electric signal. The AE technique allows the growth of damage to be sensed since stress waves are emitted by growing cracks or, in the case of laminates, fibre rupture. Thus the AE technique stands in contrast to many other NDE techniques which generally detect the presence, not the growth of defects. The sensing region of a piezoelectric sensor depends on its particular resonant frequency. For applications to laminates, a common sensor resonant frequency of 150 Hz is a good compromise between noise rejection and the area a sensor monitors (Lading et al., 2002).

4.2 System description and application to carbon fibre reinforced plastic coupons

The method of damage detection with ultrasonic waves should be illustrated by an experimental implementation on a CFRP (carbon fibre reinforced plastic) coupon (Lilov et al., 2010). The aim of the study is to monitor the delamination due to impact damage by acousto-ultrasonic inspection. To this end the coupon is instrumented with two piezo-ceramic transducers (PI 876-SP1). One of them is used as actuator and the other as a sensor (Fig. 18). The test set up (Fig. 19) also includes a dynamic signal analyser, which drives the actuator with a broad band swept sine signal (0..100 kHz), acquires the sensor data and calculates the structural Frequency Response Function (FRF) between actuator and sensor. A PC is connected to the analyser to control the measurements, collect the data and store them for further analysis.

To investigate the possibility of determining the severity of the impact event from the measured data, several identical coupons are instrumented and each damaged with a defined impact energy.

When directly comparing the measured FRFs from different impact tests, deviations from the original state can be observed (Fig. 20). In order to quantify these effects, a signature based analysis is applied to the FRF. To this end, FRF data H_{init} from a reference measurement, i.e. before the damage is induced, is compared to data acquired after the impact event (H_{dam}) by calculating statistical measures as damage indices. In many cases, the calculation of the

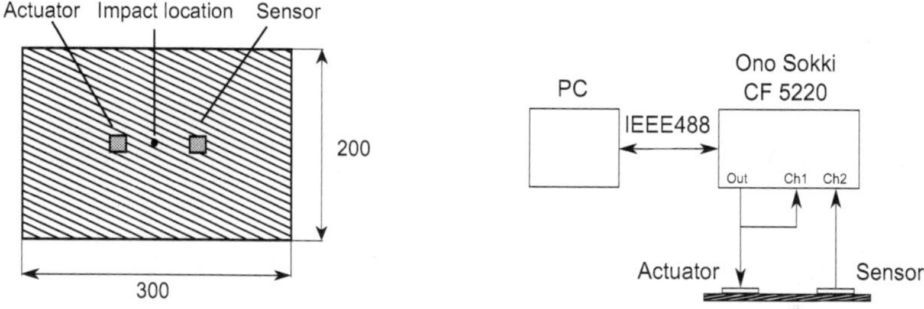

Fig. 18. Coupon instrumentation.   Fig. 19. Test set up.

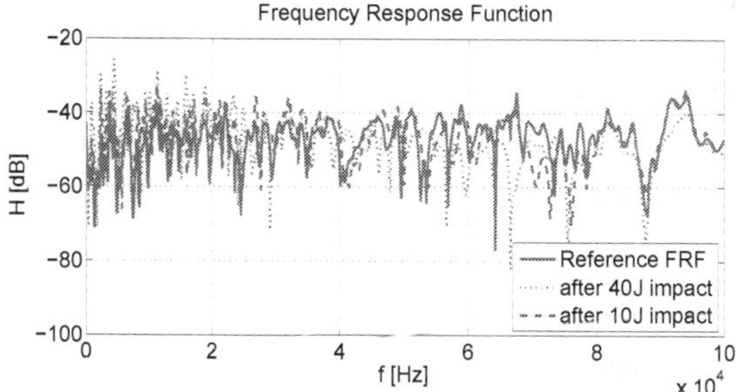

Fig. 20. Frequency response functions before and after different impacts.

correlation coefficient leads to good results:

$$Cor(H_{init}, H_{dam}) = \frac{Cov(H_{init}, H_{dam})}{\sqrt{Var(H_{init})}\sqrt{Var(H_{dam})}} \tag{12}$$

If both measurements lead to identical results, the correlation coefficient value will remain at unity. However, in lower frequency regions, noise from external sources deteriorates the FRF measurements and leads to a low coherence. At very high frequencies, the propagation of ultrasonic waves is damped due to the material properties of the CFRP coupon, this also leads to poor FRF measurements. Thus, it is advisable to choose an expedient frequency band for signature analyses. As shown in Fig. 21, regarding higher impact energies above 15 J, the correlation coefficient decreases with increasing impact energy when analysing the third-octave band around 35.6 kHz. However, for lower impact energies, a deviation of the correlation coefficient from the reference can also be observed. As stated above, this interrelation cannot be observed at a lower frequency band (e.g. 2.8 kHz). Since a broad band excitation is used, and the signals are simply analysed in the frequency domain, a localisation e.g. by time-of-flight analysis is not possible. Thus, this method is more suitable for the straightforward implementation of a damage detection system.

Certainly, to transfer this SHM method from the coupon level to an application, some challenges have to be taken up: The system has to be scaled for the application to larger

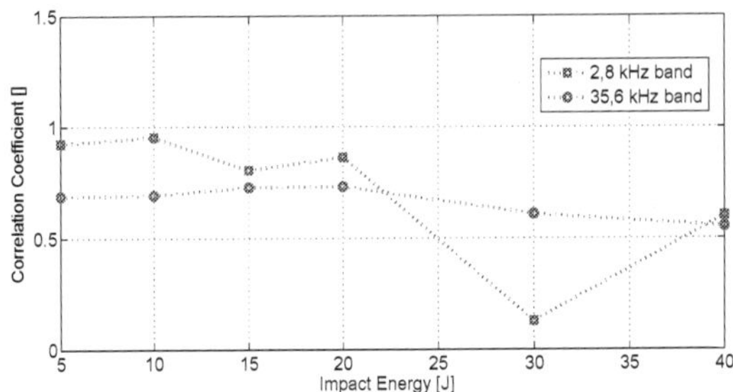

Fig. 21. Correlation coefficient for different impacts and frequency bands.

components, i.e. several actuator-sensor pairs distributed on the surface. The signal processing should be implemented on a network of smart embedded platforms instead of bulky laboratory equipment. Furthermore, a damage detection algorithm has to be implemented in order to work autonomously. This also includes differentiating between false alarms, e.g. a decrease in correlation due to temperature changes and that from actual impact damage.

5. Conclusions

This paper has given an account of the useability of different monitoring methods in wind turbines. The methods described, all serve to supervise of the integrity of structural parts.
Different projects have been undertaken at Fraunhofer LBF to develop and evaluate such methods for Structural Health Monitoring, not only in the field of wind energy. This contribution summarizes those developments, starting from the current state of the scientific and technical knowledge and demonstrates their utility for wind turbines with the help of realizations and studies of application of such systems. Three methods have been presented: Load Monitoring, vibration analysis and Acousto Ultrasonics.
Concerning Load Monitoring, the development of classical methods for assessing residual life-times and their transfer to smart sensor nodes is presented. As application example, a simple model of a wind turbine is used.
The section dealing with vibrational analysis includes a literature review of methods used for extracting health parameters from modal properties as well as the topic Operational Modal Analysis (OMA). The measurement system developed at Fraunhofer LBF makes use of one of those OMA methods and combines it with the Random Decrement method to design a data acquisition system that can be implemented on a smart sensor network. As examples, the application of this system to the model of a wind turbine as well as to a full scale pedestrian bridge is described.
As a third method, Acousto Ultrasonics is described as a method which is capable of detecting small scale damages. As an example, its application to coupons of fibre reinforced plastic is described.
Although the methods described here are not far enhanced beyond the state of the art documented in literature, their transfer to a network of sensor nodes and especially their capability of being automated is a new and invaluable development deserving discussion. It was demonstrated that, due to their relative small computational effort, all methods possess the potential to be implemented at nodes of smart sensor networks. Their most important

limitation lies in the fact that future research will be strongly connected to full scale tests and that such tests cannot be carried out by research organisations alone.

Considered as a whole, these results suggest that there are monitoring methods that can be applied to wind turbines and thereby contribute to making the process of harvesting renewable energy more reliable.

This research will serve as a base for future studies on the topic of combining the different classes of methods to develop a monitoring system that is able to predict the residual life-time of structures. Such systems are of great use for off-shore wind turbines, because they are the basis of a maintenance on demand.

6. Acknowledgements

Parts of the present and still ongoing research in this field as well as the publication of this chapter are funded by the German federal state of Hesse (project „LOEWE-Zentrum AdRIA: Adaptronik Research, Innovation, Application", grant number III L 4 - 518/14.004 (2008)) as well as in the framework of the programme „Hessen ModellProjekte" (HA-Projekt-Nr.: 214/09-44). This financial support is gratefully acknowledged.

7. References

Andersen, P. (1997). *Identification of Civil Engineering Structures using Vector ARMA models*, PhD-Thesis, University of Aalborg, Aalborg.

Andersen, P. & Brincker, R. (2000). *The Stochastic Subspace Identification Techniques*, www.svibs.com.

Andersen, P., Brincker, R., Goursat, M. & Mevel, L. (2007). Automated Modal Parameter Estimation For Operational Modal Analysis of Large Systems, *Proceedings of the 2nd International Operational Modal Analysis Conference*, ed. Brincker R. & Møller N., Copenhagen, pp. 299–308.

Asmussen, J. C. (1998). *Modal Analysis Based on the Random Decrement Technique, Application to Civil Engineering Structures*, PhD-Thesis, University of Aalborg, Aalborg.

Ballah, S. & Soh, C. K. (2003). Structural impedance based diagnosis by piezo-transducers. *Earthquake Engineering and Structural Dynamics*, No. 32, pp. 1897-1916.

Bartel, T., Atzrodt, H., Herold, S. & Melz, T. (2010). Modelling of an Active Mounted Plate by means of the Superposition of a Rigid Body and an Elastic Model, *Proceedings of ISMA2010. International Conference on Noise and Vibration Engineering*, pp. 511-523.

Bendat, J. S. & Piersol, A. G. (2000). *Random Data: Analysis and Measurement Procedures*, John Wiley & Sons, New York.

Bodeux, J. B. & Colinval, J. C. (2001). Application of ARMAV models to the identification and damage detection of mechanical and civil engineering structures. *Smart Materials and Structures*, Vol. 10, pp. 479-489.

Boller, C. & Buderath, M. (2006). Fatigue in aerostructures – where structural health monitoirng can contribute to a complex subject. *Phil. Trans. R. Soc. A*, No. 365, Dec 2006.

Brincker, R., Krenk, S. & Jensen, J. L. (1991). Estimation Of Correlation Functions By The Random Decrement, *Proceedings of The Florence Modal Analysis Conference*, Florence, pp. 783–788.

Brincker, R., Zhang, L. & Andersen, P. (2000). Modal Identification from Ambient Responses using Frequency Domain Decomposition, *Proceedings of the 18th IMAC*, San Antonio, TX, USA, pp. 625–630.

Buff, H., Friedmann, A., Koch, M., Bartel, T. & Kauba, M. (2011). Systematic preparation of Random Decrement Method based Structural Health Monitoring, *Proceedings of the ICEDyn, International Conference on Structural Engineering Dynamics*, Tavira, 2011. Accepted for presentation.

Carden, E. P. & Fanning, P.(2004). Vibration Based Condition Monitoring: A Review. *Structural Health Monitoring*, Vol. 3(4), pp. 355–377.

Chesné, S.& Deraemaeker, A. (2010). Sur l'utilisation des transmissibilités pour la localisation des défauts dans les systèmes non dispersifs. *10ème Congrès Francais d'Acoustique*, Lyon.

Ciang, C. C., Lee, J.-R. & Bang, H.-J. (2008). Structural health monitoring for a wind turbine system: a review of damage detection methods. *Measurement Science and Technology*, No. 19, DOI: 10.1088/0957-0233/19/12/122001.

Cole, H. (1973). *On-line failure detection and damping measurement of aerospace structures by random decrement signatures*, Report NASA CR-2205.

DIN 45 667 (1969). *Klassierverfahren für das Erfassen regelloser Schwingungen*, Beuth-Vertrieb.

DIN 50 100 (1978). *Dauerschwingversuch – Begriffe, Zeichen, Durchführung, Auswertung*, Beuth-Vertrieb.

Ewins, D. J. (2000). *Modal Testing: Theory, Practice and Application*, Research Studies Press, Ltd., Baldock.

Fan, W. & Qiao, P. (2010). Vibration-based Damage Identification Methods: A Review and Comparative Study, *Structural Health Monitoring*, Vol. 10(1), pp. 83-111.

Farrar, C. R. & Doebling, S. W. (1997). An Overview of Modal-Based Damage Identification Methods. *EUROMECH 365 International Workshop: DAMAS 97, Structural Damage Assessment Using Advanced Signal Processing Procedures*.

Friedmann, A., Koch, M. & Mayer, D. (2010). Using the Random Decrement Method for the Decentralized Acquisition of Modal Data, *Proceedings of ISMA2010. International Conference on Noise and Vibration Engineering*, Leuven, September 2010, pp. 3275-3286.

Fritzen, C. P., Kraemer, P. & Klinkov, M. (2008). Structural Health Monitoring of Offshore Wind Energy Plants, *Proc. 4th Europ. Workshop on Structural Health Monitoring*, Krakow, Poland, pp.3-20.

Giurgiutiu, V.& Cuc, A. (2005). Embedded Non-destructive Evaluation for Structural Health Monitoring, Damage Detection, and Failur Prevention. *The Shock and vibration Digest*, Vol. 37(2), pp. 83-105.

Giurgiutiu, V. (2008). Structural Health Monitoring with Piezoelectric Wafer Active Sensors. Elsevier, Amsterdam, ISBN: 978-0-12-088760-6.

Haibach, E. (1971). Probleme der Betriebsfestigkeit von metallischen Konstruktionsteilen. *VDI-Berichte Nr. 155*, pp. 51-57.

Haibach, E. (2006). *Betriebsfestigkeit – Verfahren und Daten zur Bauteilberechnung*, Springer-Verlag, ISBN-10 3-540-29363-9, Berlin, Heidelberg.

Hameed, Z., Hong, Y. S., Cho, Y. M., Ahn, S. H. & Song, C. K. (2007). Condition monitoring and fault detection of wind turbines and related algorithms: A review, *Renewable and Sustainable Energy Reviews*, Vol. 13, pp. 1–39.

Have, A. A. ten (1992) *Wisper and Wisperx – Final definition of two standardised fatigue loading sequencesa for wind turbine blades*, National Aerospace Laboratory NLR, Amsterdam, Netherlands, NLR TP 91476 U.

Herold, S. (2003). *Simulation des dynamischen und akustischen Verhaltens aktiver Systeme im Zeitbereich*, Dissertation, TU Darmstadt, 2003.

Herlufsen, H., Gade, S. & Møller, N. (2005). Identification Techniques for Operational Modal Analysis – An Overview and Practical Experiences, *Proceedings of the 1st International Operational Modal Analysis Conference*, Copenhagen.

Heuler, P. & Klätschke, H. (2004). Generation and use of standardized load spectra and load-time histories. *International Journal of Fatigue*, No. 27, pp. 974-990.

Humar, J., Bagchi, A. & Xu, H. (2006). Performance of Vibration-based Techniques for the Identification of Structural Damage. *Structural Health Monitoring*, Vol. 5(3), pp. 215–241.

Kammer, D. C. (1991). Sensor Placement on-orbit modal identification and correlation of large structures, *Journal of Guidance, Control and Dynamics* 14 (1991), pp. 251-259.

Kessler, S. S., Spearing, S. M. & Soutis, C. (2002). Damage detection in composite materials using Lamb waves methods. *Smart Materials and Structures 11*, pp. 269-278.

Kuo, S. M. & Morgan, D. R. (1996). *Active Noise Control Systems: Algorithms and DSP Implementations*, John Wiley & Sons, New York.

Lading, L., McGugan, M., Sendrup, P., Rheinländer, J. & Rusborg, J. (2002). Fundamentals for Remote Structural Health Monitoring of Wind Turbine Blades - a Preproject, Annex B - Sensors and Non-Destructive Testing Methods for Damage Detection in Wind Turbine Blades. Risø National Laboratory, Roskilde, Denmark, Risø-R-1341(EN).

Lee, U. & Shin, J. (2001). A frequency response function-based structural damage identification method. *Computers and Structures*, Vol. 80, pp. 117–132.

Liang, C., Sun, F. P. & Rogers, C. A. (1994). Coupled Electro-Mechanical Analysis of Adaptive Material Systems – Deteremination of the Actuator Power Consumption and System Energy Transfer. *Journal of Intelligent Materials and Structures*, Vol. 5, pp. 12-20.

Lilov, M., Kauba, M. & Mayer, D. (2010) Structural monitoring and damage detection on CFRP specimens by using broadband acousto ultrasonic and electromechanical impedance measures, *Proc. 5th Europ. Workshop on Structural Health Monitoring*, Sorrento, Italy, p. G28.

Lunze, J. (2010). *Regelungstechnik 1*, Springer-Verlag, ISBN 978-3-642-13807-2, Berlin, Heidelberg.

Lynch, J., Wang, Y., Kenneth, J. L. , Yi, J. H. & Yun, C.-B. (2006). Performance Monitoring of the Geumdang Bridge using a Dense Network of High-Resolution Wireless Sensors, *Smart Materials and Structures* Vol. 15(6), pp. 1561–1575.

McConnell, K. G. (1995). *Vibration Testing – Theory and Practice*, John Wiley & Sons, Inc., New York.

Montalvão, D., Maia, N. M. M. & Ribeiro, A. M. R. (2006). A Review of Vibration-based Structural Health Monitoring with Special Emphasis on Composite Materials. *The Shock and Vibration Digest*, Vol. 38, No. 4, pp. 295-324.

Park, G., Sohn, H., Farrar, C. R. & Inman, D. J. (2003). Overview of Piezoelectric Impedance-Based Health Monitoring and Path Forward. *The Shock and Vibration Digest*, Vol. 35(6), pp. 451-463.

Peeters, B., van der Auweraer, H. & Deblauwe, F. (2006). 10 Years Of Industrial Operational Modal Analysis: Evolution In Technology And Applications, *Proceedings of the IOMAC Workshop 2006.*

Radaj, D. (2003). *Ermüdungsfestigkeit - Grundlagen für Leichtbau, Maschinen- und Stahlbau,* Springer-Verlag, ISBN 3-540-44063-1, Berlin, Heidelberg.

Rodrigues, J. & Brincker, R. (2005). Application of the Random Decrement Technique in Operational Modal Analysis, *Proceedings of the 1st International Operational Modal Analysis Conference,* ed. Brincker, R. & Møller, N., Copenhagen, pp. 191-200.

Rytter, A. (1993). *Vibration based inspection of civil engineering structures,* PhD-Thesis, Department of Building Technology and Structural Engineering, University of Aalborg, Aalborg.

Schütz, W., Klätschke, H., Hück, M. & Sonsino, C. M. (1989). Standardised Load Sequences for Offshore Structures – WASH I. *Fatigue Fract. Engng. Struct.,* Vol. 13(1), pp. 15-29.

Siebel, T. & Mayer, D. (2011). Damage Detection on a Truss Structure using Transmissibility Functions, *Eurodyn 2011,* Leuven, Belgium. Accepted for presentation.

Silva, C., Rocha, B. & Suleman, A. (2010). Guided Lamb Waves Based Structural Health Monitoring Through a PZT Network System. *2nd International Symposium on NDT in Aerospace 2010 - We.1.B.4.*

Sohn, H., Farrar, C. R., Hemez, F. M., Shunk, D. D., Stinemates, D. W., Nadler, B. R. & Czarnecki, J. J. (2004). *A Review of Structural Health Monitoring Literature: 1996 – 2001,* Los Alamos National Laboratory, Report No. *LA – 13976 – MS.*

Sonsino, C. M. (2004). Principles of Variable Amplitude Fatigue Design and Testing. *Journal of ASTM International,* Vol. 1, No. 9, Paper ID JAI19018.

Sonsino, C. M. (2005). Dauerfestigkeit – Eine Fiktion. *Konstruktion,* No. 4, pp. 87-92, ISSN: 0373-3300.

Uhl, T. (2007). The inverse identification problem and its technical application. *Archive of Applied Mechanics,* No. 77, 2007, pp. 325-337.

Westermann-Friedrich, A. & Zenner, H. (1988). *Zählverfahren zur Bildung von Kollektiven aus Zeitfunktionen – Vergleich der verschiedenen Verfahren und Beispiele,* FVA-Merkblatt Nr. 0/14, Forschungsvereinigung Antriebstechnik e.V..

Viberg, M. (1995). Subspace-based Methods for the Identification of Linear Time-invariant Systems. *Automatica,* Vol. 31(12), pp. 1835-1851.

Zagrai, A. N. & Giurgiutiu, V. (2002). Electro-Mechanical Impedance Method for Crack Detection in Thin Plates. *Journal of Intelligent Material Systems and Structures,* Vol. 12, pp. 709-718.

Zhang, L. (2004). An Overview of Major Developments and Issues in Modal Identification. *Proc. IMAC XXII,* Detroit.

Zhang, L., Brincker, R. & Andersen, P. (2005). An Overview of Operational Modal Analysis: Major Development and Issues, *Proceedings of the 1st International Operational Modal Analysis Conference,* Copenhagen.

Zimmerman, A. T., Shiraishi, M., Swartz, R. A. & Lynch, J. P. (2008). Automated Modal Parameter Estimation by Parallel Processing within Wireless Monitoring Systems, *Journal of Infrastructure Systems* Vol. 14(1), pp. 102-113.

10

Magnetic Suspension and Self-pitch for Vertical-axis Wind Turbines

Liu Shuqin
*School of Electrical Engineering,
Shandong University,
China*

1. Introduction

Energy is important for the development of human civilization. As conventional energy exhausts, the development of clean and renewable energy, such as wind and solar becomes ever important to people's live. The wind power has been harnessed by mankind for a long time and the associated technology is more advanced than other clean energies. Nowadays wind power increasingly attracts interests and its utilization has entered a rapid development stage.

There are two types of wind turbines, namely horizontal-axis wind turbine (HAWT) and vertical-axis wind turbine (VAWT). The latter has many advantages, such as low cost, simple-structured blades, convenient installation and maintenance, and the ability to utilize wind from all directions without the need of a steering mechanism. A special VAWT implemented with magnetic suspension and self-pitch blade design will be introduced in this chapter. It not only has the advantages of traditional VAWT, but also the advantages of low threshold starting wind-speed, high wind power efficiency, etc.

Discussion in this chapter includes the benefits of applying magnetic bearing technology to the wind turbine. Specifically, the entire wind turbine rotor weight can be supported by magnetic bearings. The friction of the bearings is essentially non-existence. There is no need for bearing lubrication, and the maintenance cost can be reduced. Furthermore, the magnetic suspension technology can eliminate mechanical vibration and reduce noise. Since low friction also reduces starting torque, the magnetic bearings enable producing power at lower wind speed than conventional bearings.

The discussion also includes a blade-pitch adjusting technique. Conventional VAWT applies special blade adjusting mechanism which is complicated in structure, costly to fabricate and wastes power. Herein, the blades in a magnetic suspended VAWT are designed to adjust the pitch automatically and do not require any dedicated special devices. The blade pitch is adjusted naturally during rotation for the best windward angle. As a result the blades always produce the maximum thrust wind force improving the wind turbine efficiency. Thus, the magnetically suspended and self-pitched vertical-axis wind turbine will be designed with uncomplicated structure, high efficiency and low cost.

2. Vertical-axis wind turbine

2.1 Basic principle of vertical axis-machine

Wind machine is a kind of energy conversion device, which converts wind energy into mechanical, electric or heat energy. According to their rotor layout, wind turbines can be categorized into horizontal-axis wind turbines (HAWT) and vertical-axis wind turbines. The rotor of a HAWT is horizontal and must point to the wind. The rotating plane of HAWT blades is perpendicular to the wind direction during operation. The turbine blades are radial and their number is commonly 2~3. The shape of a turbine blade is always similar to a wing. It is perpendicular to the rotating shaft and there is an angle between the blade and rotating plane. Since the HAWT works on the principle of airfoil lift, its torque tends to be large and efficiency high in utilizing wind energy. Yaw device is necessary to turn the blades rotating plane in line with the wind direction. The HAWTs have been investigated thoroughly in theory and have the advantages of high wind energy utilization efficiency. They are the mainstream commercial products of the current wind turbine technology.

The shaft of a VAWT is perpendicular to the ground and the wind direction. VAWT accepts the wind from all horizontal directions; there is no need of a yaw control devices, making the structure design simple and reducing precession force on blades. Comparing to HAWT, a VAWT has the advantages of low noise and less adverse effect on environment.

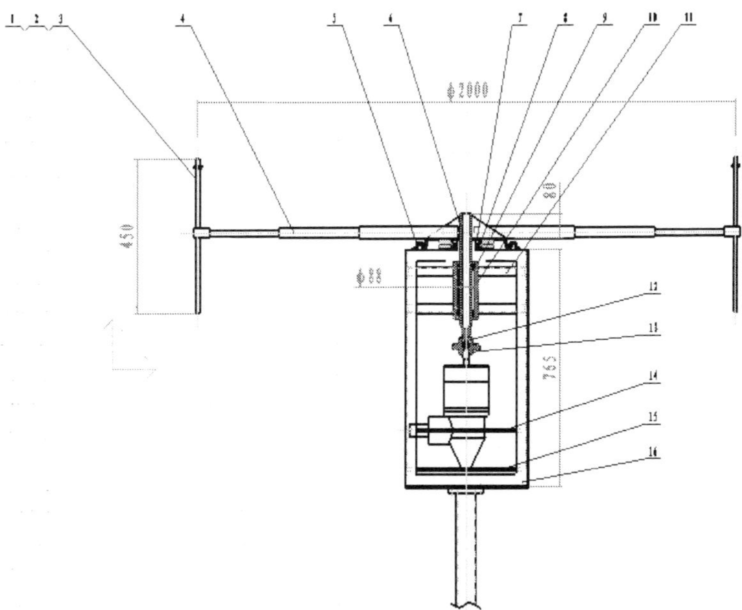

1 nose clip, 2 blade nose, 3 blade support pole, 4 wheel arm, 5 upper cover, 6 baffle, 7 position disk, 8 bearing cover, 9 upper ratchet, 10 lower ratchet, 11 generator support, 12 turbine frame, 13 ventral shield, 14 support shelf.

Fig. 1. Components of a vertical-axis wind turbine

There are two kinds of VAWTs. The first kind works on the principle of wind drag. Its typical structure is the S-type (Savonius) wind turbine, such as the cup-shaped wind wheel blades for wind-speed measurement. The S-type wind turbine consists of two axis - staggered half cylinders, and the starting torque is large. Unsymmetrical airflows exist around the rotor producing the lateral thrust to the turbine blades. The second kind of VAWT works on the principle of airfoil lift. Its typical application is Darrieus wind turbines. Darrieus wind turbines have various forms, including H- type and ϕ-type. An H-type wind turbine has a simple structure, but centrifugal force may generate serious bending stress on its turbine wheel connections. The flexural blades in ϕ-type wind turbine only bear the tension without centrifugal load; the bending stress in blades of ϕ-type wind turbine is therefore mitigated.

The VAWT wheel rotates in a horizontal plane and there is no vertical motion of the blades. Researchers used to believe that the tip-speed ratio (the ratio of blade tip rotational speed to wind speed) of VAWT cannot be greater than 1, and therefore the associated wind energy efficiency is lower than that of a HAWT. The VAWT blades were designed by using blade element momentum method. But the airflow through a VAWT, typically separated unstable flows, is more complicated than those of HAWT. The blades elements moment theory is not suitable for its analysis and design, and this is one reason for less development in VAWT. But as technology progressing, researchers have realized that only S-type VAWTs are limited by the tip-speed ratio less than 1. For the lift-type wind wheel (Darrieus-type) the tip-speed ratio can reach as high as 6. Therefore, its wind energy utilization efficiency is not lower than HAWT. Recognizing the advantage, many institutions have started investigation on VAWTs and achieved considerable advancement in recent years.

2.2 Different vertical axis wind turbines

As the wind power technology develops, the unique advantages of VAWTs have been unveiled and appreciated, especially for those small wind turbine applications. The recent progress in VAWT research has enabled many commercializations of small VAWTs as follows.

2.2.1 Sail S-type roof VAWT

Figure 2 shows the VAWT manufactured by Enviro-Energies Holdings, a Canadian company. The main line of its production is a 10KW wind turbine, with advantages of low-speed, high power output, quiet operation, and maintenance free. It can be installed on roof for domestic power need.

2.2.2 Light type VAWT

As shown in Figure 3, Urban Green Energy (http://www.urbangreenenergy.com/products/uge-4k) has developed and patented a revolutionary new dual axis design that eliminates the main concern of other vertical axis wind turbines that is premature bearing failure. Its vertical axis machines include those rated for 600W, 1KW and 4KW. They can be installed on top of a tower, roof and other suitable places. These wind turbines can be connected to utility grid. They are made of glass or carbon fibers. The 4KW model weighs about 461Kg, cuts in wind speed at 3.5m/s and is rated at wind speed 12m/s.

Fig. 2. VAWT made by Enviro-Energies and installed on the company's roof

Fig. 3. VAWT made by Urban Green Energy

2.2.3 H type VAWT

Figure 4 shows a small VAWT developed by Ropatec of Italy. The Ropatec products are sold to more than thirty countries worldwide, supplying reliable electrical power for families, farms, remote pastoral areas, communications companies and enterprise groups. Its VAWTs have four ratings, i.e., 1KW, 3KW, 6KW and 20KW. The 20KW wind turbine has 5 blades; the others have 3 blades.

Fig. 4. H-type VAWT made by Ropatec of Italy

Fig. 5. Aesthetic VAWT made by Oy Windside in Finland

2.2.4 Aesthetic VAWT

A new type of VAWT as shown in Figure 5 has been developed by Oy Windside in Finland. The wind turbine as a derivative of marine engineering, can be used for charging batteries and provide an environment-friendly image. One of the applications of Oy Windside wind turbines is about "wind art". The concept is to integrate the wind turbine into art and provide lighting. The aesthetics function and ecological concept are both considered in the turbine design.

2.2.5 Drag-type and lift-type combined VAWT

Figure 6 shows the VAWT produced by Green Giant Tech, Taiwan. Green Giant Tech has combined the benefits of the Savonius-type blade and the Darrieus-type blade, applying both the drag and lift forces of wind power. Its VAWT includes three models rated at 400W, 3.6KW and 5KW. It also manufactures street lighting devices for wind and solar power.

Fig. 6. VAWT made by Green Giant Tech's in Taiwan

2.3 Magnetic suspension and self-pitch for vertical axis wind turbine

Maglev Engineering Research Center, Shandong University, China has committed to the magnetic bearing research and related product development. Recently, magnetic suspension technology has been applied to the vertical axis wind turbine, in which the entire rotor weight of a VAWT was suspended by magnetic bearing. The turbine friction was greatly reduced, and start-up wind speed decreased. Figure 7 shows the magnetically suspended and self-pitched VAWT for street lighting.

Since the self-pitch technique and magnetic suspension were applied to the Shandong University VAWT, the wind power efficiency and system performance of the wind turbine have been greatly improved. The manufacturing cost and operational cost were also effectively reduced. This new magnetically suspended vertical-axis wind generator has irreplaceable advantages compared with other VAWTs in the market. Due to its low cost and suitability for high-power single devices, this new VAWT design will have broad commercial potential.

Fig. 7. Magnetically suspended and self-pitched VAWT from Maglev Engineering Center, Shandong University

3. Application of magnetic suspension in wind turbine

3.1 Principle and types of magnetic suspension

Magnetic suspension means that an object is suspended by magnetic attraction and/or repulsion forces to achieve non-contact support and low-friction in motion. The magnetic suspension as a branch in mechatronics technology, involves disciplines in electromagnetism, power electronics, signal processing, control engineering, statics and dynamics. Therefore, the development of the magnetic suspension technology has been closely related to and relied on the development of these fields of disciplines.

Because of no mechanical contact in the magnetic bearing, it has many advantages, including no wear, no contamination, suitable for long-term use in vacuum and corrosive environment, no mechanical friction, low noise, low power loss and no need of lubrication or sealing. Therefore, magnetic suspension technology can be used for high-speed applications to eliminate mechanical problems related to lubrication and power loss.

There are many applications of the magnetic suspension technology, including maglev train and magnetic bearing. Many countries have developed different types of maglev trains; Germany, Japan and China are among those having the most mature maglev technology. The fastest speed of the maglev train can reach 500km/h. On the other hand, the magnetic bearing technology has been widely applied in the aerospace industry, medical health field and new energy power. Some magnetic bearings have been tested in space shuttles and rockets in USA and Japan and achieved satisfactory performances. There were magnetic bearings successfully implemented in artificial heart pump developed by University of Virginia and University of Utah.

Base on the source of the magnetic field, magnetic bearings can be classified into three kinds as follows:

1. Active Magnetic Bearing (AMB) - The magnetic field is controllable. The control system detects the position of the rotor and actively controls the suspension of the rotor. The bearing stiffness and damping are electronically tuneable and the load capacity is large.

The principle of active magnetic suspension is depicted by a simple magnetic suspension system in Figure 8, which shows a displacement sensor, controller and actuator. The actuator includes electromagnets and power amplifiers. Assuming there is a downward perturbation, the position change of the suspended object, such as the rotor in Figure 8, can be detected by the sensor and the displacement signal immediately is transmitted to the controller. The signal will be transformed into a command by the microprocessor of the controller, which in turn produces a change or increase of control current in the electromagnets. The suspended object will be then pulled back up to its original position by the changing magnetic field generated in the electromagnet. Therefore, the rotor will always be kept at a preset position regardless the perturbation is down or up.

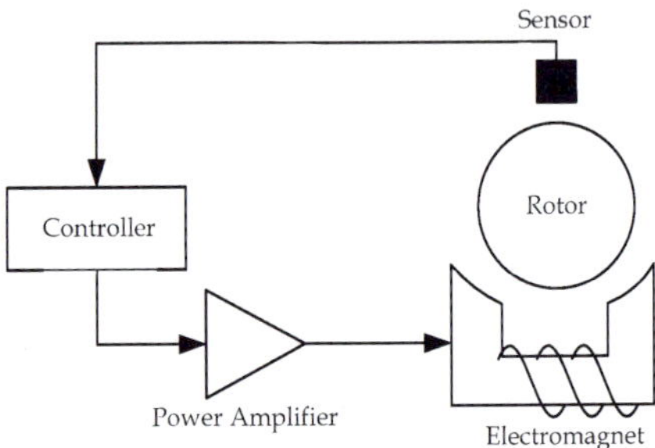

Fig. 8. Diagram of active magnetic suspension system

2. Passive Magnetic Bearing (PMB) - As showed in Figure 9, the PMB magnetic field is created using permanent magnets or superconductors. The rotor is suspended by passive magnetic forces. The advantages of PMB are its simple structure, low cost, low power loss, etc., but the load capacity of PMB is small and so is the stiffness.

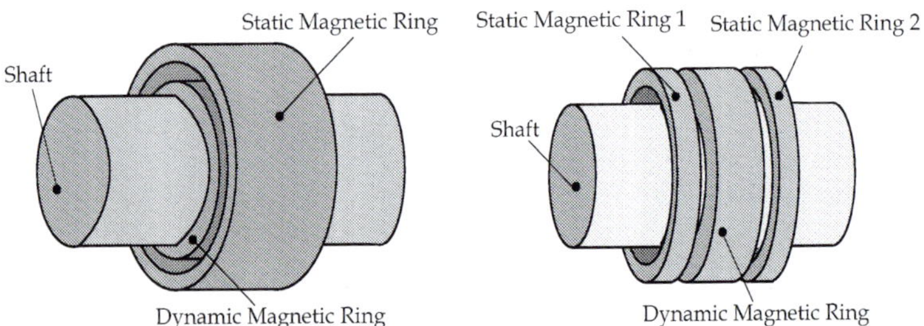

Fig. 9. Permanent magnetic bearings

3. Hybrid magnetic bearing (HMB) - As showed in figure 10, the mechanical structure of HMB includes electromagnet and permanent magnet (or superconductor); the magnetic force is generated by both the permanent magnet and electromagnet. Its structure complexity, cost and performance are average between AMB and PMB.

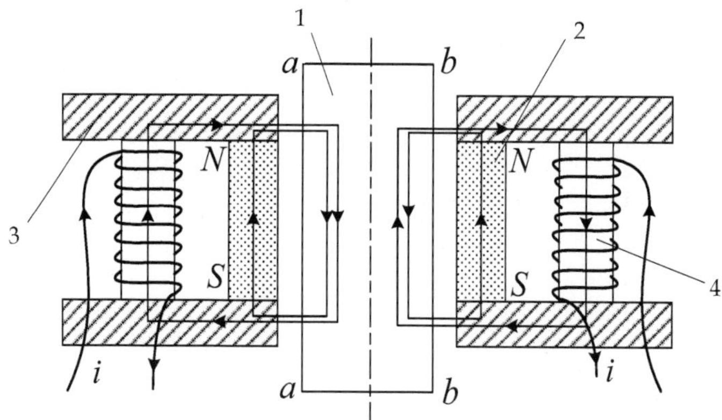

1 rotor, 2 permanent magnets, 3 stator, 4 electromagnet coils.

Fig. 10. Hybrid magnetic bearing

Based on the functions of magnet bearings, magnetic bearing can be further classified as axial (thrust) magnetic bearing and radial magnetic bearing.

3.2 Magnetic suspension technology in wind turbine

Permanent magnet bearing has the advantages of low power consumption, no mechanical contact and suitable for severe adverse environment. In recent years, with the rapid development of permanent magnet material, the technology of permanent magnetic suspension has been expanded to wind turbine applications. This has greatly reduced the cost and stringency of wind power. Specifically, the application of magnetic suspension to wind turbines has achieved the following advantages:

1. Starting wind speed is reduced by magnetic suspension due to reduced bearing friction and power output of wind turbine is increased for the same wind speed.
2. Magnetic suspension has largely changed the traditional wind turbine rotor system design using special rolling-element or oil-film bearings. These traditional bearings depend on careful lubrication and sealing for long service life, impact resistance, and high reliability. The magnetic suspension not only reduces the cost of the bearings and their maintenance, but also reduces the downtime of the wind turbine and therefore, improves the over-all efficiency of the system.

Magnetic suspension technology applied to wind turbine is an emerging technology; the development of maglev magnetic wind turbine is just beginning. Although there claimed many maglev wind turbine products have been developed, relevant published studies are rare.

Currently, the studies on magnetic suspension of wind turbines have been focused on the HAWTs. A typical magnetic suspension for an HAWT is to use PMB in radial directions and a mechanical bearing or a ball in axial direction.

Figure 11 shows a magnetic suspension system of wind turbine with five degrees of freedom using radial and axial magnetic bearings.

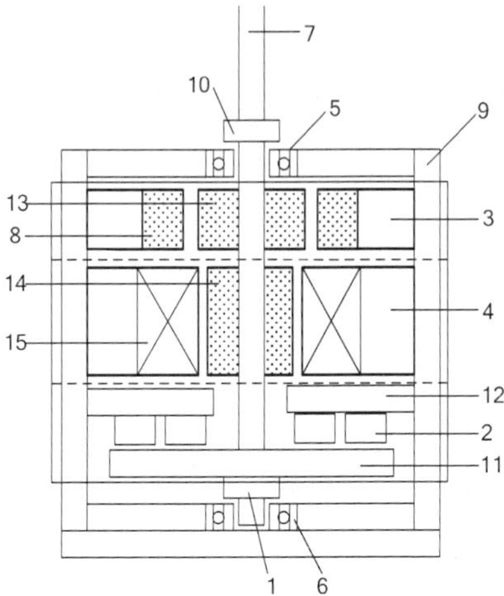

1 lower backup thrust ring, 2 axial magnetic bearing, 3 radial magnetic bearing, 4 generator, 5 upper backup bearing, 6 lower backup bearing, 7 rotor shaft, 8 radial magnetic bearing stator, 9 shell, 10 upper backup thrust ring, 11 axial magnetic bearing rotor disk, 12 axial magnetic bearing stator, 13 radial magnetic bearing rotor, 14 generator rotor, 15 generator stator

Fig. 11. Magnetic suspension system of vertical axis wind turbine

The thrust (axial) bearing is the most important part of the VAWT magnetic suspension system. It supports the weight of the blades and generator rotor. An implementation of the thrust bearing using permanent magnets is presented in Figure 12(a). A passive radial bearing structure resisting radial disturbance is shown in Figure 12(b).

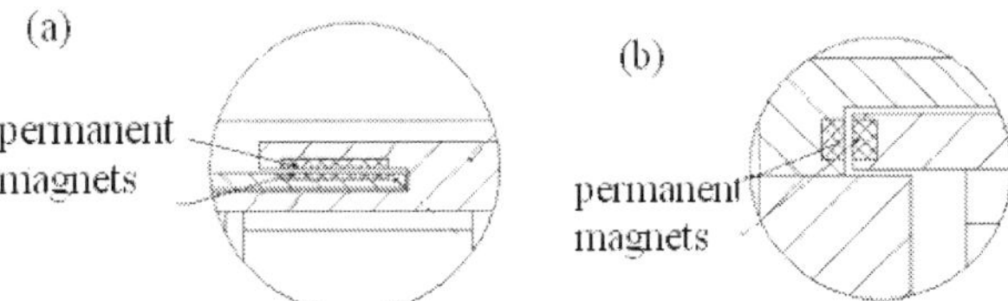

Fig. 12. Axial (a) and radial (b) magnetic bearing in VAWT

PMBs are normally made of NdFeB (neodymium iron boron) magnets. These magnets are currently the best choice for magnetic bearings because of its high magnetic energy product and low cost. Also its process-ability is better than other permanent materials, such as SmCo alloy. But the temperature stability and erosion resistance are needed to improve for NdFeB magnets. Inferior to SmCo, typical NdFeB Curie temperature is 312°C and reversible temperature coefficient of remanence is -0.13/°C. Typical NdFeB working temperature is below 80°C. For better temperature stability and erosion resistance, the NdFeB magnets chosen for wind turbines have been manufactured by hot-rolling process. The process yields good tension strength, and oxidation resistance.

4. Self-pitch technique in wind turbine

Pitch control is one of the key technologies in wind turbine and its development has progressed from fixed pitch to controllable pitch. The fixed-pitch design has the blades fixed on the arms so that the windward angle will not change with wind speed in operation. The controllable pitch design is to change the blade pitch angle according to the wind speed in order to control the generator output power. Moreover, it is necessary to control blade pitch to lower starting torque during wind turbine start-ups. Comparing with the fixed-pitch wind turbines, the controlled-pitch designs can produce more wind power, because it can vary blade pitch according to variable wind speed.

For stable loading control, security and high efficiency, blade-pitch control is employed in most large HAWTs. The pitch-controlled system consists of pitch bearing and pitch gear. When blade pitch adjustment is needed, the control device will drive a small gear through the pitch bearing to achieve the desired the blade angle. Currently on commercial market, the pitch control devices are powered by hydraulic or electrical variable pitch systems. The pitch control is generally achieved by applying variable-speed constant-frequency control. And the main control strategies applied to wind turbines include the classical PD (proportion-integral) control, PID (proportional-integral-derivative) control, fuzzy control, neural networks, and the optimal control theory.

In recent years, the development of VAWT was mainly concentrated in the fixed-pitch type, and most of the machines presented in Chapter 2 are of this type. However, the fixed-pitch turbines have low utilization of wind power and poor start-up performance. If VAWT is combined with a pitch-controlled device, the self-starting performance and the wind power utilization will improve because adjusting the blade pitch angle changes the aerodynamic characteristic of the blades. At present, research on the pitch-controlled VAWT has made some success. However, those pitch-controlled devices were complicated and costly. With high power consumption and low reliability in operation, these progresses have not yet lead to inexpensive wind power.

Overall, the traditional VAWTs have disadvantages of poor start-up performance, complicated pitch control system, high cost, low utilization of wind power and inability to keep the optimal blade pitch angle all the time in operation. And all these disadvantages severely limit its applications in the wind power field. To overcome these shortcomings, presented in the subsequent section is another kind of self-pitch vertical axis wind turbine with high efficiency and structure simplicity.

4.1 The structure and principle of self-adjusting pitch vertical axis wind turbine

The straight wing VAWT is the lift-type wind turbine. During its rotation, the lift force perpendicular to the blade is constantly changing. When a blade is at positions parallel to the wind direction, the lift on the blade is very small. At positions perpendicular to the wind, the lift would be large.

The angle between the incoming air flow velocity at the blade nose and the chord of the blade (connection between blade leading edge and trailing edge) is called the attack angle, also known as the pitch angle, represented by "α". For the lift-type VAWT, according to aero-foil lift and drag coefficient curves, good aerodynamic performance occurs within a small range of attack angle. Out of this range, the wind machine will be subjected to large aerodynamic drag resulting in lower utilization of wind energy or even the braking effect. Therefore, one can change the blade attack angle so that blades always generate large aerodynamic torque; improve the wind energy utilization and the self-starting capability of the wind turbine.

The self-adjusting blade pitch mechanism described herein is based on the centrifugal positioning and the lift adjustment during operation of wind turbine blades. The blade CG (center of gravity) is at the nose, and it is hinged to the round-arm of the turbine wheel. The centrifugal force of the blade is balanced by the centripetal force generated by the round-arm. Since the blade nose and round-arm is on a straight line, the chord would be perpendicular to the round-arm. The action that the blade maintains itself to be perpendicular to the round-arm by the centrifugal force is called centrifugal positioning. When the angle β between wind direction and blade is larger than the best pitch angle α (see Figure 13), the lift will pull the blade to a counter-clockwise rotation, and reduce β to α. The combined action of the centrifugal position and lift always maintains the attack angle α in the optimum range. The figure shows, the blade's center of gravity must be located before the point where the lift applies. Otherwise the self-pitching will not work.

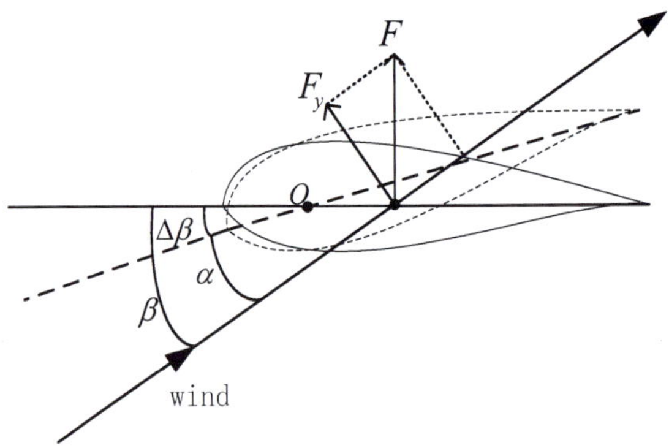

Fig. 13. The principle of self-adjusting pitch

Thus, by properly-designing the nose CG position and the lift force bearing point on blade, one can achieve the automatic adjustment of the pitch angle and maintain it in the optimum range. The working process of a self-pitched VAWT is described as follows.

When the wind turbine is stationary, some blades will automatically adjust in parallel to wind and some perpendicular to wind; it is essentially a drag-type wind turbine. The blade parallel to wind direction bears large wind thrust and the blade perpendicular to wind direction bears small wind drag. The difference between the thrust and drag starts the wind turbine in low wind speed. At the moment of starting, the turbine blades will automatically adjust the attack angle and work under lift type mode. When the wind is relative stable, the blade pitches will be adjusted periodically during each rotation. Those blades in parallel with the wind will have zero pitch angles and experience no pull or pushing force. Those blades at any other positions will have pushing force acting on them. The whole process is automatic without using any devices or driving force. The scheme of self-pitch VAWT is shown in figure 14.

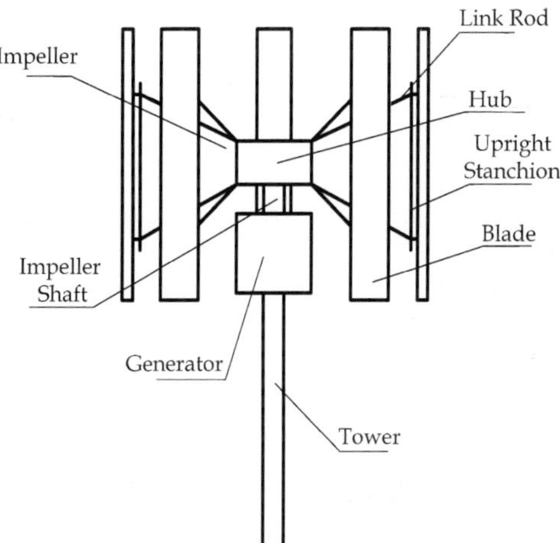

Fig. 14. The scheme of self-pitch vertical axis wind turbine

4.2 The design of the blade of self-pitch

The blades of self-pitch VAWT are designed in straight wing shape without bends or variable cross section; they are easy to manufacture with low-cost. Arm design is based on aerodynamic principles and triangular vector connections, which reduce the threat of opposite wind pressure on the blades. Therefore, the new design improves deformation resistance and reduces blade cost.

The blades of the self-pitch VAWT are uniformly distributed along the circumference. Since fewer blades may lead to low power output and too many blades may adversely affect the aerodynamic characteristic, a VAWT with straight wing blades in general have 3 to 6 blades. In order for self-pitch adjustment, the blade is designed with nose CG located before the maximum thickness position. This ensures that the lift force acting point is behind or after the CG. Therefore, the blade pitch angles will be kept in the best range under the centrifugal force and lift torque.

Furthermore, the rapid development of blade material and unique structure for typhoon resistance provides a potential opportunity to increase the power capacity of VAWT.

4.3 The experiments of the self-pith vertical axis turbine

In the above two sections, the principle of self-pitch VAWTs has been presented. In this section, their good performance will be verified through comparative experiments with other type wind turbines. Experiments process is as follows:

Wind field is produced by 2*3 array wind generators using variable-frequency drive. Wind speed is controlled in the range of 1~11m/s. A self-pitch VAWT with streamline symmetrical blades (type A), a fixed-pitch VAWT with streamline symmetrical blades (type B) and a self-pitch VAWT with arc blades (type C) have been tested. Their start-up properties and power outputs were measured and compared. The parameters of these tested wind turbines are listed in table 1.

Parameters	wind turbine A	wind turbine B	wind turbine C
Pitch type	Self-pitch	Fixed-pitch	Self-pitch
Number of blades	5	5	5
Blade shape	Streamline symmetrical blades	Streamline symmetrical blades	Arc blades
Sizes of blade	120cm(L)×20cm(W) ×2.5 cm(T)	120cm(L)×20cm(W) ×2.5 cm(T)	120cm(L)×20cm(W) ×1.75 cm(T)
Weight of each bade	1.65kg	1.65kg	1.15kg
Diameter of arm rotation	2m	2m	2m
Generator	300W permanent magnet generator	300W permanent magnet generator	300W permanent magnet generator

Table 1. The parameters of wind turbine A, B and C

The wind turbine rotational speed and output power as functions of wind speed were recorded from the experiments, and they are plotted in Figure 15 and 16, respectively.

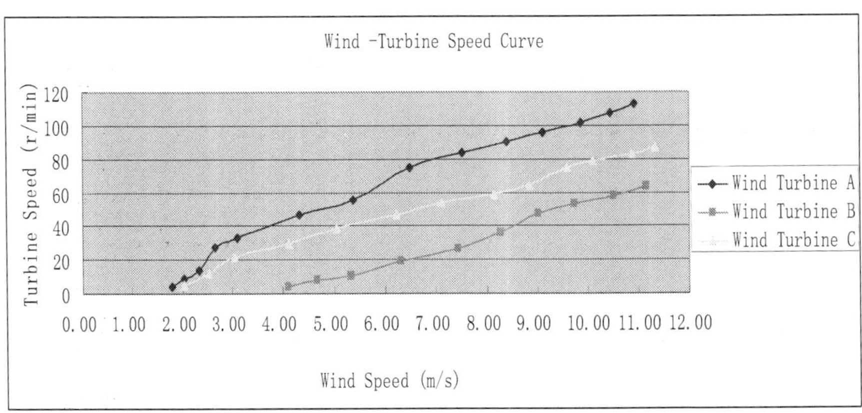

Fig. 15. Turbine speed vs. wind speed

It can be seen from these two figures that wind turbines A and with self pitch had good start-up properties and starting wind speed was about 1.8m/s. However, starting wind speed of wind turbine B with fixed pitch was above 4m/s. At the same wind speed, the turbine rotational speed of turbine A was greater than turbine C, and the turbine speed of C is in turn, greater than B. As the wind speed increased, the output power of turbine A increases rapidly and reaches the rated value, 300W at 9m/s. At 11m/s wind speed, output power of wind turbine B is only a quarter of turbine A, and output power of wind turbine C reaches a half of turbine A.

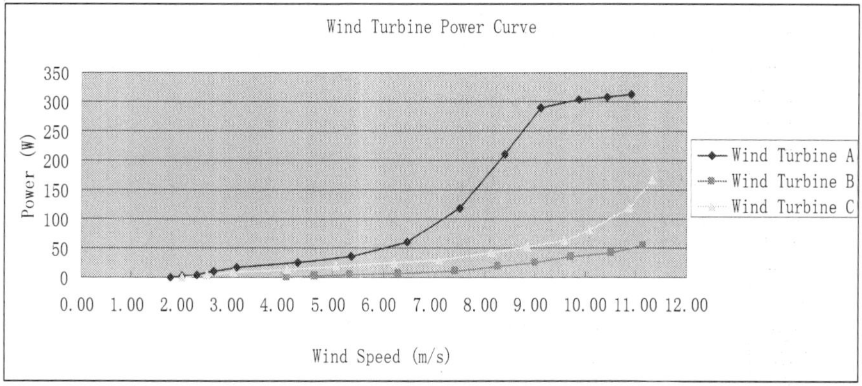

Fig. 16. Output power vs. wind speed

Experiment results show that the start-up properties and efficiency of wind utilization were enhanced in the self-pitch VAWTs due to the implemented self-pitch technology and optimal design of blades. Magnetic suspended self-pitch VAWT is a new and high-efficiency wind turbine.

5. Conclusion

In this chapter, the basic principle and main features of a magnetic suspended self-pitch vertical axis wind turbine have been presented. Compared with the horizontal axis wind turbine and the traditional vertical axis wind turbine, this new VAWT has the following advantages:

- The magnetic suspension in all degrees of freedom has applied permanent magnets. The bearing friction and mechanical noise are low. The starting torque, thus the starting wind speed is low. As a result, the wind turbine consumes little power and has a high efficiency using wind power.
- During rotation, the blades automatically adjust their attack angles to the optimal values and therefore, generate large lifts.
- The wind turbine structure is designed based on aerodynamics and triangular vector connection method. As a result, the threat of normal blade pressure was mitigated, and the blade manufacturing cost reduced.
- Blades can be made with different sizes as spare parts, which can be used for the same wind turbine but according to local wind power resource needs. This is a unique feature not available for other wind turbines.
- In order to lower wind power investment, the power capacity of any single wind turbine should be increased as much as possible. The VAWT presented herein can easily increase the unit power by simply elongating the turbine wheel arms and blades.
- This VAWT design has a simple structure and low manufacturing cost. It needs no complicated control mechanism to adjust the blade attack angles. It has a direct-drive permanent magnet generator with no need of gear box or yawing wind direction control.

- The blades automatically adjust attack angles to reduce wind thrust, and therefore, can survive in Typhoon.
- This VAWT can start up by itself.

In summary, the new vertical axis wind turbines have adopted magnetic bearings and self-pitch blade design. Consequently their utilization rate of wind power is improved, overall efficiency of wind turbine is enhanced, and the production cost as well as maintenance cost is reduced. The magnetically-suspended self-pitch vertical-axis wind turbines have irreplaceable advantages and demonstrated the superiority of vertical axis wind turbines.

6. References

TIAN Hai-jiao; WANG Tie-long; WANG Ying. Summarize of the development of the vertical-axis wind turbine. In : Applied Energy Technology, Vol.6, pp. 22-27, ISSN 1009-3230.

Steinbuch, M. Dynamic modeling and robust control of a wind energy conversion system. Ph., D. Thesis, University of Delft, Holland 1989.

H. Camblong. Digital robust control of a variable speed pitch regulated wind turbine for above rated wind speeds. In: Control Engineering Practice, Vol.16, No. 8, pp. 946-958, ISSN 0967-0661.

Ahmet Serdar Yilmaz a, Zafer Ozer. Pitch angle control in wind turbines above the rated windspeed by multi-layer perception and radial basis function neural networks. In : Expert Systems with Applications, Vol. 36, pp. 9767-9775, ISSN 0957-4174.

YANG Hui-jie, YANG Wen-tong. The new development of foreign small vertical axis wind generators. In : Power Demand Side Management, Vol. 9, No. 2, pp. 68-70, ISSN 1009-1831.

QU Jian-jun, XU Ming-wei, LI Zhong-jie, ZHI Chao. Renewable Energy Resources, Vol. 28, No. 1, pp. 101-104.

Shuqin Liu, Zhongguo Bian, Deguang Li, Wen Zhao.A Magnetic Suspended Self-pitch Vertical Axis Wind Generator. Proceedings of Power and Energy Engineering Conference (APPEEC), ISBN: 978-1-4244-4812-8, chengdu, March 2010.

HU Ye-faa; XU Kai-guob, et. al. Analysis and Design of Magnetic Bearings Used in Magnetic Suspending Wind Power Generator. In : Bearing, Vol. 7, pp. 6-10, ISSN 1000-3762.

Global Wind Technology INC, Watkins Philip G. Omni-Directional Wind Turbine Electric Generation System. America, WO2005108785,2005.

11

The Analysis and Modelling of a Self-excited Induction Generator Driven by a Variable Speed Wind Turbine

Ofualagba, G and Ubeku, E.U
Federal University of Petroleum Resources, Effurun
Nigeria

1. Introduction

Induction machine is used in a wide variety of applications as a means of converting electric power to mechanical work. The primary advantage of the induction machine is its rugged brushless construction and no need for separate DC field power. These machines are very economical, reliable, and are available in the ranges of fractional horse power (FHP) to multi –megawatt capacity. Also, unlike synchronous machines, induction machines can be operated at variable speeds. For economy and reliability many wind power systems use induction machines, driven by a wind turbine through a gear box, as an electrical generator. The need for gearbox arises from the fact that lower rotational speeds on the wind turbine side should be converted to high rotor speeds, on the electrical generator side, for electrical energy production.

There are two types of induction machine based on the rotor construction namely, squirrel cage type and wound rotor type. Squirrel cage rotor construction is popular because of its ruggedness, lower cost and simplicity of construction and is widely used in stand-alone wind power generation schemes. Wound rotor machine can produce high starting torque and is the preferred choice in grid-connected wind generation scheme. Another advantage with wound rotor is its ability to extract rotor power at the added cost of power electronics in the rotor circuit.

This chapter focuses on the electrical generation part of a wind energy conversion system. After a brief introduction of the induction machine, the electrical generator used in this chapter, a detailed analysis of the induction machine operated in stand-alone mode is presented. As a generator, induction machines have the drawback of requiring reactive power for excitation. This necessitates the use of shunt capacitors in the circuit. The effect of magnetization inductance on self-excitation of the induction generator is discussed. Also, this chapter presents the two existing methods to analyze the process of self-excitation in induction machine and the role of excitation-capacitors in its initiation.

Simulation results of the self-excited induction generator driven by the variable speed wind turbine are presented in the last section of this chapter. The process of voltage build up and the effect of saturation characteristics are also explained in the same section.

2. Induction machine

In the electromagnetic structure of the Induction machine, the stator is made of numerous coils with three groups (phases), and is supplied with three phase current. The three coils are physically spread around the stator periphery (space-phase), and carry currents which are out of time-phase. This combination produces a rotating magnetic field, which is a key feature of the working of the induction machine. Induction machines are asynchronous speed machines, operating below synchronous speed when motoring and above synchronous speed when generating. The presence of negative resistance (i.e., when slip is negative), implies that during the generating mode, power flows from the rotor to the stator in the induction machine.

2.1 Equivalent electrical circuit of induction machine

The theory of operation of induction machine is represented by the per phase equivalent circuit shown in Figure 1 (Krause et al., 1994).

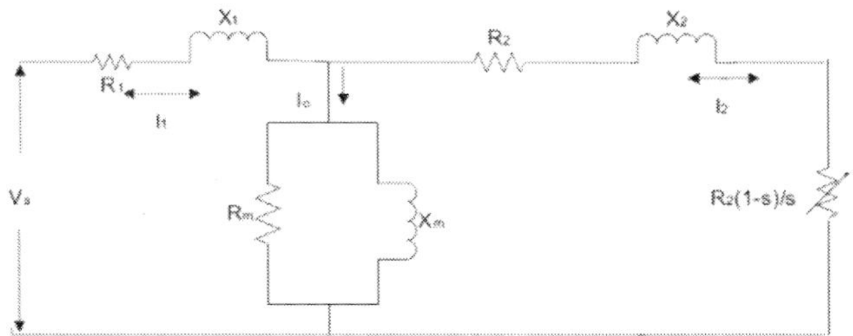

Fig. 1. Per-phase equivalent circuit of the induction machine referred to the stator.

In the above figure, R and X refer to the resistance and inductive reactance respectively. Subscripts 1, 2 and m represent stator, rotor values referred to the stator side and magnetizing components, respectively.

Induction machine needs AC excitation current for its running. The machine is either self-excited or externally excited. Since the excitation current is mainly reactive, a stand-alone system is self-excited by shunt capacitors. In grid-connected operation, it draws excitation power from the network, and its output frequency and voltage values are dictated by the grid. Where the grid capacity of supplying the reactive power is limited, local capacitors can be used to partly supply the needed reactive power (Patel, 1999).

3. Self-Excited Induction Generator (SEIG)

Self-excited induction generator (SEIG) works just like an induction machine in the saturation region except the fact that it has excitation capacitors connected across its stator terminals. These machines are ideal choice for electricity generation in stand-alone variable speed wind energy systems, where reactive power from the grid is not available. The induction generator will self-excite, using the external capacitor, only if the rotor has an adequate remnant magnetic field. In the self-excited mode, the generator output frequency and voltage are affected by the speed, the load, and the capacitance value in farads (Patel, 1999). The steady-state per-phase equivalent circuit of a self-excited induction generator is shown in the Figure 2.

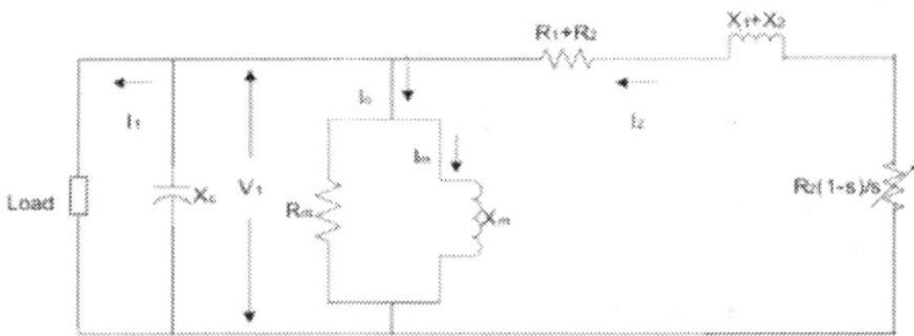

Fig. 2. Self-excited induction generator with external capacitor.

The process of self-excitation in induction machines has been known for many decades (Basset & Potter, 1935). When capacitors are connected across the stator terminals of an induction machine, driven by an external prime mover, voltage will be induced at its terminals. The induced electromotive force (EMF) and current in the stator windings will continue to rise until the steady-state condition is reached, influenced by the magnetic saturation of the machine. At this operating point the voltage and the current will be stabilized at a given peak value and frequency. In order for the self-excitation to occur, for a particular capacitance value there is a corresponding minimum speed (Wagner, 1935). So, in stand-alone mode of operation, it is necessary for the induction generator to be operated in the saturation region. This guarantees one and only one intersection between the magnetization curve and the capacitor reactance line, as well as output voltage stability under load as seen in the Figure 3:

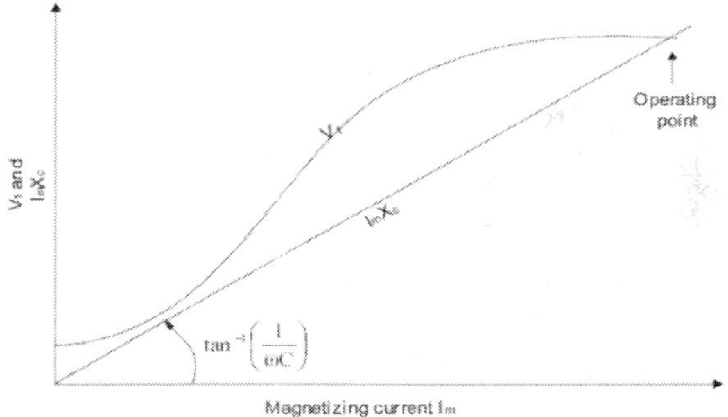

Fig. 3. Determination of stable operation of self-excited induction generator.

At no-load, the capacitor current $I_c = V_1/X_c$ must be equal to the magnetizing current $I_m = V_1/X_m$. The voltage V_1 is a function of I_m, linearly rising until the saturation point of the magnetic core is reached. The output frequency of the self-excited generator is, $f = 1/(2\pi C X_m)$ and $\omega = 2\pi f$ where C is self-exciting capacitance.

4. Methods of analysis

There are two fundamental circuit models employed for examining the characteristics of a SEIG. One is the per-phase equivalent circuit which includes the loop-impedance method adopted by (Murthy *et al*, 1982) and (Malik & Al-Bahrani, 1990), and the nodal admittance method proposed by (Ouazene & Mcpherson, 1983) and (Chan, 1993). This method is suitable for studying the machine's steady-state characteristics. The other method is the *dq*-axis model based on the generalized machine theory proposed by (Elder *et al.*, 1984) and (Grantham *et al.*, 1989), and is employed to analyze the machine's transient state as well as steady-state.

4.1 Steady-state model

Steady-state analysis of induction generators is of interest both from the design and operational points of view. By knowing the parameters of the machine, it is possible to determine the performance of the machine at a given speed, capacitance and load conditions.

Loop impedance and nodal admittance methods used for the analysis of SEIG are both based on per-phase steady-state equivalent circuit of the induction machine (Figure 4), modified for the self-excitation case. They make use of the principle of conservation of active and reactive powers, by writing loop equations (Murthy *et al*, 1982], (Malik & Al-Bahrani, 1990), (Al-Jabri & Alolah, 1990) or nodal equations (Ouazene & Mcpherson, 1983), (Chan, 1993), for the equivalent circuit. These methods are very effective in calculating the minimum value of capacitance needed for guaranteeing self-excitation of the induction generator. For stable operation, *excitation capacitance* must be slightly higher than the minimum value. Also there is a speed threshold, below which no excitation is possible, called as the cutoff speed of the machine. In the following paragraph, of loop impedance method is given for better understanding.

The per-unit per-phase steady-state circuit of a self-excited induction generator under RL load is shown in Figure 4 (Murthy *et al.*, 1982), (Ouazene & Mcpherson, 1983).

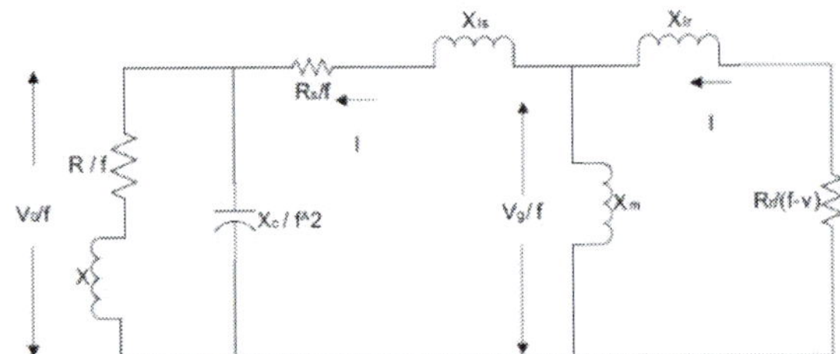

Fig. 4. Equivalent circuit of self-excited induction generator with R-L Load.

Where:

R_s, R_r, R : p.u. per-phase stator, rotor (referred to stator) and load resistance respectively.

Xls, Xlr, X, Xm : p.u. per-phase stator leakage, rotor leakage (referred to stator), load and magnetizing reactances (at base frequency), respectively.

$Xsmax$: p.u. maximum saturated magnetizing reactance.

C : per-phase terminal-excitation capacitance.

Xc : p.u. per-phase capacitive reactance (at base frequency) of the terminal excitation capacitor.

f, v : p.u.frequency and speed, respectively.

N : base speed in rev/min

Z_b : per-phase base impedance

fb : base frequency

V_g, V_0 : per-phase air gap and output voltages, respectively.

In the analysis of SEIG the following assumptions were made (Murthy *et al.*, 1982):

1. Only the magnetizing reactance Xm is assumed to be affected by magnetic saturation, and all other parameters of the equivalent circuit are assumed to be constant. Self-excitation results in the saturation of the main flux and the value of X_m reflect the magnitude of the main flux. Leakage flux passes mainly in the air, and thus these fluxes are not affected to any large extent by the saturation of the main flux.
2. Stator and rotor leakage reactance, in per-unit are taken to be equal. This assumption is normally valid in induction machine analysis.
3. Core loss in the machine is neglected.

For the circuit shown in Figure 4, the loop equation for the current can be written as:

$$I\,Z = 0 \tag{1}$$

Where Z is the net loop impedance given by

$$Z = \left(\left(\frac{Rr}{f-v}\right) + jX_{lr} \,\|\, jX_m\right) + \frac{Rs}{f} + jX_{ls} + \left(\frac{-jX_C}{f^2} \,\|\, \left(\frac{R}{f} + jX\right)\right) \tag{2}$$

Since at steady-state excitation $I \neq 0$, it follows from (equation 1) that $Z = 0$, which implies that both the real and imaginary parts of Z are zeros. These two equations can be solved simultaneously for any two unknowns (usually voltage and frequency). For successful voltage-buildup, the load-capacitance combination and the rotor speed should result in a value such that $X_m = X_{smax}$, which yields the minimum value of excitation capacitance below which the SEIG fails to self-excite.

4.2 Steady-state and transient model (abc-dq0 transformation)

The process of self-excitation is a transient phenomenon and is better understood if analyzed using a transient model. To arrive at transient model of an induction generator, *abc-dq0* transformation is used.

4.2.1 abc-dq0 transformation

The *abc-dq0* transformation transfers an *abc* (in any reference frame) system to a rotating *dq0* system. (Krause *et al.*, 1994) noted that, all time varying inductances can be eliminated by referring the stator and rotor variables to a frame of reference rotating at any angular velocity or remaining stationary. All transformations are then obtained by assigning the appropriate speed of rotation to this (*arbitrary*) *reference frame*. Also, if the system is balanced the zero component will be equal to zero (Krause *et al.*, 1994).

A change of variables which formulates a transformation of the 3-phase variables of stationary circuit elements to the arbitrary reference frame may be expressed as (Krause *et al.*, 1994):

$$F_{qdos} = K_s f_{abcs} \tag{3}$$

Where:

$$F_{qdo}s = \begin{bmatrix} f_{qs} \\ f_{ds} \\ f_{os} \end{bmatrix}; f_{abc}s = \begin{bmatrix} f_{as} \\ f_{bs} \\ f_{cs} \end{bmatrix}; K_s = \frac{2}{3} \begin{bmatrix} \cos\theta & \cos\left(\theta - \frac{2\pi}{3}\right) & \cos\left(\theta + \frac{2\pi}{3}\right) \\ \sin\theta & \sin\left(\theta - \frac{2\pi}{3}\right) & \sin\left(\theta + \frac{2\pi}{3}\right) \\ \frac{1}{2} & \frac{1}{2} & \frac{1}{2} \end{bmatrix}; \theta = \int_0^t \omega(\xi)d\xi + \theta(0);$$

ξ is the dummy variable of integration

For the inverses transformation:

$$(K_s)^{-1} = \begin{bmatrix} \cos\theta & \sin\theta & 1 \\ \cos\left(\theta - \frac{2\pi}{3}\right) & \sin\left(\theta - \frac{2\pi}{3}\right) & 1 \\ \cos\left(\theta + \frac{2\pi}{3}\right) & \sin\left(\theta + \frac{2\pi}{3}\right) & 1 \end{bmatrix}$$

In (equation 3), f can represent voltage, current, flux linkage, or electric charge. The subscript s indicates the variables, parameters and transformation associated with stationary circuits. This above transformation could also be used to transform the time-varying rotor windings of the induction machine. It is convenient to visualize the transformation equations as trigonometric relationships between variables as shown in Figure 5.

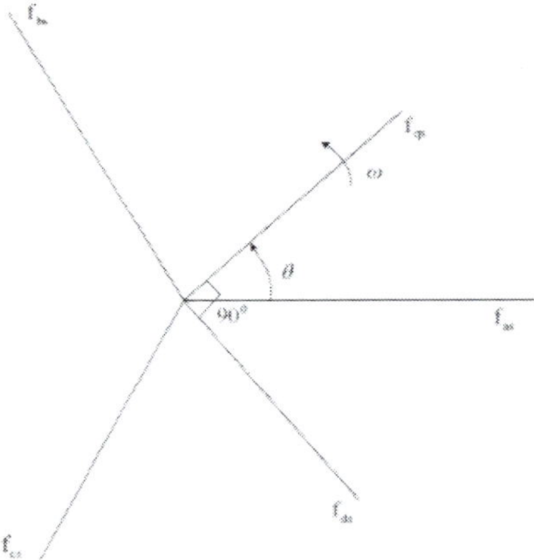

Fig. 5. Transformation for stationary circuits portrayed bytrigonometric relationships.

The equations of transformation may be thought of as if the f_{qs} and f_{ds} variables are directed along axes orthogonal to each other and rotating at an angular velocity of ω, where upon f_{as}, f_{bs}, and f_{cs} (instantaneous quantities which may be any function of time), considered as variables directed along stationary paths each displaced by 120^0. Although the waveforms of the qs and ds voltages, currents and flux linkages, and electric charges are dependent upon the angular velocity of the frame of reference, the waveform of the total power is same regardless of the reference frame in which it is evaluated (Krause *et al.*, 1994).

4.2.2 Voltage equations in arbitrary reference-frame variables

The winding arrangement for a 2-pole, 3-phase, wye-connected, symmetrical induction machine is shown in Figure 6.

The stator windings are identical sinusoidally distributed windings, displaced 120^0, with N_s equivalent turns and resistance r_s. The rotor is consists of three identical sinusoidally distributed windings, with N_r equivalent turns and resistance r_r. Note that positive a, b, c sequence is used in both in Figures 5 and 6.

The voltage equations in machine variables can be expressed as (Krause *et al.*, 1994):

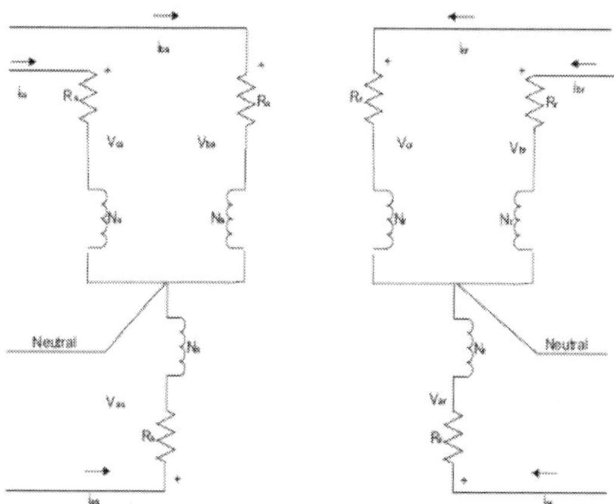

Fig. 6. Two-pole, 3-phase, wye connected symmetrical induction machine.

$$V_{abcs} = r_s i_{abcs} + p\lambda_{abcs} \tag{4}$$

$$V_{abcr} = r_r i_{abcr} + p\lambda_{abcr} \tag{5}$$

Where:
Subscript s denotes parameters and variables associated with the stator.
Subscript r denotes parameters and variables associated with the rotor.
V_{abcs}, V_{abcr} are phase voltages.
I_{abcs}, I_{abcr} are phase currents.
λ_{abcs}, λ_{abcr} are the flux linkages and $p = d/dt$.
By using the *abc-dq0* transformation and expressing flux linkages as product of currents and winding inductances, we obtain the following expressions for voltage in arbitrary reference frame (Krause *et al.*, 1994):

$$V_{qdos} = r_s i_{qdos} + \omega\lambda_{dqs} + p\lambda_{qdos} \tag{6}$$

$$V'_{qdor} = r'_r i'_{qdor} + (\omega - \omega_r)\,\lambda'_{dqr} + p\lambda'_{qdor} \tag{7}$$

Where:
ω is the electrical angular velocity of the arbitrary frame.
ω_r is the electrical angular velocity of the rotor.

$$(\lambda_{dqs})^T = [\lambda_{ds} \quad -\lambda_{qs} \quad 0]\; ; (\lambda'_{dqr})^T = [\lambda'_{dr} \quad -\lambda'_{qr} \quad 0]$$

"'" denotes rotor values referred to the stator side.
Using the relations between the flux linkages and currents in the arbitrary reference frame and substituting them in (equations 6) & (equations 7), the voltage and flux equations are expressed as follows:

$$V_{qs} = r_s i_{qs} + \omega\lambda_{ds} + p\lambda_{qs} \tag{8}$$

$$V_{ds} = r_s i_{ds} - \omega\lambda_{qs} + p\lambda_{ds} \tag{9}$$

$$V'_{qr} = r'_r i'_{qr} (\omega - \omega_r) \lambda'_{dr} + p\lambda'_{qr} \tag{10}$$

$$V'_{dr} = r'_r i'_{dr} - (\omega - \omega_r) \lambda'_{qr} + p\,\lambda'_{dr} \tag{11}$$

$$\lambda_{qs} = L_{ls}i_{qs} + L_m (i_{qs} + i'_{qr}) \tag{12}$$

$$\lambda_{ds} = L_{ls}i_{ds} + L_m (i_{ds} + i'_{dr}) \tag{13}$$

$$\lambda'_{qr} = L'_{lr}i_{qr} + L_m (i_{qs} + i'_{qr}) \tag{14}$$

$$\lambda'_{dr} = L'_{lr}i'_{dr} + L_m (i_{ds} + i'_{dr}) \tag{15}$$

Where:

L_{ls} and L_{ms} are leakage and magnetizing inductances of the stator respectively.

L_{lr} and $L_{m}r$ are leakage and magnetizing inductances of the rotor respectively.

Magnetizing inductance, $L_m = \dfrac{3}{2} L_{ms}$

The voltage and flux linkage equations suggest the following equivalent circuits for the induction machine:

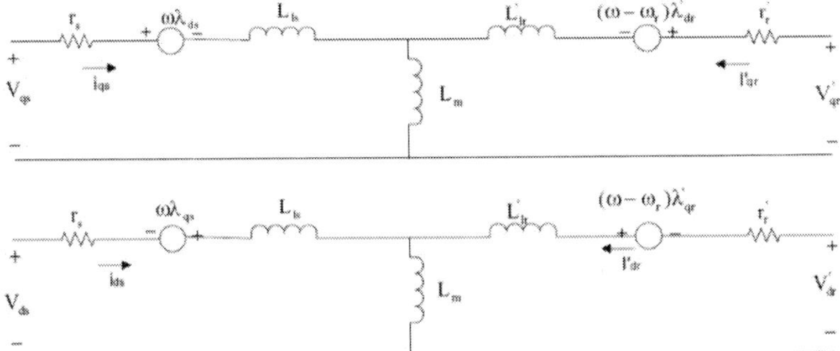

Fig. 7. Arbitrary reference-frame equivalent circuits for a 3-phase, symmetrical induction machine

4.2.2.1 Torque equations

The expression for electromagnetic torque, positive for motor operation and negative for generator operation, in terms of the arbitrary reference variables can be expressed as (Krause *et al.*, 1994):

$$\text{For motor action, } T_e = \left(\frac{3}{2}\right)\left(\frac{P}{2}\right) L_m(i_{qs}i'_{dr} - i_{ds}i'_{qr}) \tag{16}$$

$$\text{For generator action, } T_e = \left(\frac{3}{2}\right)\left(\frac{P}{2}\right) L_m(i_{ds}i'_{qr} - i_{qs}i'_{dr}) \tag{17}$$

The torque and speed are related by the following expressions:

$$\text{For the motor operation, } T_{e\text{-}motor} = J\left(\frac{2}{P}\right) p\omega_r + T_D \tag{18}$$

$$\text{For the generator operation, } T_D = J\left(\frac{2}{P}\right) p\omega_r + T_{e\text{-}gen} \tag{19}$$

Where:

P: Number of poles.

J: Inertia of the rotor in (Kg m^2).

T_D: Drive torque in (Nm).

4.2.3 Stationary reference frame

Although the behavior of the induction machine may be described by any frame of reference, there are three which are commonly used (Krause *et al.*, 1994). The voltage equations for each of these reference frames can be obtained from the voltage equations in the arbitrary reference frame by assigning the appropriate speed to ω. That is, for the stationary reference frame, $\omega = 0$, for the rotor reference frame, $\omega = \omega_r$, and for the synchronous reference frame, $\omega = \omega_e$.

Generally, the conditions of operation will determine the most convenient reference frame for analysis and/or simulation purposes. The stator reference frame is used when the stator voltages are unbalanced or discontinuous and the rotor applied voltages are balanced or zero. The rotor reference frame is used when the rotor voltages are unbalanced or discontinuous and the stator applied voltages are balanced. The stationary frame is used when all (stator and rotor) voltages are balanced and continuous. In this thesis, the *stationary reference frame(ω=0)* is used for simulating the model of the self-excited induction generator (SEIG).

In all asynchronously rotating reference frames($\omega \neq \omega_r$) with $\theta(0)=0$ (see (equation 3)), the phasor representing phase a variables (with subscript as) is equal to phasor representing qs variables. In other terms, for the rotor reference frame and the stationary frame, $f_{as} = f_{qs}$, $f_{bs} = f_{as} < 120^0$ and $f_{cs} = f_{as} < 240^0$

4.2.4 SEIG model

As discussed above, the *dq* model of the SEIG in the stationary reference frame is obtained by substituting ω=0 in the arbitrary reference frame equivalent of the induction machine shown in Figure 7. Figure 8 shows a complete *dq*-axis model, of the SEIG with load, in the stationary reference frame. Capacitor is connected at the stator terminals for the self-excitation. For convenience, all values are assumed to be referred to the stator side and here after " ' " is neglected while expressing rotor parameters referred to the stator (Seyoum *et al.*, 2003).

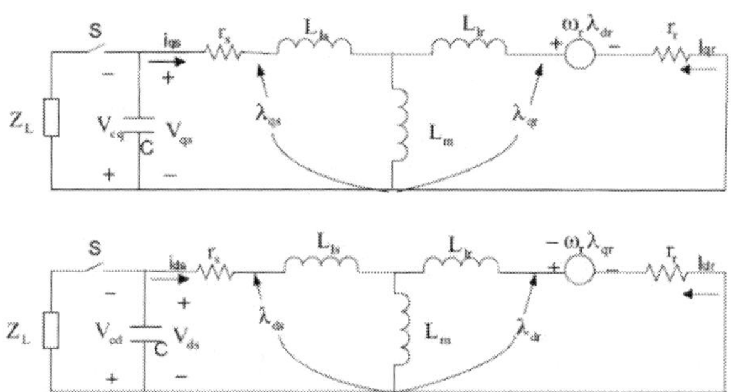

Fig. 8. *dq* model of SEIG in stationary reference frame (All values referred to stator)

For no-load condition, rearranging the terms after writing loop equations for Figure 8, we obtain the following voltage equations expressed in the form of a matrix (Krause *et al.*, 1994), (Seyoum *et al.*, 2003):

$$\begin{bmatrix} 0 \\ 0 \\ 0 \\ 0 \end{bmatrix} = \begin{bmatrix} r_s + pL_s + 1/pC & 0 & pL_m & 0 \\ 0 & r_s + pL_s + 1/pC & 0 & pL_m \\ pL_m & -\omega_r L_m & r_r + pL_r & -\omega_r L_r \\ \omega_r L_m & pL_m & \omega_r L_r & r_r + pL_r \end{bmatrix} \begin{bmatrix} i_{qs} \\ i_{ds} \\ i_{qr} \\ i_{dr} \end{bmatrix} + \begin{bmatrix} V_{cqo} \\ V_{cdo} \\ K_q \\ K_d \end{bmatrix} \tag{20}$$

Where:

K_d and K_q are constants representing initial induced voltages along the *d*-axis and *q*-axis respectively, due to the remaining magnetic flux in the core.
V_{cqo} and V_{cdo} are initial voltages in the capacitors.

$$L_s = L_{ls} + L_m \text{ and } L_r = L_{lr} + L_m.$$

The above equations can further be simplified in the following manner using (equations 8-15):
In the stationary reference frame (equations 8) can be written as,

$$V_{qs} = -V_{cq} = r_s i_{qs} + 0 \times \lambda_{ds} + p\lambda_{qs} \tag{21}$$

Substituting (equation 12) in (equation 21), will result

$$0 = V_{cq} + r_s i_{qs} + L_s p i_{qs} + L_m p i_{qr} \tag{22}$$

Solving for pi_{qr} by substituting (equation 14) and (equation 15) in (equation 10) yields:

$$Pi_{qr} = \frac{1}{L_{lr}}(V_{qr} - r_r i_{qr} + \omega_r L_r i_{dr} + \omega_r L_m i_{ds} - L_m L_m i_{qs}) \tag{23}$$

Substituting (equation 23) in (equation 22) results in the final expression for i_{qs} as:

$$pi_{qs} = \frac{1}{L}(-r_r L_r i_{qs} - L_m^2 \omega_r i_{ds} + L_m r_r i_{qr} - L_m \omega_r L_r i_{dr} + L_m K_q - L_r V_{cq}) \tag{24}$$

Where:

$$L = L_s L_r - L_m^2$$

Similarly, the expressions for other current components are obtained and the SEIG can be represented in a matrix form as:

$$pI = AI + B \tag{25}$$

Where:

$$A = \frac{1}{L}\begin{bmatrix} -L_r r_s & -L_m^2 \omega_r & L_m r_r & -L_m \omega_r L_r \\ L_m^2 \omega_r & -L_s r_s & L_m \omega_r L_r & L_m r_r \\ L_m r_s & L_s \omega_r L_m & -L_s r_r & L_s \omega_r L_r \\ -L_s \omega_r L_m & L_m r_s & -L_s \omega_r L_r & -L_s r_r \end{bmatrix} ; \quad B = \frac{1}{L}\begin{bmatrix} L_m K_q - L_r V_{cq} \\ L_m K_d - L_r V_{cd} \\ L_m V_{cq} - L_s K_q \\ L_m V_{cd} - L_s K_d \end{bmatrix}$$

$$I = \begin{bmatrix} i_{qs} \\ i_{ds} \\ i_{qr} \\ i_{dr} \end{bmatrix} ; V_{cq} = \frac{1}{C} \int i_{qs} dt + V_{cq} \,|t = 0 \; ; V_{cq} = \frac{1}{C} \int i_{ds} dt + V_{cd} \,|t = 0$$

Any combination of R, L and C can be added in parallel with the self-excitation capacitance to act as load. For example, if resistance R is added in parallel with the self-excitation capacitance, then the term $1/pC$ in (equation 20) becomes $R/(1+RpC)$. The load can be connected across the capacitors, once the voltage reaches a steady-state value (Grantham *et al.*, 1989), (Seyoum *et al.*, 2003).

The type of load connected to the SEIG is a real concern for voltage regulation. In general, large resistive and inductive loads can vary the terminal voltage over a wide range. For example, the effect of an inductive load in parallel with the excitation capacitor will reduce the resulting effective load impedance (Z_{eff}) (Simoes & Farret, 2004).

$$Z_{eff} = R + j\left(\omega L - \frac{1}{\omega C}\right) \tag{26}$$

This change in the effective self-excitation increases the slope of the straight line of the capacitive reactance (Figure 3), reducing the terminal voltage. This phenomenon is more pronounced when the load becomes highly inductive.

5. Simulation results

A model based on the first order differential equation (equation 25) has been built in the MATLAB/Simulink to observe the behavior of the self-excited induction generator. The parameters used, obtained from (Krause *et al.*, 1994), are as follows.

Machine Rating			IB (abc)	r_r	r_s	X|s	X|r	Xm	J
Hp	Volts	Rpm	Amps	Ohms	Ohms	Ohms	Ohms	Ohms	Kg.m^2
500	2300	1773	93.6	0.187	0.262	1.206	1.206	54.02	11.06

Table 1. Induction Machine Parameters

All the above mentioned values are referred to the stator side of the induction machine and the value of self-exciting capacitance used is 90 micro farads.

From the previous subsection, it can be said that with inductive loads the value of excitation capacitance value should be increased to satisfy the reactive power requirements of the SEIG as well as the load. This can be achieved by connecting a bank of capacitors, across the load meeting its reactive power requirements thereby, presenting unity power factor characteristics to the SEIG. It is assumed in this thesis that, such a reactive compensation is provided to the inductive load, and the SEIG always operates with unity power factor.

5.1 Saturation curve

As explained in the previous section, the magnetizing inductance is the main factor for voltage build up and stabilization of generated voltage for theunloaded and loaded conditions of the induction generator (Figure 3). Reference (Simoes & Farret, 2004) presents a method to determine the magnetizing inductance curve from lab tests performed on a machine. The saturation curve used for the simulation purposes is, obtained from (Wildi,

1997) by making use of the B-H saturation curve of the magnetic material (silicon iron 1%), shown in Figure 9.

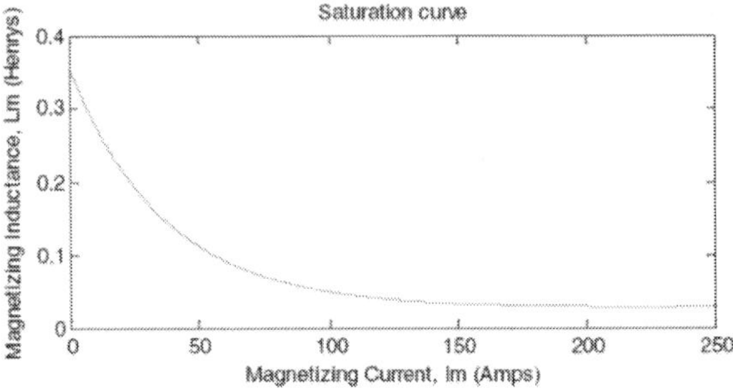

Fig. 9. Variation of magnetizing inductance with magnetizing current.

Using least square curve fit, the magnetizing inductance Lm can be expressed as a function of the magnetizing current I_m as follows:

$$L_m = 1.1*(0.025+0.2974*exp(-0.00271*I_m)) \tag{27}$$

Where,

$$I_m = \sqrt{[\,(\,i_{ds} + i_{dr}\,)^2 + (\,i_{qs} + i_{qr}\,)^2\,]}$$

It must be emphasized that the machine needs residual magnetism so that the self-excitation process can be started. Reference (Simoes & Farret, 2004) gives different methods to recover the residual magnetism in case it is lost completely. For numerical integration, the residual magnetism cannot be zero at the beginning; its role fades away as soon as the first iterative step for solving (equation 25) has started.

5.2 Process of self-excitation

The process of self-excitation can be compared with the resonance phenomenon in an RLC circuit whose transient solution is of the exponential form $Ke^{p_1 t}$ (Elder *et al.*, 1984), (Grantham *et al.*, 1989). In the solution, K is a constant, and root p_1 is a complex quantity, whose real part represents the rate at which the transient decays, and the imaginary part is proportional to the frequency of oscillation. In real circuits, the real part of p_1 is negative, meaning that the transient vanishes with time. With the real part of p_1 positive, the transient (voltage) build-up continues until it reaches a stable value with saturation of iron circuit. In other terms, the effect of this saturation is to modify the magnetization reactance X_m, such that the real part of the root p_1 becomes zero in which case the response is sinusoidal steady-state corresponding to continuous self-excitation of SEIG.

Any current (resulting from the voltage) flowing in a circuit dissipates power in the circuit resistance, and an increasing current dissipates increasing power, which implies some energy source is available to supply the power. The energy source, referred to above is provided by the kinetic energy of the rotor (Grantham *et al.*, 1989).

With time varying loads, new steady-state value of the voltage is determined by the self-excitation capacitance value, rotor speed and load. These values should be such that they guarantee an intersection of magnetization curve and the capacitor reactance line (Figure 3), which becomes the new operating point.

The following figures show the process of self-excitation in an induction machine under no-load condition.

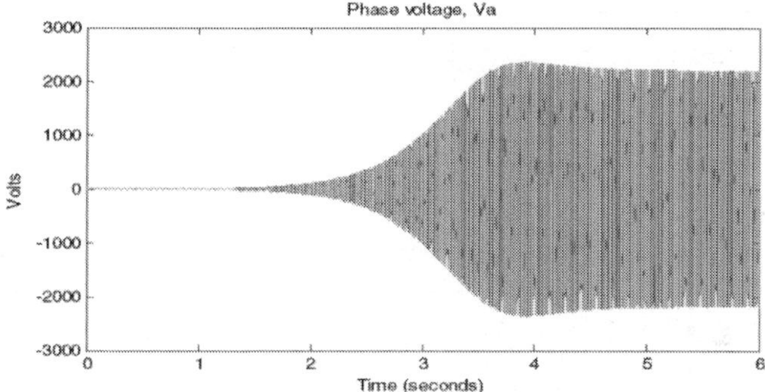

Fig. 10. Voltage build up in a self-excited induction generator.

From Figures 10 and 11, it can be observed that the phase voltage slowly starts building up and reaches a steady-state value as the magnetization current I_m starts from zero and reaches a steady-state value. The value of magnetization current is calculated from the instantaneous values of stator and rotor components of currents (see (equation 27)). The magnetization current influences the value of magnetization inductance Lm as per (3.27), and also capacitance reactance line (Figure 3). From Figures 10-12, we can say that the self-excitation follows the process of magnetic saturation of the core, and a stable output is reached only when the machine core is saturated.

In physical terms the self-excitation process could also be explained in the following way. The residual magnetism in the core induces a voltage across the self-exciting capacitor that produces a capacitive current (a delayed current). This current produces an increased voltage that in turn produces an increased value of capacitor current. This procedure goes on until the saturation of the magnetic filed occurs as observed in the simulation results shown in Figures 10 and 11.

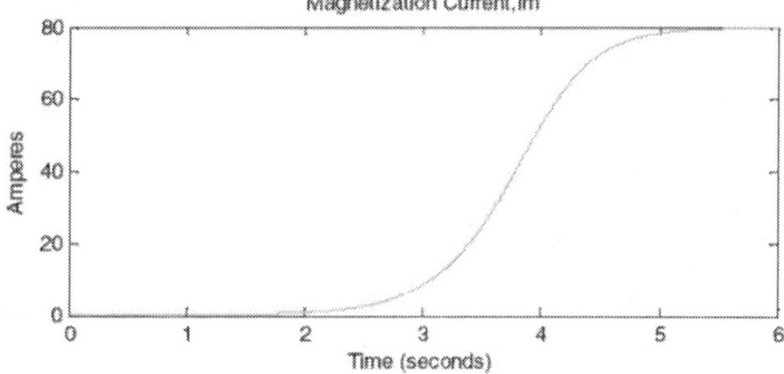

Fig. 11. Variation of magnetizing current with voltage buildup.

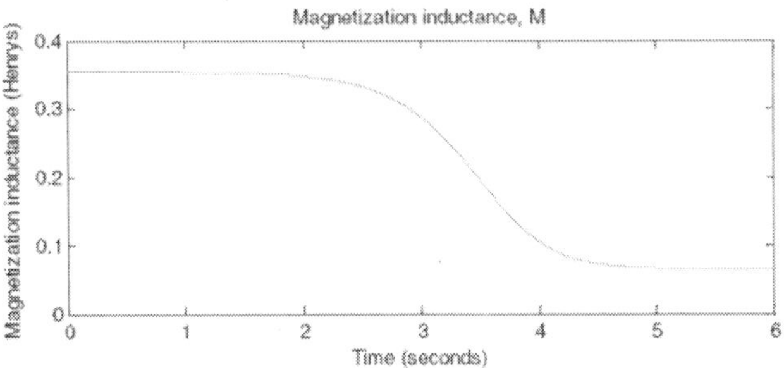

Fig. 12. Variation of magnetizing inductance with voltage buildup.

For the following simulation results the WECS consisting of the SEIG and the wind turbine is driven by wind with velocity of 6 m/s, at no-load. At this wind velocity it can only supply a load of approximately 15 kW. At t=10 seconds a 200 kW load is applied on the WECS. This excess loading of the self-excited induction generator causes the loss of excitation as shown in the Figure 13.

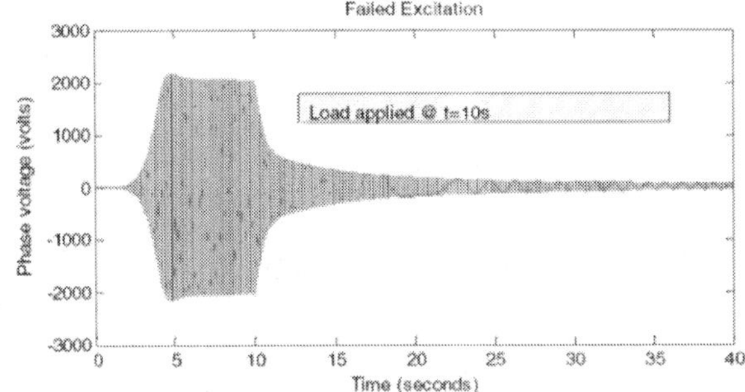

Fig. 13. Failed excitation due to heavy load.

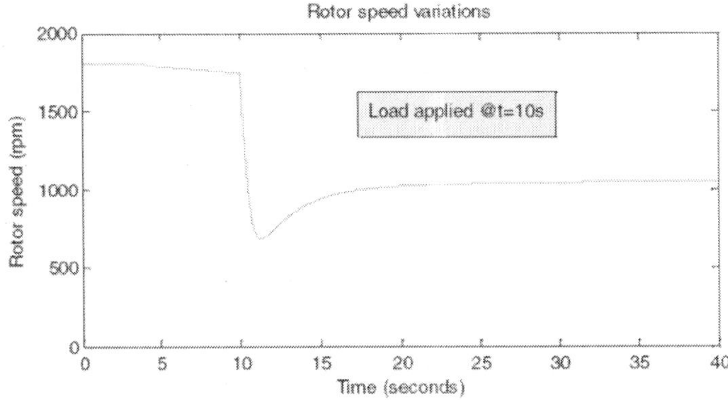

Fig. 14. Generator speed (For failed excitation case)

Figure 14 shows the rotor speed variations with load during the loss of excitation. The increase in load current should be compensated either by increasing the energy input (drive torque) thereby increasing the rotor speed or by an increase in the reactive power to the generator. None of these conditions were met here which resulted in the loss of excitation. It should also be noted from the previous section that there exists a minimum limit for speed (about 1300 rpm for the simulated machine with the self-excitation capacitance equal to 90 micro-farads), below which the SEIG fails to excite.

In a SEIG when load resistance is too small (drawing high load currents), the self-excitation capacitor discharges more quickly, taking the generator to the de-excitation process. This is a natural protection against high currents and short circuits.

For the simulation results shown below, the SEIG-wind turbine combination is driven with an initial wind velocity of 11m/s at no-load, and load was applied on the machine at t=10 seconds. At t = 15 seconds there was a step input change in the wind velocity reaching a final value of 14 m/s. In both cases the load reference (full load) remained at 370 kW. The simulation results obtained for these operating conditions are as follows:

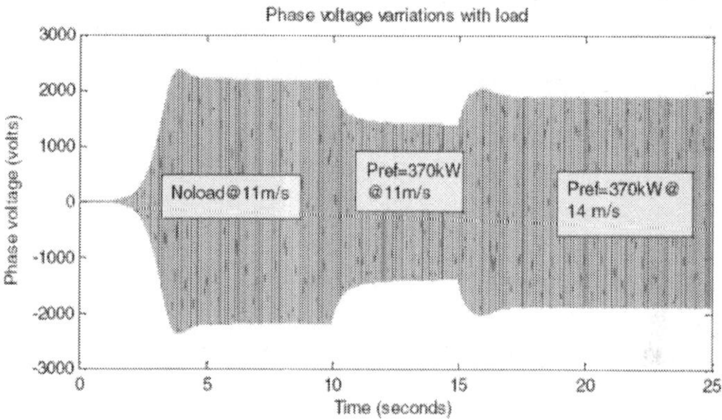

Fig. 15. SEIG phase voltage variations with load.

For the voltage waveform shown in Figure 15, the machine reaches a steady-state voltage of about 2200 volts around 5 seconds at no-load. When load is applied at t=10seconds, there is a drop in the stator phase voltage and rotational speed of the rotor (shown in Figure 18) for the following reasons.

We know that the voltage and frequency are dependent on load (Seyoum *et al.*, 2003). Loading decreases the magnetizing current I_m, as seen in Figure 16, which results in the reduced flux. Reduced flux implies reduced voltage (Figure 15). The new steady-state values of voltage is determined (Figure 3.3) by intersection of magnetization curve and the capacitor reactance line. While the magnitude of the capacitor reactance line (in Figure 3) is influenced by the magnitude of I_m, slope of the line is determined by angular frequency which varies proportional to rotor speed. If the rotor speed decreases then the slope increases, and the new intersection point will be lower to the earlier one, resulting in the reduced stator voltage. Therefore, it can be said that the voltage variation is proportional to the rotor speed variation (Figure 18). The variation of magnetizing current and magnetizing inductance are shown in the Figures 16 and 17 respectively.

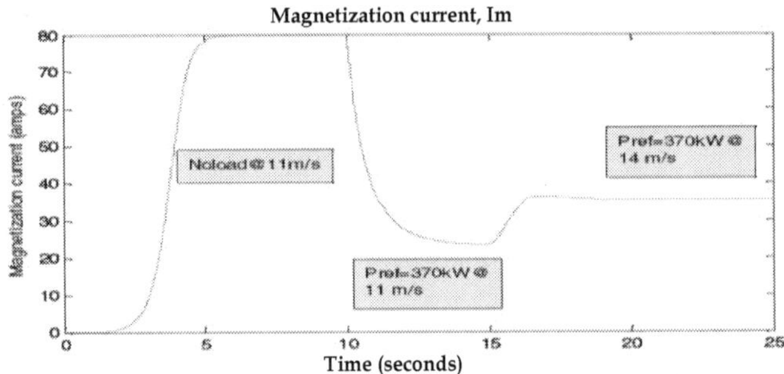

Fig. 16. Magnetizing current variations with load.

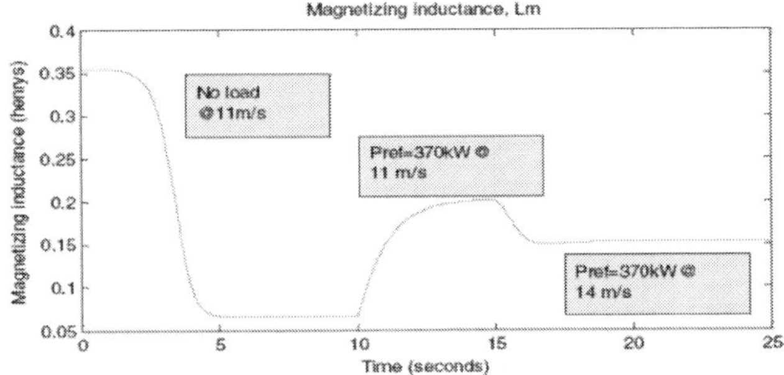

Fig. 17. Magnetizing inductance variations with load.

Figures 16 and 17 verify that the voltage is a function of the magnetizing current, and as a result the magnetizing inductance (see (equation 27)), which determines the steady-state value of the stator voltage.

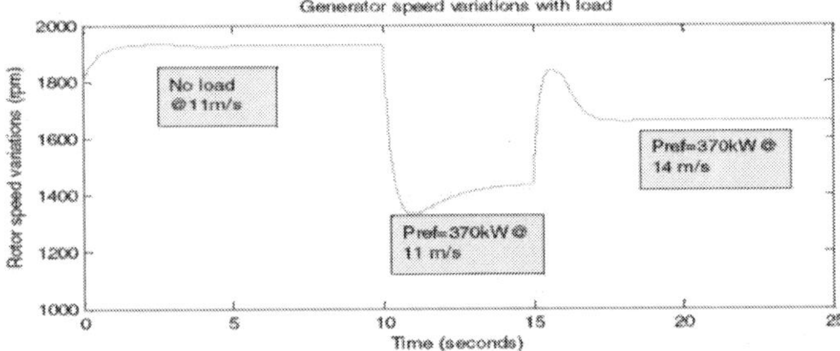

Fig. 18. Rotor speed variations with load

Figure 18, shows the variations of the rotor speed for different wind and load conditions. For the same wind speed, as load increases, the frequency and correspondingly synchronous speed of the machine decrease. As a result the rotor speed of the generator, which is slightly above the synchronous speed, also decreases to produce the required amount of slip at each operating point.

As the wind velocity increases from 11m/s to 14m/s, the mechanical input from the wind turbine increases. This results in the increased rotor speed causing an increase in the stator phase voltage, as faster turning rotor produces higher values of stator voltage. The following figures show the corresponding changes in the SEIG currents, WECS torque and power outputs.

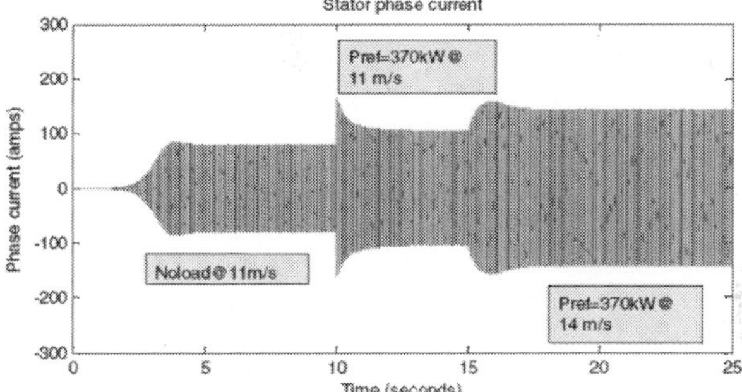

Fig. 19. Stator current variations with load.

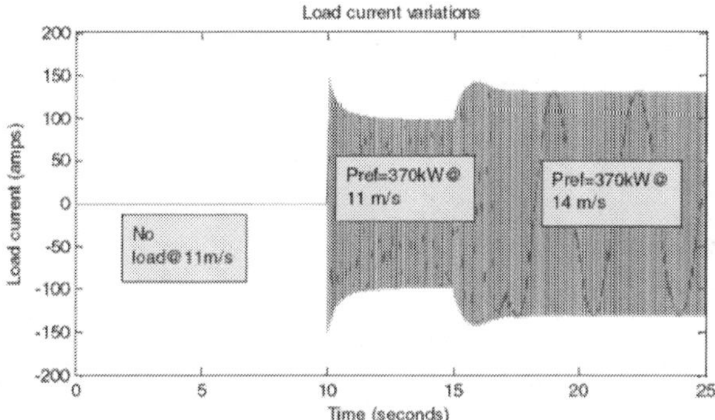

Fig. 20. Load current variations with load.

From Figures 19 and 20, we see that as load increases, the load current increases. When the machine is operating at no-load, the load current is zero. When the load is applied on the machine, the load current reaches a steady-state value of 100 amperes (peak amplitude). With an increase in the prime mover power input, the load current further increases and reaches the maximum peak amplitude of 130 amperes. Also, the stator and load currents will increase with an increase in the value of excitation capacitance. Care should be taken to keep these currents with in the rated limits. Notice that, in the case of motor operation stator windings carry the phasor sum of the rotor current and the magnetizing current. In the case of generator operation the machine stator windings carry current equal to the phasor difference of the rotor current and the magnetizing current. So, the maximum power that can be extracted as a generator is more than 100% of the motor rating (Chathurvedi & Murthy, 1989).

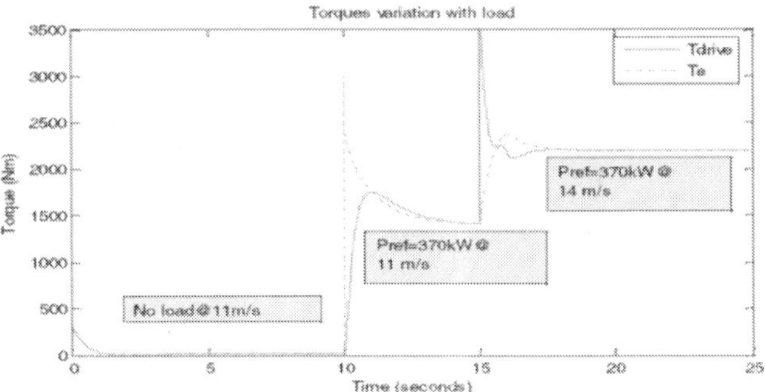

Fig. 21. Variation of torques with load.

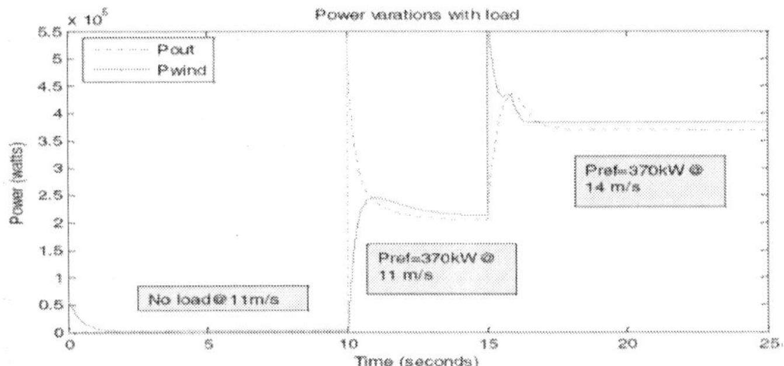

Fig. 22. Output power produced by wind turbine and SEIG.

Equation 17, has been simulated to calculate the electromagnetic torque generated in the induction generator. Figure 21 also shows the electromagnetic torque T_e and the drive torque *Tdrive* produced by the wind turbine at different wind speeds. At t=0, a small drive torque has been applied on the induction generator to avoid simulation errors in Simulink.

Figure 22 shows the electric power output of the SEIG and mechanical power output of the wind turbine. The electric power output of the SEIG (driven by the wind turbine), after t=10 seconds after a short transient because of sudden increase in the load current (Figure 20), is about 210 kW at 11 m/s and reaches the rated maximum power (370 kW) at 14 m/s. Pitch controller limits (see chapter 1) the wind turbine output power, for wind speeds above 13.5 m/s, to the maximum rated power. This places a limit on the power output of the SEIG also, preventing damage to the WECS. Since, the pitch controller has an inertia associated with the wind turbine rotor blades, at the instant t=15seconds the wind turbine output power sees a sudden rise in its value before pitch controller starts rotating the wind turbine blades out of the wind thereby reducing the value of rotor power coefficient. Note that the power loss in the SEIG is given by the difference between P_{out} and *Pwind*, shown in Figure 22.

6. Conclusion

In this chapter the electrical generation part of the wind energy conversion system has been presented. Modeling and analysis of the induction generator, the electrical generator used in this chapter, was explained in detail using dq-axis theory. The effects of excitation capacitor and magnetization inductance on the induction generator, when operating as a stand-alone generator, were explained. From the simulation results presented, it can be said that the self-excited induction generator (SEIG) is inherently capable of operating at variable speeds. The induction generator can be made to handle almost any type of load, provided that the loads are compensated to present unity power factor characteristics. SEIG as the electrical generator is an ideal choice for isolated variable-wind power generation schemes, as it has several advantages over conventional synchronous machine.

7. References

Al Jabri A. K. and Alolah A. I, (1990) "Capacitance requirements for isolated self-excited induction generator," *Proceedings, IEE*, pt. B, vol. 137, no. 3, pp. 154-159

Basset E. D and Potter F. M. (1935), "Capacitive excitation of induction generators," *Trans. Amer. Inst. Elect. Eng*, vol. 54, no.5, pp. 540-545

Bimal K. Bose (2003), *Modern Power Electronics and Ac Drives*, Pearson Education, ch. 2

Chan T. F., (1993) "Capacitance requirements of self-excited induction generators," *IEEE Trans. Energy Conversion*, vol. 8, no. 2, pp. 304-311

Dawit Seyoum, Colin Grantham and M. F. Rahman (2003), "The dynamic characteristics of an isolated self-excited induction generator driven by a wind turbine," *IEEE Trans. Industry Applications*, vol.39, no. 4, pp.936-944

Elder J. M, Boys J. T and Woodward J. L, (1984) "Self-excited induction machine as a small low-cost generator," *Proceedings, IEE*, pt. C, vol. 131, no. 2, pp. 33-41

Godoy Simoes M. and Felix A. Farret, (2004) *Renewable Energy Systems-Design and Analysis with Induction Generators*, CRC Press, 2004, ch. 3-6

Grantham C., Sutanto D. and Mismail B., (1989) "Steady-state and transient analysis of self-excited induction generators," *Proceedings, IEE*, pt. B, vol. 136, no. 2, pp. 61-68

Malik N. H. and Al-Bahrani A. H., (1990)"Influence of the terminal capacitor on the performance characterstics of a self-excited induction generator," *Proceedings, IEE*, pt. C, vol. 137, no. 2, pp. 168-173

Mukund. R. Patel (1999), *Wind Power Systems*, CRC Press, ch. 6

Murthy S. S, Malik O. P. and Tandon A. K., (1982)"Analysis of self excited induction generators," *Proceedings, IEE*, pt. C, vol. 129, no. 6, pp. 260-265

Ouazene L. and Mcpherson G. Jr, (1983) "Analysis of the isolated induction generator," *IEEE Trans. Power Apparatus and Systems*, vol. PAS-102, no. 8, pp.2793-2798

Paul.C.Krause, Oleg Wasynczuk & Scott D. Sudhoff (1994), *Analysis of Electric Machinery*, IEEE Press, ch. 3-4

Rajesh Chathurvedi and S. S. Murthy, (1989) "Use of conventional induction motor as a wind driven self-excited induction generator for autonomous applications," in *IEEE-24th Intersociety Energy Conversion Eng. Conf.*, IECEC, pp.2051-2055

Salama M. H. and Holmes P. G., (1996) "Transient and steady-state load performance of stand alone self-excited induction generator," *Proceedings, IEE-Elect. Power Applicat.*, vol. 143, no. 1, pp. 50-58

Sreedhar Reddy G. (2005), *Modeling and Power Management of a Hybrid Wind-Microturbine Power Generation System*, Masters thesis., ch. 3

Theodore Wildi, (1997) *Electrical Machines, Drives, and Power Systems*, Prentice Hall, Third
      Edition, pp. 28
Wagner C. F, (1939) "Self-excitation of induction motors," *Trans. Amer. Inst. Elect. Eng*, vol.
      58, pp. 47-51

12

Optimisation of the Association of Electric Generator and Static Converter for a Medium Power Wind Turbine

Daniel Matt[1], Philippe Enrici[1], Florian Dumas[1] and Julien Jac[2]
[1]Institut d'Electronique du Sud, Université Montpellier 2
[2]Société ERNEO SAS
France

1. Introduction

This chapter shows the ways of optimising a medium power wind power electromechanical system, generating anything up to several tens of kilowatt electric power. The optimisation criteria are based on the cost of the electromechanical generator associated with a power electronic converter; on the power efficiency; and also on a fundamental parameter, often neglected in smaller installations, which is torque ripple. This can cause severe noise pollution. For a wind turbine generating several kW of electric power, the best solution, without a shadow of a doubt, is to use a permanent magnet electromagnetic generator. This type of generator has obvious advantages in terms of reliability, ease of operation and above all, efficiency. Despite problems concerning the cost of magnets, almost all manufacturers of small or medium power wind turbines use permanent magnet generators (Gergaud et al. 2001). This chapter deals with this type of system. The objective is to demonstrate that only a judicious choice of the configuration of the permanent magnet synchronous generator, amongst the different options, will allow us to satisfy the criteria required for optimal performance. We will study examples of a conventional permanent magnet generator with distributed windings, a permanent magnet generator with concentrated windings (Magnusson & Sandrangani, 2003) and a non-conventional Vernier machine (Matt & Enrici, 2005). How these different machines work will be detailed in the following paragraphs.

2. Description of the electromechanical conversion system

The chosen electromechanical conversion system is represented in Fig. 1. The principle of a turbine directly driving a generator has been chosen in preference to adding a speed multiplier gearbox between the turbine and the generator.
There are many advantages to using a mechanical drive without a gearbox, which requires regular maintenance and which has a pronounced rate of breakdown. These devices are also a significant source of noise pollution when sited near housing. Noise pollution is one of the principal factors in the chosen optimisation criteria. Finally, the gearbox can cause chemical pollution due to the lubricant which it contains. However, the omission of a gearbox means an increase in both size and cost of the generator, which then operates at a very low speed.

For this reason, a balance between size, cost and performance of the system must be considered. More and more wind turbine manufacturers are using the direct drive concept.

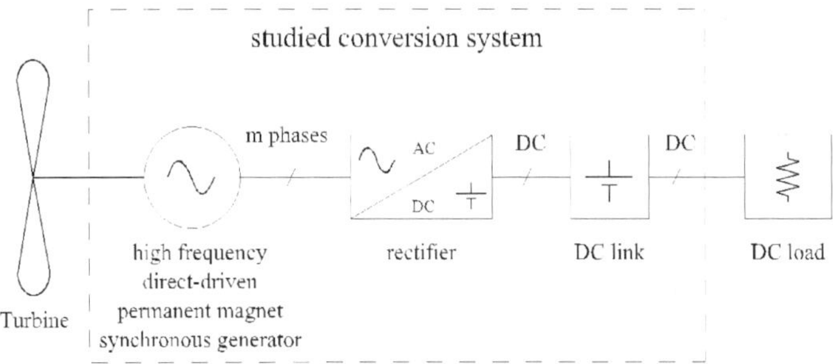

Fig. 1. Structure of the conversion system

As stated, one of the major design difficulties is sizing the generator, which whilst operating at low speed, must supply high torque. The size is proportional to the torque so the mass or volume power ratio of a generator tends to be low.

The main deciding parameter for the size of the generator is the electrical conversion frequency. These energy systems are therefore all sized on the same basis: the frequency of the completed conversion cycle (electric, thermal, mechanical). The optimal solution chosen for the generator will have the characteristics of a "high frequency" machine, typically between 100 and 200 Hz or even more in certain cases, for a rotation speed generally in the order of 100 to 200 rpm. In this context, optimised direct drive gives a mass-power ratio close to that obtained by an indirect drive, with increased efficiency and reliability. This quest for high conversion frequencies is beneficial to noise pollution, high frequency vibrations being more easily filtered by the mechanical structure of the wind turbine.

A second design difficulty concerns the choice of static converter associated with the machine, in order to fulfil the generating requirements of the end user. This is a difficult choice, because the behaviour of the converter can have serious repercussions on the behaviour of the generator with regard to the chosen performance criteria.

Whether the turbine is on an isolated site or is connected to the grid, most power electronic converters have a DC bus like that in Fig. 1. The study presented in this chapter will be limited mainly to DC bus systems i.e. combined with a permanent magnet synchronous generator and rectifier.

It should be noted that direct AC to AC conversion solutions, like that in Fig. 2, adapted for linking the generator to the grid, exist (Barakati, 2008), but while these solutions are appealing on paper, they haven't really been put into practice. They conflict with the design of the matrix converter which uses bidirectional switches for coupling (Thyristor solutions also exist).

We return to diagram on Fig. 1 which corresponds to the system under study. Different solutions exist for the rectifier. They are shown in Fig. 3.

Two of these are based on the concept of active rectifiers. The structure of these rectifiers is that of an inverted PWM inverter, the energy flowing from an AC connection to a DC connection (Mirecki, 2005; Kharitonov, 2010). A variation of this structure, called Vienna, also uses the notion of a bidirectional switch (Kolar et al, 1998).

The interest of an active rectifier lies in the fact that the driver gives complete control of the current waveform produced by the generator, the rectifier itself imposes no specific stress on

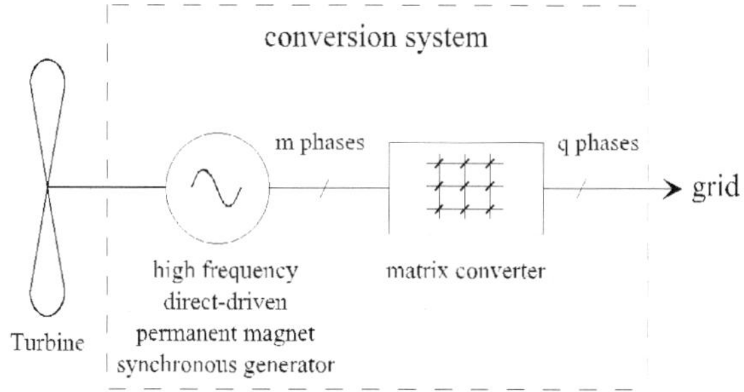

Fig. 2. Connection of generator to the grid using a matrix converter

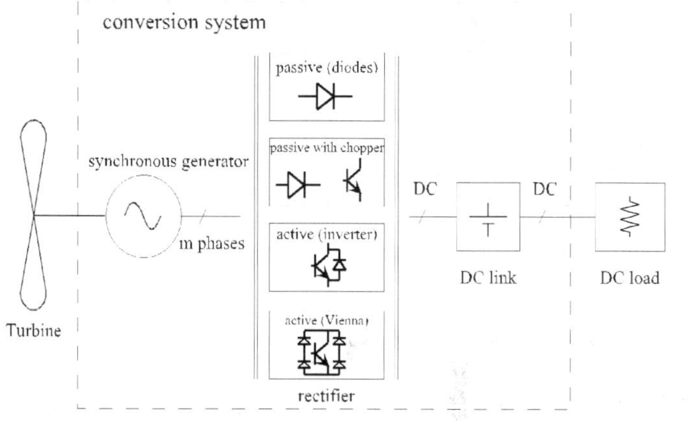

Fig. 3. Different configurations of rectifiers

the machine. If the EMF of the generator is sinusoidal, control of the rectifier will give a sinusoidal current in phase with the EMF, the ohmic loss will be minimised, the sizing optimal. This configuration and ideal operating mode will serve as a reference in the following paragraphs for comparing different generator configurations.

The disadvantage of using active rectifiers is essentially economic. The structure of the power electronic used, although classic, is complex, which makes for high costs and poor reliability, especially in comparison with the solutions which we are soon going to present. In the case of medium power, which is the focus of this chapter, the active rectifier, despite its drawbacks, is the most common solution.

Despite the undeniable advantages of active rectifiers, the conventional structure of passive diode rectifiers can be preferable in wind turbine systems because they are robust in most conditions. These are the two other solutions illustrated in Fig. 3. The rectifier can be used alone, or with a chopper when a degree of fine tuning is required in maintaining optimal performance of the system. The chopper is not always indispensable and only slightly improves the performance of the wind turbine. (Gergaud, 2001; Mirecki, 2005).

We will confine ourselves therefore to the study of a permanent magnet synchronous generator with a passive diode rectifier. We will show that by careful selection of the configuration of the generator, highly satisfactory operation of the conversion system is achieved, with minimal drop in performance (efficiency, torque ripple) compared to a system using an active rectifier.

3. Choosing the structure of the synchronous generator

To satisfy the conditions which we have imposed, we will study the behaviour of three distinct permanent magnet synchronous generators: a conventional structure widely used; a "Vernier" structure (Toba & Lipo, 2000; Matt & Enrici, 2005), less well known, but perfectly adapted to operating at very low speed; and a harmonic coupling structure (Magnussen & Sadarangani, 2003), nowadays a classic, but little used in the field of wind turbines. Their operating modes are reviewed in the following paragraphs. The conventional structure will serve as a reference by which to compare the other two structures, which are better adapted to running at high frequencies for low rotation speeds.

The electrical system under study will be modelled on the diagram in Fig. 4. The electrical generator is represented by a simplified Behn-Eschenburg model, which is sufficiently precise for this general comparison. This model is particularly pertinent, since the 3 machines studied generate sinusoidal EMF with almost no harmonics.

The addition of the "DC model" gives an accurate estimation of the reduction in average rectified voltage, E_s, due on the one hand to the overlap engendered by the synchronous inductance, Ls, of the generator (Δ_{Ee}), and on the other hand to resistive voltage drop, (Δ_{Er}), (Mirecki, 2005). This demonstrates the power limitation associated with synchronous inductance. Thus, conforming to the rules of impedance matching, the maximum power, P_{max}, transferred to the continuous load is obtained when $E_b = E_s/2$, which allows us to express the following:

$$P_{max} = \frac{E_s^2}{4(Y+R)} \tag{1}$$

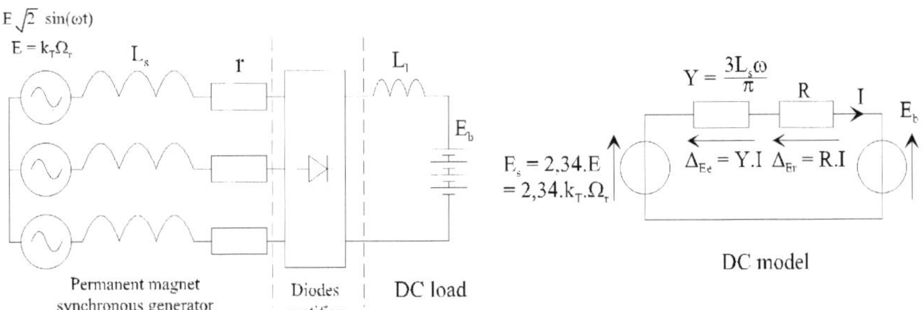

Fig. 4. Modelling of the Electrical System

This phenomenon of power limitation can be used to advantage in a wind turbine with a passive conversion system, without regulating power. It is then possible, with the appropriate value of parameter Y (see Table 1), to obtain close to maximum power (MPPT), at variable speed, without any control mechanism (Matt et al.; 2008). However, the optimisation of the inductance L_s on which Y depends, will be dictated by the compromises reached, which we will explain in the descriptions of the studied generators (Abdelli, 2007). The following table shows the main notations used for the study of the electrical system.

Electrical parameters	Notations	Remarks
Electric frequency, pulsation	f_e, ω	-
Rotation speed, pulsation of rotation	N, Ω_r	-
Coefficient of torque or EMF	k_T	In steady state
Phase EMF	E	In steady state
Synchronous inductance	L_s	In steady state
Phase resistor	r	In steady state
Number of pole pair	p	-
Filtering inductance	L_1	Optional
DC bus voltage	E_b	-
Average rectified voltage	E_s	3 phases Graetz rectifier
Overlap "resistor"	Y	Non dissipative
DC model resistor	R	Dissipative

Table 1. Variables of the electrical system

The comparative study of the following three generators was done using a CAD power electronics tool (PSIM, Powersim Inc.), based on the diagrams in Figs. 1 and 4.
The three structures compared are of a similar cylindrical design and overall size. They are designed to supply an electrical output of 10 KW for a rotation speed of 150 rpm.

4. Operation using a conventional synchronous generator

The first permanent magnet generator studied is a classic design. Its general structure is represented in Fig. 5. The armature of this machine has a three-phase pole pitch winding with a large number of poles of which we will list the precise characteristics. The field system magnets are fixed along the rotor rim and form an almost continuous layer.

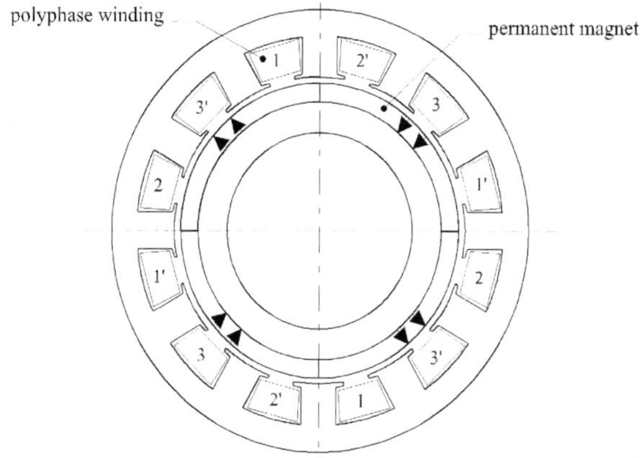

Fig. 5. Conventional Permanent magnet generator

Rather than designing a generator specifically for this comparison, we have chosen to adopt the characteristics of a commercial machine, currently used in medium power wind turbines. The useful characteristics for the model are summed up in Table 2 (refer to Table 1 for the notations).

Characteristics	Values
Nominal rotation speed (rpm)	150
f_e at nominal speed (Hz)	45
Nominal power with resistive load (W)	9740
E at nominal rotation speed (steady state) (V)	160
r (steady state) (Ω)	1
L_s (mH)	7,4
p	18
Joule losses at nominal current (W)	1600
Iron and mechanical losses (W)	200
Torque ripple without load (cogging torque) (Nm)	8
Nominal efficiency with resistive load (%)	84

Table 2. Characteristics of the reference generator

The principal characteristic of this generator lies in the angle of the slots which greatly reduce cogging torque ripple (8 Nm). Another consequence of this angle is the limitation of the harmonics of the electromotive forces which, as a result, are practically sinusoidal.

The optimisation of the mass-power ratio of this machine is achieved by using as many poles as possible in order to obtain a high electrical operating frequency. The winding, however, cannot have more than 36 poles because of the high number of slots (108). The working frequency is therefore equal to 45 Hz at the speed of 150 rpm. This phenomenon is a major structural drawback for conventional low speed designs.

This generator is sold as a kit, only the active parts are supplied. The stator comprises the armature, magnetic circuit and windings, inside an aluminium tube; the rotor is made up of a steel tube to which are attached the magnets. The dimensions are shown in Table 3.

Dimensions	Values
External diameter (mm)	500
Airgap diameter (mm)	400
Stator length (mm)	175
Rotor length (mm)	110
Internal diameter of the rotor (mm)	350
Mass, rotor and stator (kg)	73

Table 3. Principal dimensions of a conventional generator

The electrical parameters in Table 2 are given for operation with the armature giving directly onto a resistive load. The power factor is then unitary but the current is slightly out of phase in relation to the electromotive force (dephasing of an angle $\Psi \approx 20°$).

Operating with an active rectifier as mentioned in paragraph 2 allows minimisation of the armature current for any given power due to the phasing of the current with the electromotive force. This ideal mode of operation will give us a reference efficiency of 10 kW at 150 rpm. With sinusoidal current, the electromotive force being sinusoidal, torque ripple is negligible; only cogging torque remains, measured at 8 Nm by the manufacturer. Simulated study of this operating mode is pointless, given the simplicity of the waveforms produced through the use of this rectifier.

With the data in Table 2 we get the following characteristics:

Characteristics	Values
Output power (kW)	10
EMF, E (V)	160
Armature RMS current (A)	20,8
Joule losses (W)	1300
Iron and mechanical losses (W)	200
Efficiency (%)	87
Torque ripple (%)	1

Table 4. Operating in $\cos\Psi = 1$ (active rectifier)

The slight gain in efficiency obtained here, as we have already mentioned, is at the cost of an increase in complexity of the conversion system.

We are now going to study the consequences of operating with a strictly passive rectifier, which we recommend for this kind of application.

Digital simulation of the above gives the following waveforms for armature current (set against the electromotive force) and the electromagnetic power.

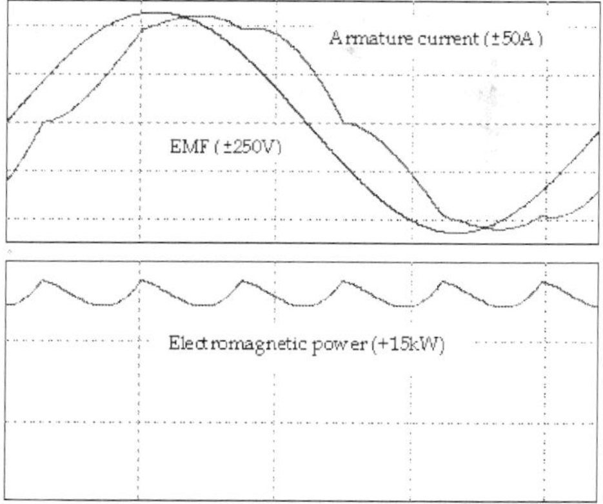

Fig. 6. Waveforms with passive rectifier

Table 5 shows the results obtained.

The deterioration of the current form engendered by the rectifier has two important consequences. In the first place, the appearance of harmonics and the dephasing of the current with the EMF leads to a significant increase in the RMS value of the current for any given power, the output passes from 87% with an active rectifier to 76% with a diode rectifier. There is significant derating of the generator. Secondly, the current harmonics increase significantly the rate of torque ripple which goes from a value of almost zero to

Characteristics	Values
Output power (kW)	10
EMF, E (V)	160
Armature RMS current (A)	31,5
Joule losses (W)	2976
Iron and mechanical losses (W)	200
Efficiency (%)	76
Torque ripple (%)	13

Table 5. Operating with a diode rectifier

close to 13%. This phenomenon is far from being insignificant: it causes operating noise, one of the main disadvantages of wind turbines.

Torque ripple produces a resonant frequency that is audible and unpleasant, often close to the natural frequency of the structure. In our example, this frequency is equal to six times the first harmonic frequency, i.e. 270 Hz.

Medium power wind turbines are often situated near to residential areas so this torque ripple problem can be very disturbing. An effective way of remedying the problem would be to augment the parameter L_s of the generator, but this would reduce efficiency even further. Therefore, the type of generator presented does not work well with a passive rectifier.

The result presented is obtained with a voltage source load, E_b. This is possible thanks to the synchronous inductance of the generator which smooths out the output current. Further filtering, through the inductance L_1, is often added, if only to limit ripple current load when E_b is an accumulator, or to smooth out the output voltage in the case of a load on a capacitive bus.

The waveforms obtained with an inductance L_1 of 20mH are shown in Fig. 7.

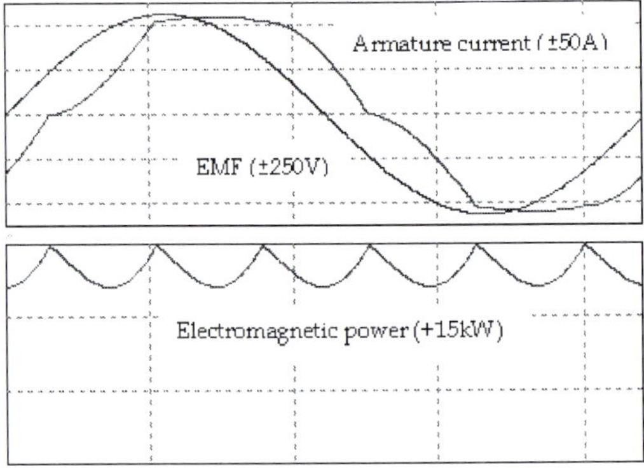

Fig. 7. Waveforms with a passive rectifier and filter choke

The introduction of a filter choke L_1, has the notorious consequence of increasing the rate of electromechanical power ripple, which is precisely what we wish to reduce, as shown in the results in Table 6.

From here on, we will no longer factor in the inductance L_1 which is detrimental to operation.

Characteristics	Values
Output power (kW)	10
EMF, E (V)	160
Armature current (RMS) (A)	32,7
Joule losses (W)	3207
Iron and mechanical losses (W)	200
Efficiency (%)	75
Torque ripple (%)	21

Table 6. Operating with diode rectifier and filter choke

To recap, two intrinsic characteristics of the conventional structure of a generator limit performance in a passive rectifier configuration: the operating frequency, which is difficult to increase because of the way the armature is designed, and ohmic loss which remains high for this application.

5. Operation of a Vernier permanent magnet synchronous generator

The Vernier synchronous generator, using magnets, is an interesting and viable alternative to the last configuration. Despite being the subject of numerous studies (Toba & Lipo, 2000; Matt & Enrici, 2005; Matt et al, 2007) it is less well known. It is represented in Fig. 8.

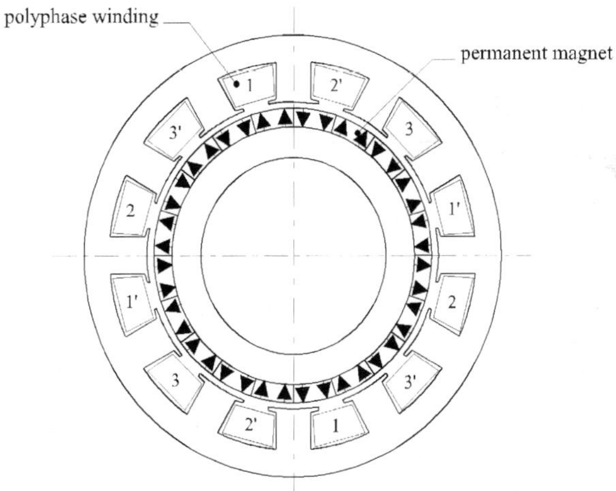

Fig. 8. Vernier permanent magnet synchronous generator

The design of the Vernier generator is similar to the example previously studied (Fig 5), but with the Vernier machine the component used in the magnetic field results from the coupling between the field system magnets and variation of reluctance due to the armature slots. Operation is modified as follows.

The armature of the Vernier machine has exactly the same structure as a conventional machine, the polyphase winding with p pole pairs spread out through the N_s slots which are wide open.

The structure of the field system is also similar to that of a conventional machine, but the number of pairs of magnets, N_r, along the rotor rim is not at all related to the number of pole-pairs: it can be much larger. This is what makes the Vernier generator unique. The working electrical frequency of the machine is now uniquely linked to N_r, as seen in the following formula:

$$\Omega_r.T_e = \frac{2\pi}{N_r} \Rightarrow f_e = \frac{1}{T_e} = \frac{N_r.\Omega_r}{2\pi} \tag{2}$$

This frequency can be high, even though the number of pole-pairs may be small. The limitations on frequency increase at low rotation speeds are generally lower than for the preceding configuration.

In closing this descriptive summary, we can summarise the coupling relations between the magnetic fields, expressed in terms of the only spatial variable, θ, which is the azimuthal coordinate in the airgap.

The N_s slots airgap permeance has a periodicity equal to $2\pi/N_s$. The airgap magnetomotive force created by the magnets, having a remanent flux density equal to M, has a periodicity equal to $2\pi/N_r$. Consequently, the field component, b_{1an}, created by the magnets, and having the periodicity $2\pi/|Ns-Nr|$, comes into play. This can be expressed thus:

$$b_{1an} = k_1.M.\cos((N_s-N_r).\theta) \tag{3}$$

The coefficient k_1, which defines the field amplitude $b_{1an}(\theta)$, is deduced using the finite elements method (of an elementary domain) (Matt, & Enrici, 2005), as in Fig. 9. It's value, which depends on the ratios of the dimensional proportion parameters, is generally between 0,1 et 0,2. This is not the most precise of methods since the slot pitch is slightly different from the magnet pitch, but in most cases it is sufficiently accurate.

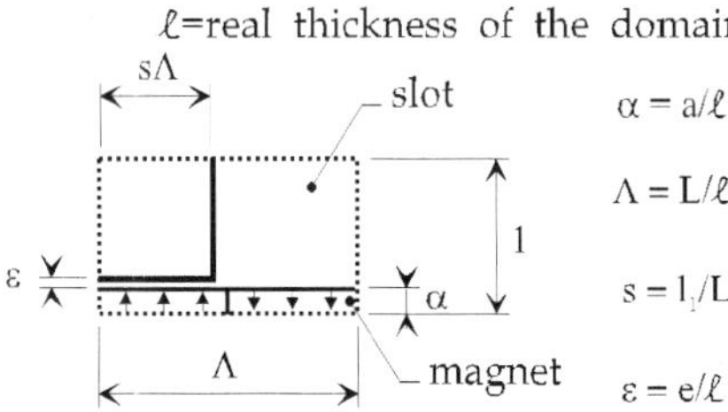

Fig. 9. Magnet-slot interaction in the elementary domain

The flux density field component, b_{1an}, is the main component used in the Vernier machine.

The armature currents can be likened to a thin coating (Matt et al, 2011) of current, conventionally called electric loading, λ_1, with a periodicity equal to $2\pi/p$, which we will express thus:

$$\lambda_1 = A_1.\cos(p\theta) \tag{4}$$

The amplitude, A_1, of the electric loading, is obtained by looking at the ratio of the amount of current at the core of the slot and that at the slot pitch, taking into account the winding factor.

The magnetic field, b_{1an}, and the magnetomotive force created by the electric loading, λ_1, combine to generate electromagnetic torque, C_{em}, provided that the spatial periodicity of b_{1an} and λ_1 are identical. The following formula must therefore be verified:

$$|Ns\text{-}Nr| = p \tag{5}$$

Once this first condition is met, and the functions (3) and (4) are in phase, the electromagnetic torque will be maximised.

The formula (4) shows that the number of pole-pairs, p, is dissociated from the number of magnet pairs, N_r, since the choice of number of slots is relatively large.

For the electromechanical conversion to take place, it is necessary to verify a second condition: the rotation speeds of b_{1an} and the magnetomotive force produced by λ_1 must be identical, which, taking into account the spatial periodicity of the two functions, leads to the following expression:

$$\frac{N_r.\Omega_r}{|N_s - N_r|} = \frac{N_r.\Omega_r}{p} = \frac{\omega}{p} = \Omega_c \tag{6}$$

The expression (6) obtained being identical to the expression (2), the main criteria for optimal operation are met.

The formula (7) above demonstrates the ratio of the field speed, Ω_c, to the rotor speed, Ω_r:

$$\frac{\Omega_c}{\Omega_r} = K_v = \frac{N_r}{p} \tag{7}$$

The coefficient K_v, the speed ratio, is called the Vernier Ratio, and is characteristic of the eponymous machine.

We shall conclude this explanatory part by expressing the electromagnetic torque, C_{em}, of the Vernier machine, which refers to the principal elements that we have just cited:

$$C_{em} = K_v.R^2.L. \int_{2\pi} b_{1an}.\lambda_1.d\theta \tag{8}$$

The dimensions R and L of the expression (8) represent the airgap radius and the iron length respectively. This expression can be misleading: the coefficient K_v seems to be a torque-multiplying coefficient if we refer to traditional expressions, whereas here, this coefficient compensates the low flux density b_{1an} (see (3)).

In practice, direct comparison of the performance levels of the Vernier configuration and that of a conventional configuration is delicate (Matt & Enrici, 2005). We simply note here that increasing the operating frequency, which the design of the Vernier machine allows, gives a gain of 50 to 100% in the mass-power ratio at very low speed, at the price of a substantial increase in rotor manufacturing costs.

The Fig. 10 shows an example of industrial production for a small electric car with a Vernier engine.

Unfortunately, at present, there is no commercialised Vernier generator specific to the field of wind turbines. We will therefore base our comparison on a theoretically scaled model. The sizing calculations are not within the scope of the summary that we are presenting and will not be detailed, but the references given in the prior explanations cover the main elements.

For this theoretical sizing, we will use a maximum number of the characteristics of the preceding generator in order to ensure the most precise comparison, notably when discussing the thermic aspects, which are always difficult to comprehend in a scaled model. The size is similar (even the external dimensions), and the configuration of the armature winding will be identical (same number of slots, same number of poles).

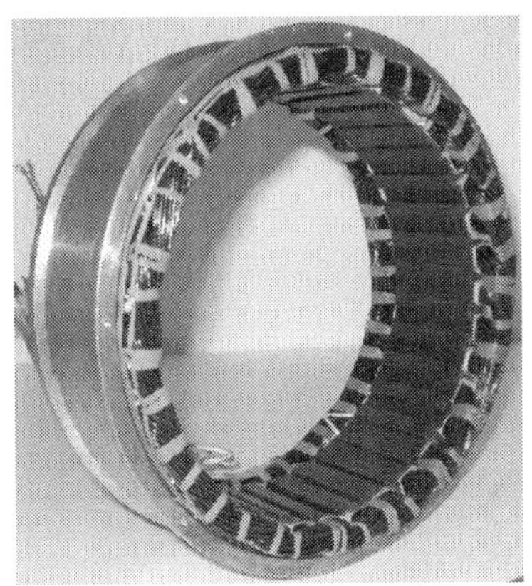

Fig. 10. Vernier engine for an electric vehicle (photography ERNEO)

Unfortunately, at present, there is no commercialised Vernier generator specific to the field of wind turbines. We will therefore base our comparison on a theoretically scaled model. The sizing calculations are not within the scope of the summary that we are presenting and will not be detailed, but the references given in the prior explanations cover the main elements.

For this theoretical sizing, we will use a maximum number of the characteristics of the preceding generator in order to ensure the most precise comparison, notably when discussing the thermic aspects, which are always difficult to comprehend in a scaled model. The size is similar (even the external dimensions), and the configuration of the armature winding will be identical (same number of slots, same number of poles).

The following table presents the principal characteristics obtained with a Vernier generator operating in association with an active rectifier.

The main dimensions of this generator are shown in Table 8.

The advantages of this configuration in the context of wind turbines, which impose an operating point with high torque for a very low speed of rotation, are shown in the folowing two tables.

Given that the mode of interaction between the magnets and slots results in a continuous and gradual shift of the magnets relative to the slots, which increases as the numbers N_s and N_r increase, the Vernier structure is a machine of naturally sinusoidal electromotive force with almost zero cogging torque and no slot tilt.

We observe that the high operating frequency at low speed, 225 Hz instead of 45 Hz, leads to a mass-power ratio of more than two times that obtained previously. We go from 140 W/kg to around 380 W/kg (without the housing) taking into account only the weight of the active parts.

Finally, the efficiency of the sized Vernier machine is comparable to a conventional machine, in the same operating conditions, with three times more iron loss, but with half the Joule

Part 3

Wind Turbine Control and System Integration

13

Advanced Control of Wind Turbines

Abdellatif Khamlichi, Brahim Ayyat
Mohammed Bezzazi and Carlos Vivas
University Abdelmalek Essaâdi
Morocco

1. Introduction

Wind energy technology has experienced huge progress during the last decade. This was encouraged by the need to develop ambient friendly clean and renewable forms of energy and the continuously rise of oil price. Sophisticated designs of wind turbines were performed. Large-size wind turbine farms are nowadays producing electricity at great scale throughout the world. Wind energy represents actually the most growing renewable energy; the rate of growth reaches actually 30% in Europe.

The cost of wind energy was not always cheaper than that of the other energy resources if the impact on environment and the risk linked to the classical forms of energy is not considered. The cost has however experienced a regular drop since the early 1970s. Cost reduction continues to constitute a main concern in the field of wind energy and research and development programs are considering it as a top priority. The objective is to extract optimal electric energy from wind with high quality specifications and with reduced installation and servicing expenses (Ackermann & Söder, 2002; Gardner et al., 2003; Sahin, 2004; AWEA, 2005).

Among the most important issues that allow to deal with cost reduction and its stabilisation in the field of wind turbines, one finds controller design for these installations. The objective is to deserve better use of the available energy in wind by providing through intelligent control its optimal extraction. Three main goals are generally pursued during designing of wind turbine controllers. The first one is to optimize use of the wind turbine capacity (optimal extraction of electric energy from the kinetic energy contained in the incident wind). The second is to alleviate mechanical loads in order to increase life of wind turbine components (fatigue loads should be reduced during operation). The third one is to improve power quality to approach the habitual performances met in the classical forms of energy, this is to assess compatibility of wind energy with the common standards about consumption of electricity (Ackermann, 2005).

Control must take into account variability of wind resource and should also cope with the intermittent nature of wind energy. The idea is to exploit optimally the wind resource when it is available and to limit overloading at risky high wind speeds. For this raison modern wind turbines are variable speed. They function by seeking optimal orientation of the wind turbine rotor and by pitching the blades to limit the captured energy from wind when wind speed exceeds the cut-off limit.

Knowing that if the electric generator is directly connected to the grid, then only one rotational speed can be used in order to synchronize with the grid frequency, modern wind turbines have incorporated electronic converters as an interface between the generator and the grid. This enables to decouple the rotational speed of the electric generator from that of the grid. The electric generator speed can in this way be varied in order to track the optimum tip-speed-ratio which is function of the instantaneous wind speed. Many configurations of active controllers were developed for this purpose (Burton et al., 2001; Hansen et al., 2005; Thiringer & Petersson, 2005; AWEA, 2005). But in practice, pitch-controlled wind turbines are the most performant ones, especially in the mid to high power range. Flexibility is the main advantage of pitching wind turbine blades. This arrangement has enabled to deal with the various concerns intervening in control of wind turbines and was recognized to recompense for the investments cost needed during the research and development operations or those associated to their realization and servicing.

Classical controls have used gain scheduling techniques. These consist in selecting a set of operating points and designing linear controllers for each linearized system near a given operating point. A set of linear time-invariant plants is then considered. The gain-scheduled controller is constructed from this family of linear controllers by operating a switching strategy by means of interpolations (Rugh & Shamma, 2000).

Gain scheduling techniques using roughly interpolation suffer however from lack to guarantee stability and robustness. For this raison, control based on linear parameter varying systems was introduced (Shamma & Athans, 1991; Leith & Leithead, 1996; Ekelund, 1997). This control consists first in describing the wind turbine dynamics by reformulating the nonlinear system as a linear system whose dynamics depend on a vector of time-varying exogenous parameters, the scheduling parameters. The advantage is that the controller design can be achieved through solution of a convex optimisation problem with linear matrix inequalities (Packard, 1994; Becker & Packard, 1994; Apkarian & Gahinet, 1995; Apkarian & Adams, 1998). The existence of efficient numerical methods that enable to solve this optimisation problem has enabled designing high performant controllers based on linear parameter varying and gain scheduling techniques.

Robustness appears to be a main feature in the controller design of wind turbine systems since they are so complex and work in the presence of many uncertainties affecting system parameters and inputs. For instance, the system is elastic in reality and vibrates according to complex patterns. The aerodynamic forces generated by the wind passing through the rotor plane are highly nonlinear. Wind speed varies stochastically and can not be measured through the whole rotor plane to use this information in control. These nonlinearities lead to huge variations in the dynamics of the wind turbine through the whole operating range of useful wind speeds.

To deal with control purposes, a simplified dynamic model for the wind turbine is usually considered. To represent reasonably wind turbine behavior, this model is obtained through an identification process. Identification can be performed conventionally without specifying the order of the model and without making assumptions regarding the wind turbine dynamics. Another more enhanced identification procedure relies on lumped representation of the mechanical system. This last is assumed to be a multi-body system consisting of rigid bodies linked together by flexible joints. The components of the model are adjusted by identification so as the parameters match as close as possible the real dynamic behaviour. In both approaches, the model is subject to parameter uncertainties and lack in general to be valid at high frequencies.

A large number of wind turbine control systems have been developed without taking into account modelling errors in the design process. Few contributions were dedicated to this crucial features of controllers and robust gain scheduling techniques existing nowadays are far from being robust and optimal (Bongers et al., 1993; Bianchi et al., 2004; Bianchi et al., 2005).

In order to optimize conversion efficiency of kinetic energy contained in wind to electric power, advanced strategies of control were introduced without using wind speed measurement. This was performed at first in the context of linearized wind turbine models (Boukhezzar et al., 2006; Boukhezzar et al., 2007). More advanced controllers were presented later (Vivas et al., 2008; Khamlichi et al., 2008; Khamlichi et al., 2009; Bezzazi et al., 2010). While not using wind speed measurement, these last are based on nonlinear observers that are built by using the extended Kalman filter. They were found to provide reliable information about wind speed, enabling to design control in continuous time without the need to make linearization of the system dynamics. These observers were implemented in well known controllers such as aerodynamic feed forward torque control (Vihriälä et al., 2001) and indirect speed control (Leithead & Connor, 2000).

A study regarding performance evaluation of the extended Kalman filter based controllers has been carried out in the below-rated power zone. This was performed through comparison between these controllers and some of the classical ones, chosen as reference. Comparison has focused on the accuracy of tracking the optimal rotor speed, the aerodynamic capture efficiency, control signal characteristics and the generated mechanical forces. The obtained results have shown that the advanced controls are quite pertinent. They are robust; they yield satisfactory results and give better enhancement of power conversion efficiency.

2. Control strategy statement of wind turbines

The details of the control systems used in wind turbines may vary largely from one installation to another, but they all have common elements that are considered in any controller design. This will be illustrated in the following through using a simple wind turbine model which permits to display the intervening turbine components and to review the ordinary basic functional elements that are used to build the controllers.

A wind turbine can be typically modelled, in first approximation, as a rigid mass-less shaft linked to rotor inertia at one side and to the drive train inertia at the other side, figure 1. The captured aerodynamic torque acts on the rotor and the generator electrical torque acts on the drive train.

The aerodynamic torque results from the local action of wind on blades. It is given by the sum of all elementary contributions related to the local wind speed that apply to a given element of a blade and which depend on the rotor speed, the actual blade pitch, the yaw error, the drag error, and any other motion due to elasticity of the wind turbine structure. Except from wind speed and aeroelastic effects, each of the other contribution inputs to aerodynamic torque (rotor speed, pitch, yaw and drag) may be monitored by specific control systems.

All wind turbines are equipped with yaw drives that monitor yaw error and with supplementary devices that are used to modify rotor drag. In the particular case of variable speed wind turbines, these installations can operate at different speeds or equivalently variable tip-speed ratios. Pitch-regulated wind turbines are controlled by modifying the blade orientation with respect to the direction of incident wind.

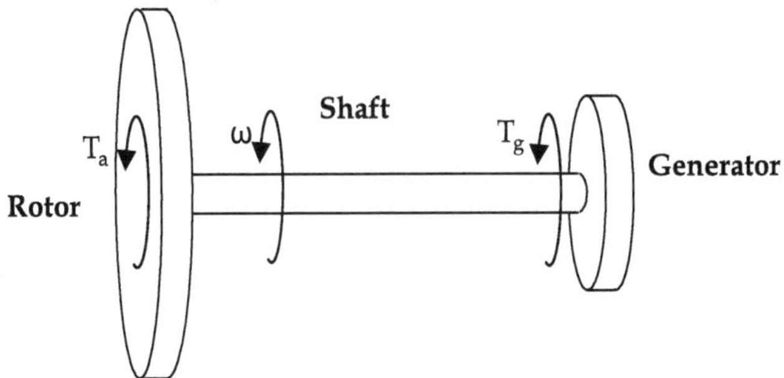

Fig. 1. Simplified model of a wind turbine

Neglecting elastic and aeroelastic effects, dynamics of the wind turbine rotor can be described by a one degree-of-freedom rigid body model (Wilkie et al., 1990) as

$$J\dot{\omega} + D\omega = T_a - T_g \tag{1}$$

where ω is the rotor speed, J the equivalent inertia of power train, D the equivalent damping coefficient, T_g the applied generator torque as seen from the rotor and T_a the aerodynamic torque.

Denoting $C_p(\lambda,\beta)$ the wind turbine power coefficient which is function of the pitch angle β and the tip-speed ratio λ, the aerodynamic torque acting on the rotor writes

$$T_a = \frac{1}{2}\rho\pi R^3 v^2 \frac{C_p(\lambda,\beta)}{\lambda} \tag{2}$$

where ρ is the air density, R the rotor radius and v the effective wind speed.

The tip-speed ratio is defined as $\lambda = \omega R / v$. Because of the speed multiplication resulting from the gear box, the high-speed shaft rotates with the rate $\omega_g = n\omega$, where n is the gear box multiplication factor.

It should be noted here that the effective wind speed appearing in equation (2) is not the average wind speed that acts at large on the rotor plane, but some hypothetical wind speed that have to be identified. This can be performed for instance if one fixes the pitch angle and the rotor speed, then measures the aerodynamic torque and solves after that the nonlinear equation (2) to compute the effective wind speed v. Using a reference wind speed in equation (2) instead of the unknown effective wind speed will result in wind speed error and consequently aerodynamic torque error. These two errors are however not perfectly correlated since in case of the aerodynamic torque, air density and rotor blades aerodynamic coefficients may also vary as function of the ambient conditions or because of wear affecting the blades.

Surface defining $C_p(\lambda,\beta)$ depends on the geometric configuration of the wind turbine blades and the aerofoils composing them. This surface admits a unique maximum denoted $C_{p,opt}$ which is obtained for $\beta = \beta_{opt}$ and $\lambda = \lambda_{opt}$. As the extracted power is given by $P_a = \rho\pi R^2 v^3 C_p(\lambda,\beta) / 2$, energy extraction from the kinetic energy of wind is optimal for $C_p(\lambda,\beta) = C_{p,opt}$.

Control strategy is usually defined by indicating the desired variations of wind turbine velocity and torque in the (ω, T_a) plane. Among the common strategies used in practice one finds that one depicted in figure 2.

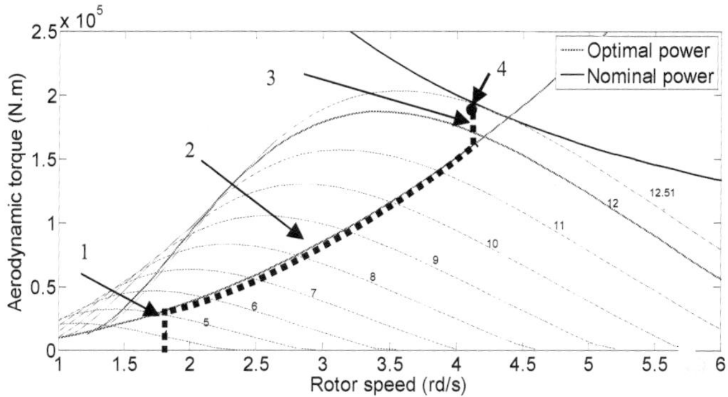

Fig. 2. Strategy of control illustrated in the (ω, T_a) plane as the dashed curve

In the (ω, T_a) plane, the red curves are obtained for different wind speeds by imposing the constant pitch angle $\beta = \beta_{opt}$.

Zone 1 corresponds to the segment between the starting wind speed v_{start} and the optimum lower wind speed $v_{min,opt} = \omega_{min} R / \lambda_{opt}$. In zone 1 the rotor speed is maintained constant at the value ω_{min}. This zone serves to reach at constant pitch angle $\beta = \beta_{opt}$ and constant rotor speed ω_{min} the operating point located on the maximum efficiency curve, blue curve. In zone 1, the tip-speed ratio varies from $\lambda = \omega_{min} R / v_{start}$ to λ_{opt}. The starting aerodynamic torque and the starting extracted power are given respectively as $T_{start} = \rho \pi R^2 v_{start}^3 C_p (\omega_{min} R / v_{start}, \beta_{opt}) / (2\omega_{min})$ and $p_{start} = T_{start} \omega_{min}$. The last point situated in the branch associated to zone 1 and located on the maximum efficiency curve has the following coordinates $\left(\omega_{min}, T_{min,opt} = \rho \pi R^5 C_{p,opt} \omega_{min}^2 / (2\lambda_{opt}^3)\right)$ in the (ω, T_a) plane. The extracted power varies in this zone in the interval $\left[p_{start}, p_{min,opt} \rho \pi R^5 C_{p,opt} \omega_{min}^3 / (2\lambda_{opt}^3)\right]$.

Zone 2 corresponds to tracking the maximum efficiency curve where the objective is to adjust the rotor speed to wind speed such that the captured aerodynamic torque is always optimal, the pitch angle as well as the tip-speed ratio are kept constant at their optimal values β_{opt} and λ_{opt}. For a given wind speed v, the optimal rotor speed is defined by $\omega_{opt} = \lambda_{opt} v / R$. The maximum rotor speed is fixed at the value ω_{max} which is slightly below the rated rotor speed corresponding to the intersection between the rated (nominal) power curve, black curve, and the optimum efficiency curve, blue curve. The rated rotor speed is given by

$$\omega_{rated} = \lambda_{opt} \left(\frac{2 p_{rated}}{\rho \pi R^5 C_{p,opt}} \right)^{\frac{1}{3}} \tag{3}$$

where p_{rated} is the rated generator power.

The maximum wind speed corresponding to zone 2 is given by $v_{max,opt} = \omega_{max} R / \lambda_{opt}$. The last point located on the maximum efficiency curve for zone 2 has the following coordinates

$\left(\omega_{\max}, T_{\max,opt} = \rho\pi R^5 C_{p,opt}\omega_{\max}^2 / (2\lambda_{opt}^3)\right)$ in the (ω, T_a) plane. The extracted power varies in

the interval $\left[p_{\min,opt} = \rho\pi R^5 C_{p,opt}\omega_{\min}^3 / (2\lambda_{opt}^3), p_{\max,opt} = \rho\pi R^5 C_{p,opt}\omega_{\max}^3 / (2\lambda_{opt}^3)\right]$.

Zone 3 constitutes a transition phase between zone 2 and the rated power zone, zone 4 corresponding to the circle point on the black curve. In zone 3 rotor speed is maintained constant at the value $\omega_{\max}$. The wind speed varies in this zone form $v_{\max,opt}$ to $v_{\max,rated}$ which is obtained as solution of the following nonlinear equation

$$v^3 C_p(\omega_{\max}R / v, \beta_{opt}) = \frac{2p_{rated}}{\rho\pi R^2} \tag{4}$$

The maximum aerodynamic torque in zone 3 is $T_{rated} = p_{rated} / \omega_{\max}$. The extracted power varies in this zone in the interval $\left[p_{\max,opt} = \rho\pi R^5 C_{p,opt}\omega_{\max}^3 / (2\lambda_{opt}^3), p_{rated}\right]$.

The first three zones are termed below-rated power region as the extracted power is always smaller than the rated power p_{rated}.
Finally, zone 4 corresponds to the maximum load zone (above-rated power region). In this zone pitch angle is permanently adjusted in order to reduce the captured aerodynamic torque, assuring continuous rated power generation. Giving a wind speed $v \geq v_{\max,rated}$, the pitch angle in zone 4 is obtained as solution of the following nonlinear equation

$$v^3 C_p(\omega_{\max}R / v, \beta) = \frac{2p_{rated}}{\rho\pi R^2} \tag{5}$$

The extracted power is constant in zone 4 and is equal to p_{rated}.
Table 1 recalls the wind speed limits corresponding to each zone and the associated extracted power limits.

Zone	Minimum wind speed limit	Maximum wind speed limit	Minimum extracted power	Maximum extracted power	Pitch angle
1	v_{start}	$v_{\min,opt}$	p_{start}	$p_{\min,opt}$	fixe
2	$v_{\min,opt}$	$v_{\max,opt}$	$p_{\min,opt}$	$p_{\max,opt}$	fixe
3	$v_{\max,opt}$	$v_{\max,rated}$	$p_{\max,opt}$	p_{rated}	fixe
4	$v_{\max,rated}$	v_{end}	p_{rated}	p_{rated}	variable

Table 1. Limits of the different control zones in terms of wind speed and extracted power

For control purposes power can be used as control variable in the transition zones 1 and 3 as well as in the above-rated power zone 4. In these zones rotor speed is maintained constant, while in addition for zone 4 pitch is controlled to maintain the power constant.
In the below-rated zone 2, the rotor speed is varied as function of the actual wind speed to optimize permanently power extraction. The optimal power that can be extracted varies then as function of the wind speed. For a given wind speed the pursued reference in terms of extracted power, aerodynamic torque and rotor speed writes

$$p_{opt} = \frac{1}{2}\rho\pi R^2 C_{p,opt} v^3$$

$$T_{opt} = \frac{1}{2}\rho\pi R^3 v^2 \frac{C_{p,opt}}{\lambda_{opt}} \tag{6}$$

$$\omega_{opt} = \lambda_{opt} v / R$$

The optimal extracted power can be used as reference for control, but because it is proportional to v^3, error on the effective wind speed will have an important effect on the reference power to be tracked and hence on control efficiency. This error is proportional to v^2. The aerodynamic torque can not be used as control input because this quantity is not easy to measure in practice. So, the control variable that is usually used is the rotor speed. Control should track the reference $\omega_{opt} = \lambda_{opt} v / R$ by changing the generator torque in order to change the rotor speed. The effective wind speed v which can not be measured may be estimated by using power and rotor speed measurements, p_{mes} and ω_{mes}, and solving the following non linear equation

$$C_p\left(\frac{R\omega_{mes}}{v}, \beta_{opt}\right)v^3 - \frac{2p_{mes}}{\rho\pi R^2} = 0 \tag{7}$$

Using equation (7) to extract the estimated effective wind speed is however numerically costly. Moreover, the process of solving this equation is not robust because of noise that could affect the measurements p_{mes} and ω_{mes}, and the delays that are inherent to any measurement system and which yield always late information on the actual wind speed. Variations could result also from ambient conditions such as for example air density ρ which is temperature dependent or the power coefficient which is sensitive to gusts, wear and debris impacting the blades. Since these perturbations affecting the ideal model are not straightforward to take into account, control operates inefficiently and loss of extracted power occurs systematically. In order to emphasize the requisite for developing intelligent controls that can handle more effectively wind turbine system uncertainties, a review is performed in the next section about the standard methods of control that have so far been proposed for these installations.

3. Review of classical controllers for wind turbines

3.1 Standard proportional integral control

Since it is simple to design and easy to implement, the classical proportional integral (PI) control is widely used in industry applications. This controller which requires little feedback information can be employed over most plants for which a dynamical model can be derived. PI controller can be used alone or in conjunction with other control and modelling techniques such as linearization or gain scheduling.

For fixed pitch wind turbines operating in the below-rated power zone 2, capture of maximum energy that is available in the wind can be achieved if the turbine rotor operates such that the tip-speed ratio is made equal to the optimal value λ_{opt}. This regime can be obtained by tracking the optimal rotor speed. Useful details about PI controlling of wind turbines are given in (Bossanyi, 2000; Muljadi et al., 2000; Burton et al., 2001). One can find three kinds of control loops for tracking the optimal rotor speed. In all of them the target rotor speed is given by $\omega_{opt} = \lambda_{opt} v / R$ where the wind speed v is assumed to be known from measurements and the generator torque is synthesized as

$$T_g = K\omega^2 \tag{8}$$

with

$$K = \frac{1}{2}\rho\pi R^5 \frac{C_{p,opt}}{\lambda_{opt}^3} \tag{9}$$

Based on the rotor speed measurement and a generator torque control loop, a control loop on λ is built. The PI controller zeroes the difference between the target and the measured rotor speed and imposes the generator torque reference. This control is known as the Indirect Speed Control (ISC). One can expect large torque variations, as the torque demand varies rapidly in this configuration.

Based only on the rotor speed feedback, a torque control loop can be built using as reference T_g given by equation (8) (Pierce, 1999). A variant of this control, known as the Aerodynamic Torque Feed forward (ATF) where the aerodynamic torque and the rotor speed are estimated using a Kalman filter was presented in (Vihriälä et al., 2001). An advantage of this control structure is the increased mechanical compliance of the system, but rotor speed variations result in general to be high.

An active power loop can also be built using once more the measured rotor speed, in conjunction with the captured power and the inner torque control loop. The target power is defined by the first equation in (6). By zeroing the power error, the operating point is driven to move to the maximum power point (Burton et al., 2001). The drawback of this control is sensitivity of the reference to error measurement of wind speed.

To show how the first variant of PI control using the λ loop can be derived, let us notice that equations (1), (2), (8) and (9) yield the following ordinary differential equation

$$\left(\dot{v} + \frac{D}{J}v\right)\lambda + v\dot{\lambda} = \frac{\rho\pi R^4}{2J}\left(\frac{C_p(\lambda,\beta)}{\lambda^3} - \frac{C_{p,opt}}{\lambda_{opt}^3}\right)\lambda^2 v^2 \tag{10}$$

To apprehend how this standard control works, let us assume that v is constant. Since in reality damping is small so that $(D/J)\lambda \ll 1$ is satisfied, the sign of $\dot{\lambda}$ depends on the sign of the difference in the right hand side of (10). Taking into account that $C_p(\lambda,\beta) < C_{p,opt}$, it follows from (10) that $\dot{\lambda} < 0$ when $\lambda > \lambda_{opt}$, and the rotor decelerates towards λ_{opt}. When $\lambda < \lambda_{opt}$, $\dot{\lambda} > 0$ happens and the rotor accelerates towards λ_{opt}. Thus, the control defined by equations (8), (9) and (10) causes the rotor speed, for a well definite wind turbine, to approach the optimal tip speed ratio enabling to track always the optimal extraction power curve. This control is easier to understand under constant wind conditions, but this behavior occurs only in an averaged sense under time-varying wind conditions.

It was assumed so far that the wind speed is measured and that this value is equal to the effective wind speed appearing in equation (2). It was assumed also that turbine properties used to calculate the gain K in equation (9) are accurate. These conditions are rarely met in practice, because air density varies as function of temperature and, over time, debris build up and blade erosion change the $C_p(\lambda,\beta)$ surface and thus $C_{p,opt}$. Consequently an inaccurate gain K is used in control. Knowing that a small error on the tip-speed ratio of about 5%, causes a significant energy loss of about 2% (Johnson et al., 2006), the potential improvement for energy capture and cost savings has motivated investigation on adaptive control approaches that could enhance control efficiency.

3.2 Adaptive control

The dynamic behavior of a wind turbine is highly dependent on wind speed due to the non-linear relationship between wind speed, turbine torque and pitch angle. System parameter variations can be tackled by designing a controller for minimum sensitivity to changes in these parameters. Adaptive control schemes continuously measure the value of system parameters and then change the control system dynamics in order to make sure that the desired performance criteria are always met. In (Simoes et al., 1997; Bhowmi et al., 1999; Song et al., 2000) use is made of adaptive control to compensate for unknown and time-varying parameters in the below-rated power zone 2. In (Johnson et al., 2006) an adaptive control scheme is developed for this region. The authors have addressed also the question of theoretical stability of the torque controller, showing that the rotor speed is asymptotically stable under the adaptive generator torque control law in the constant wind speed input case and L2 stable with respect to time-varying wind input. Furthermore, the authors have derived a method for selecting the adaptive gain to ensure convergence to its optimal value.
In the context of gain-scheduling, adaptive control structures which allow different control goals to be formulated; depending on the considered operating points were also introduced. To decide which controller to apply to the plant varies from simply switching between the controllers associated to the various operating points to quite sophisticated interpolation strategies (Shamma, 1996). In addition to its complexity, the adaptive structure has a major drawback in that the rotor characteristics should be known quite accurately.

3.3 Search algorithms control

Based on measurement of the captured power, heuristic search algorithms have been proposed to perform control of wind turbines. These algorithms change constantly the rotor speed in the objective to maximize the captured power. If a reduction of rotor speed yields a decreased power then the controller would slightly increase the speed. In this manner the rotor speed could be kept permanently near the maximum power coefficient as the wind speed changes. This kind of controller does not need a wind turbine model to be implemented; it is thus insensitive to changes in operation conditions due to dirty blades, mispitched blades or ambient temperature variations.
This approach supposes that the wind turbine reacts sufficiently fast to variation of the low frequency components of wind speed; this happens in practice in case of low power wind turbines. For ensuring the optimal energy extraction, it is thus sufficient to feed the electrical generator with the torque control value corresponding to the steady state operating point placed on the optimal power extraction curve. To this end, an on-off-controller-based structure can be used in order to zeroing the difference on λ. The choice of such a structure (Munteanu et al., 2006) is characterized by its robustness to the parametric uncertainties inherent to any wind turbine.

The control law associated to the on-off-controller-based structure provides a steady state torque reference by adding two components. The first component is an equivalent control, corresponding to the optimal operating point, and depends proportionally on the low frequency wind speed squared. The second component is a high-frequency component, which switches between two values, depending on the sign of the actual error on λ. The first component is intended to drive the system to the optimal operating point, whereas the second one has the role of stabilising the system behaviour around this point.
Search algorithms based controllers do not consider mechanical loads as an objective. Wind turbines are hence vulnerable to induced torque variations which result from the alternate control input and could increase dramatically fatigue loads.

3.4 Optimal control

Optimal control theory formulates the control problem in terms of a performance index. This last is a function of the error between the references and actual system responses. Sophisticated mathematical techniques are then used to solve the constrained optimisation problem in order to determine the optimal values of design parameters. Optimal control algorithms often need a measurement of the various system state variables. If this is not possible a state estimator based on a plant model is used.

The need to consider a performance index in control of wind turbines comes from the fact that tracking the optimal rotor speed induces variations of the generator torque, thus additional cyclic mechanical stresses which reduce the lifetime of the drive train parts. Furthermore, the turbine energetic efficiency is also decreased because of the supplementary maintenance costs and reduced turbine availability time.

The above mentioned control methods have as an exclusive goal that consists in the maximization of the energy efficiency, while ignoring the possible drawbacks related to large control input efforts that could be needed. In (Ekelund, 1997), it is stated that keeping λ_{opt} in turbulent winds is possible only with large generator torque variations, thus significantly high mechanical stress. Therefore, the supplementary mechanical fatigue of the drive train should be reduced by imposing the minimization of the generator torque variations ΔT_g used as control input, around the optimal operating point. In (Ekelund, 1997) this has been expressed through a performance index consisting of the following combined optimization criterion

$$ I = E\left\{\alpha\left(\lambda - \lambda_{opt}\right)^2 + \left(\Delta T_g\right)^2\right\} \tag{11} $$

where E is the statistical expectation symbol.
The first term in equation (11) illustrates the energy efficiency maximisation, whereas the second term expresses the minimisation of the torque control variations. The trade-off between the two terms is adjusted by means of the weighting coefficient α.
Operation around optimality is ensured by minimizing only the first term in the right hand side of equation (11), but allowing important torque variations that are represented by the last term of this equation. The positive coefficient α confers flexibility to the control law in the following sense. If the wind turbulence is low, then the energy efficiency of the wind turbine will be considered as a priority and therefore α may take a large value. If the wind

turbulence is important, then through a small value of α focus will be done on reducing the mechanical stress and increasing the life service of the wind turbine components.

The mechanical loads alleviation is an issue in all the wind turbine operating modes. At high winds, when the system works in full load, an optimal linear quadratic control with Gaussian noise LQG was given in (Boukhezzar et al., 2007). A flexible drive train wind turbine was considered and the equations were linearized around the above-rated power operating point. The optimal controller was designed to determine the pitch variation $\Delta\beta$ such that to minimize the performance index

$$I = E\left\{ x^t Q x + \alpha (\Delta\beta)^2 \right\} \tag{12}$$

where x is the state vector deviation from desired reference values and Q the penalty matrix.

To derive an optimal controller, linearization of equations is needed around an operating point. The solution proposed to deal with the dependency of the linearized dynamical state system on the average wind speed was a gain-scheduling adaptive structure, together with an observer for state reconstruction, which uses the rotor speed ω as measurable output.

In (Munteanu et al., 2005) the performance index to be minimized is stochastically defined. An optimal control structure was presented in order to optimize the criterion at equation (11) without using adaptive structures. This approach, named the frequency separation principle, relies upon using a low-pass filter to separate the turbulence high frequency and the low frequency wind speed components (Nichita et al., 2002). Since the two components excite the plant dynamics in two distinct spectral ranges, the proposed structure is formed by two loops that are respectively driven by the low-frequency and the turbulence component of wind speed. While this structure desensitizes the closed loop system subject to the steady-state operating point, control cost to deal with the high-frequency components was recognized to be elevated.

4. Advanced control

In order to track the optimal rotor speed, it is required that reliable measurement of the effective wind speed v is available. But, this is impossible in reality because the effective wind speed is a rather averaged value that could only be identified by solution of equation (2) if the torque T_a was measured by means of a torquemeter for example. Therefore, it is smarter designing controls that do not require wind speed measurement. This is performed in the following through using the extended Kalman filter to derive a wind speed observer.

The control problem that states to track optimal power extraction in the below-rated power zone 2 takes the form of single-input single-output system. For a given effective wind speed v, it consists in synthesising the generator torque T_g to adjust the rotor speed ω to its optimal value ω_{opt}. The option of choosing T_g as the control variable is justified by the fact that this last is quite easy to generate through swift and effective electric machine controllers using either vector control (Pena et al., 1996) or direct torque control (Arnalte et al., 2002).

To test performance of the new introduced controllers, the well-known reference controllers ISC (Leithead & Connor, 2000; Wright, 2003) and ATF (Pena et al., 1996; Vihriälä et al., 2001; Johnson, 2004) are considered. ATF and ISC are far from being optimal and lack robustness (Boukhezzar et al., 2006; Boukhezzar et al., 2007). To overcome these deficiencies, feed back nonlinear controllers using Kalman based wind speed observer were introduced by these authors. This was performed in two steps: estimation of the aerodynamic torque and deduction of the wind speed from it by solving the nonlinear equation (2) where a Newton-Raphson like iterative scheme was used. Based on this technique of observing wind speed, a Nonlinear Static State Feedback control with Estimator (NSSFE) and a Nonlinear Dynamic State Feedback control with Estimator (NDSFE) were derived (Boukhezzar et al., 2006; Boukhezzar et al., 2007). These controllers were shown to be robust in comparison with the classic ISC and ATF. However, the major difficulty resulting from estimating wind speed by using the aerodynamic torque value is linked to the necessity to perform at each time inversion of the nonlinear equation (2). This is numerically costly. On the other hand Kalman filter requires a local linearization of system equations. Considering multiple linearizations performed over the whole below-rated zone 2 results in a conservative condition on stability which limits control performance.

As a principal contribution in this work, a direct estimation of wind speed by means of the nonlinear extended Kalman filter is proposed without using aerodynamic torque estimation. The idea consists in using a time dependant Riccati like equation to construct a robust continuous observer of the effective wind speed which is capable of rejecting system perturbations and disturbances that are acting on the generator torque T_g (Gelb, 1984; Reif et al., 1999).

Assuming that the dynamics of wind speed v is driven by a Gaussian white noise such that $\dot{v} = \xi$ and the noise affecting measurement of the rotor speed ω is ζ, denoting the state vector $x = \begin{bmatrix} \omega & v \end{bmatrix}^t$, the observable variable $y = \omega + \zeta$ and the input variable $u = T_g$, the state system equations write

$$\dot{x} = \begin{bmatrix} \dot{\omega} \\ \dot{v} \end{bmatrix} = \underbrace{\begin{bmatrix} -\dfrac{D}{J}\omega + \dfrac{1}{J}T_a(\omega, v) \\ 0 \end{bmatrix}}_{f(x,u)} + \begin{bmatrix} -\dfrac{1}{J} \\ 0 \end{bmatrix} T_g + \begin{bmatrix} 0 \\ \xi \end{bmatrix} \tag{13}$$

$$y = \begin{bmatrix} 1 & 0 \end{bmatrix} \begin{bmatrix} \omega \\ v \end{bmatrix} + \zeta = Cx + \zeta \tag{14}$$

For a dynamic system governed by equations (13) and (14) an exponential observer which is based on the extended Kalman filter can be constructed (Reif et al., 1999) as

$$\dot{\hat{x}} = f(\hat{x}, u) + K(t)(y - C\hat{x}) \tag{15}$$

with $K(t) = P(t)C^t S^{-1}$ and $\dot{P}(t) = A(t)P(t) + P(t)A^t(t) - P(t)\left(C(t)S^{-1}C - c_1^2 I\right)P(t) + c_2^2 I$ where I is the identity matrix having dimension of x, S the symmetric definite positive matrix defining noise correlations and having also dimension of x, $c_1 > 0$, $c_2 > 0$ are the weights and

$$A(t) = \frac{\partial f}{\partial x}(\hat{x}, u) = \begin{bmatrix} -\dfrac{D}{J}\omega + \dfrac{1}{J}\dfrac{\partial T_a}{\partial \omega} & \dfrac{1}{J}\dfrac{\partial T_a}{\partial v} \\ 0 & 0 \end{bmatrix} \tag{16}$$

To find the weighting constants c_1 and c_2, the following bisection algorithm is proposed:

-     - step 1: first guess of $c_1 > 0$, $c_2 > 0$ and $S = sI$ with $s > 0$;
-     - step 2: decreasing s by replacing it by $s/2$ until matrix P is positive definite;
-     - step 3: if it is impossible to find a positive definite matrix P, then $c_2 > 0$ is decreased to its half and $S = sI$ is reinitialized and steps 1 and 2 are repeated.
- The iterations are stopped if $c_2 > 0$ is lesser than a given critical value. In the present case $c_1 = 0.9$, $c_2 = 0.087$ and $s = 0.06$ were found to be the best weights. System of equations (1-2) and (13-16) is solved numerically by using Matlab command ode45.

Based on the extended Kalman filter (EKF) two kinds of robust controllers are built. The first one is a combination between (ISC) and (EKF) where the wind speed and the rotor speed are observed. The second one derives from a combination between (ATF) and (EKF).

To characterize performances of the various controllers, an efficiency ratio is defined over the period of time $\left[t_{ini}, t_{fin} \right]$ as the actual produced electric energy divided by the ideal optimal energy that could be extracted from the kinetic energy contained in the incident wind. This efficiency ratio writes

$$\eta = \frac{\int_{t_{ini}}^{t_{fin}} P_{extracted}\, dt}{\int_{t_{ini}}^{t_{fin}} P_{optimal}\, dt} = \frac{\int_{t_{ini}}^{t_{fin}} T_g \omega_g\, dt}{\int_{t_{ini}}^{t_{fin}} T_{a,opt} \omega_{opt}\, dt} \tag{17}$$

where $P_{extracted} = T_g \omega_g$ is the real electric power extracted at time t and $P_{optimal} = T_{a,opt} \omega_{opt}$ the theoretical optimal power that could be extracted.

To examine performance of the controllers, the experimental wind turbine called CART (Fingershand & Johnson, 2002) is considered. This two-bladed wind turbine has largely been studied in the literature related to control purposes. The low-speed shaft is coupled by means of a gear box to the high-speed shaft. This last constitutes the rotor of the induction electric generator. The principal characteristics of CART wind turbine are: $R = 21.38\,m$, $n = 43.165$, hub height $H = 36.6\,m$. Here the CART wind turbine is assumed to have the rated power $P_{rated} = 850\,kW$.

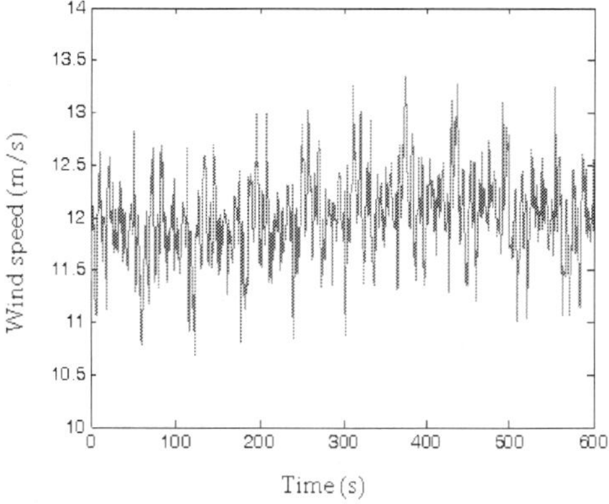

Fig. 2. Instantaneous wind speed

To illustrate results due to the introduced new controllers, a stochastic wind speed, $v(t)$, having a mean value of $12\,m.s^{-1}$, turbulence characteristic length $L = 15\,m$ and standard deviation $\sigma = 0.5$ is assumed to generate the source power of wind during the time interval of simulation $[0,600s]$, figure 2.

Reference controllers ISC, ATF, and Kalman filter based robust controllers NSSFE and NDSFE are considered in this comparative study. Their performances are evaluated through simulations carried out under Matlab environment where use was made in particular of Kalman and ode45 commands. Newton-Raphson iterations were also implemented when necessary. The obtained results are compared to the new proposed extended Kalman filter based robust controllers ISC_EKF and ATF_EKF with the same conditions of functioning regarding disturbances and measurement noise. Three kinds of results will be focused on: the generator torque which is related to the energetic cost of control, the energetic efficiency of control defined in terms of ratio η, and the torque acting on the low-speed shaft which is related to fatigue loads.

Table 2 summarizes the obtained results in terms of generator torque variations over the time interval $[0,600s]$. It gives the maximum value, the mean, value, the minimum value and the relative variation. It gives also the extracted power efficiency associated to each controller.

Table 3 gives the obtained results in terms of the mechanical torque T_{ls} variations in the low-speed shaft over the time interval $[0,600s]$. It gives the maximum value, the mean, value, the minimum value and the relative variation.

From table 2, controllers could be classified into two categories: high fluctuated generator torque (ATF, NSSFE, NDSFE) and low fluctuated generator torque (ISC, ISC_EKF, ATF-EKF).

	Max (T_g) ($10^4\ N.m$)	Mean (T_g) ($10^4\ N.m$)	Min (T_g) ($10^4\ N.m$)	Relative variation %	η (%)
ATF	16.71	13.87	11.13	40.2	80.37
ISC	15.96	13.87	11.96	28.8	80.11
NSSFE	17.12	13.87	10.49	47.8	78.91
NDSFE	17.11	13.87	10.45	48.0	78.76
ATF_EKF	15.56	14.72	13.75	12.3	81.59
ISC_EKF	16.41	14.27	12.25	29.1	80.82

Table 2. Generator torque variations and extracted power efficiency for the different controls

From table 3, it is seen that three classes of controllers are distinguishable: high fluctuated mechanical torque (NSSFE, NDSFE), intermediate fluctuated mechanical torque (ATF, ISC) and low fluctuated mechanical torque (ISC_EKF, ATF-EKF).

The obtained results show that for NSSFE and NDSFE controllers generator torque and low-speed shaft overloads are maximum while power efficiency is minimum. Although some performances of the controllers ISC and ATF are good in comparison with the previous controllers, they must be discarded as they are not robust in real operation conditions. Now comparing only the robust controllers between them shows that the proposed extended

	Max (T_{ls}) (10^5 N.m)	Mean (T_{ls}) (10^5 N.m)	Min (T_{ls}) (10^5 N.m)	Relative variation %
ATF	1.9759	1.6777	1.3900	2.9822
ISC	1.9012	1.6776	1.4666	2.2368
NSSFE	2.000	1.6774	1.3566	3.2318
NDSFE	1.9999	1.6773	1.3543	3.2251
ATF_EKF	1.8550	1.7506	1.6351	1.0433
ISC_EKF	1.8449	1.7129	1.5764	1.3201

Table 3. Mechanical torque variations in the low speed shaft for the different controls

Kalman filter based methods perform better than NSSFE and NDSFE. Comparing finally the two new proposed controllers shows that the ATF_EKF is better than ISC_EKF in terms of power efficiency, generator torque fluctuations and fatigue loads.

3. Conclusion

A review of the principal control methods that are used in the context of wind turbines was performed. Using a one-degree-of freedom mechanical model for wind turbine, focus was done on various controllers in the below-rated power zone and their performances were analyzed. These include two classic controllers that are largely used in practice, the indirect speed control and the aerodynamic torque feed forward control. Two other recent controllers using Kalman filter to make wind speed estimates were also considered. Finally two new non linear controllers were introduced. These last combine judiciously the classic controllers and an exponential continuous wind speed observer based on the extended Kalman filter.

The obtained results are promising. They have shown that the new controllers perform better than all the others since they robustify the indirect speed control and the aerodynamic torque feed forward control. They limit also to large extent generator torque fluctuations and fatigue loads, as compared with Kalman filter based methods, while maximising the extracted electric power energy.

4. Acknowledgment

Part of this research was thankfully supported by financial aid accorded to the authors by the Spanish Agency of International Cooperation AECI under project grants A/7313/06 and A/9518/07 and by French CNRS under project grant SPI03/09.

5. References

Ackermann, T. (2005). *Wind Power in Power Systems*, John Wiley & Sons Ltd, Chichester, UK.

Ackermann, T. & Söder, L. (2002). An overview of wind energy-status 2002, *Renewable and Sustainable Energy Reviews* 6(1-2), 67–127.

Apkarian, P. & Adams, R. (1998). Advanced gain-scheduling techniques for uncertain systems, *IEEE Transactions on Control Systems Technology* 6(1), 21–32.

Apkarian, P. & Gahinet, P. (1995). A convex characterization of gain-scheduled H∞ controllers, *IEEE Transactions on Automatic Control* 40(5), 853–864.

Arnalte, S., Burgos J.C. & Rodriguez-Amendeo, J.L. (2002). Direct torque control of a doubly-fed induction generator for variable speed wind turbines, *Electric Power Components and Systems*, 30(2), pp. 199-216.

AWEA (2005). *Wind energy fact sheets. Economics and cost of wind energy,* Technical report, American Wind Energy Association (AWEA). http://www.awea.org/pubs.

AWEA (2005). *Electrical guide to utility scale wind turbines,* Technical report, American Wind Energy Association (AWEA). http://www.awea.org/pubs.

Becker, G. & Packard, A. (1994). Robust performance of linear parametrically varying systems using parametrically-dependent linear feed-back, *Systems and Control Letters* 23(3), 205–215.

Bezzazi, M., Khamlichi, A., Ayyat B., Oulad Ben Zarouala, R. & El Bakkali, L., Vivas, V.C. & De Saxcé, G. (September 2010). Advanced control for wind turbine by using high performance wind speed observer, *World Renewable Energy Congress XI*, WREC2010, Abu Dhabi, UAE.

Bhowmik, S., Spée, R. & Enslin J. (1999). Performance optimization for doubly-fed wind power generation systems, *IEEE Trans. Ind. Applicat.*, vol. 35, no. 4, pp. 949–958.

Bianchi, F., Mantz, R. & Christiansen, C. (2004). Control of variable-speed wind turbines by LPV gain scheduling, *Wind Energy* 7(1), 1–8.

Bianchi, F., Mantz, R. & Christiansen, C. (2005). Gain scheduling control of variable-speed wind energy conversion systems using quasi-LPV models, *Control Engineering Practice* 13(2), 247–255.

Bongers, P., van Baars, G. & Dijkstra, S. (1993). Load reduction in a wind energy conversion system using an H∞ controller, *Proceedings of the 2nd Conference on Control Applications*, pp. 965–970. Vancouver, Canada.

Bossanyi, E.A. (2000). The design of closed loop controllers for wind turbines, *Wind Energy* 3:149-163.

Boukhezzar, B., Siguerdidjane, H. Hand, M. (2006). Nonlinear control of variable-speed wind turbines for generator torque limiting and power optimization, *ASME Transactions: Journal of Solar Energy Engineering*, 128, pp. 516-30.

Boukhezzar, B., Lupu, L., Siguerdidjane, H. & Hand, M. (2007). Multivariable control strategy for variable speed, variable pitch wind turbines, *Renewable Energy* 32:1273-1287.

Burton, T., Sharpe, D., Jenkins, N. & Bossanyi, E. (2001). *Wind Energy Handbook,* John Wiley & Sons, Ltd., Chichester, UK.

Ekelund, T. (1997). *Modeling and linear quadratic optimal control of wind turbines,* Ph.D. thesis, Chalmers University of Technology, Göteborg, Sweden.

Fingershand L.J. & Johnson K.E. (2002). *Controls Advanced Research Turbine (CART) Commissioning and Baseline Data Collection,* NREL/TP-500-32879. Golden, NREL, Colorado.

Gardner, P., Garrad, A., Jamieson, P., Snodin, H. & Tindal, A. (2003). *Wind energy, the facts,* Technical report, European Wind Energy Association (EWEA), Brussels, Belgium

Gelb A. (1984). *Applied optimal estimation.* Cambridge, MA: MIT Press.

Hansen, M., Hansen, A., Larsen, T., Øye, S., Sørensen, P. & Fuglsang, P. (2005). *Control design for a pitch-regulated, variable speed wind turbine,* Technical Report RISO-R-1500(EN), Risø National Laboratory, Roskilde, Denmark.

Johnson, K.E. (2004). *Adaptive torque control of variable speed wind turbines,* NREL Report No. NREL/TP-500-36265, National Renewable Energy Laboratory, Golden, CO.

Johnoson, K.E., Pao, L.Y., Balas M.J. & Fingersh L.J. (June 2006). Control of Variable Speed Wind Turbines, Standard and Adaptive Techniques for Maximizing Energy Capture, *IEEE Control Systems Magazine.*

Khamlichi, A., Ayyat, B., Bezzazi, M., El Bakkali, L., Vivas, V.C. & Castano, C.L.F. (2009). Modelling and control of flexible wind turbines without wind speed measurements, *Australian Journal of Basic and Applied Sciences,* 3(4), pp: 3246-3258.

Khamlichi, A., Ayyat, B., El Bakkali, L., Bezzazi, M., Vivas, C. & Castano, F. (July 2008). Modelling and control of flexible wind turbines, *Renewable Energy Congress X,* Glasgow, Scotland.

Leith, D. & Leithead, W. (1996). Appropriate realization of gain-scheduled controllers with application to wind turbine regulation, *International Journal of Control* 65(2), 223–248.

Leithead W.E. & Connor D. (2000). Control of variable speed wind turbines: Design task, *International Journal of Control,* 73(13), pp. 1173-1188.

Muljadi, E., Pierce, K. & Migliore, P. (2000). *A conservative control strategy for variable-speed stall-regulated wind turbines,* Technical Report NREL/CP-500-24791, National Renewable Energy Laboratory, Colorado, U.S.A.

Munteanu, I., Cutululis, N.A., Bratcu, A.I. & Ceang, E. (2005). Optimization of variable speed wind power systems based on a LQG approach, *Control Engineering Practice* 13(7):903-912.

Nichita, C., Luca, D., Dakyo, B. & Ceangă, E. (2002). Large band simulation of the wind speed for real time wind turbine simulators, *IEEE Transactions on Energy Conversion* 17(4):523-529.

Packard, A. (1994). Gain scheduled via linear fractional transformations, *Systems and Control Letters* 22(2), 79–92.

Pena R., Clare J.C. & Asher G.M. (1996). A doubly fed induction generator using back-to-back power converters supplying an isolated load from a variable speed wind turbine, *IEE Proc.-Electr. Power Appl,* 143(5), pp. 380-387.

Pierce, K. (1999). *Control method for improved energy capture below rated power,* Technical Report, National Renewable Energy Laboratory, Colorado, U.S.A.

Reif K., Sonnemann, F. & Unbehauen R. (1999). *Nonlinear State Observation Using H-infinity Filtering Riccati Design,* IEEE Transactions on Automatic Control, 44(1), pp. 203-208.

Rugh, W. & Shamma, J. (2000). Research on gain scheduling, *Automatica* 36(10), 1401–1425.

Sahin, A. (2004). Progress and recent trends in wind energy, *Progress in Energy and Combustion Science* 30, 501–543.

Shamma, J.S. (1996). Linearization and gain-scheduling, In: *Levine WS (ed.) The control handbook.* CRC Press, IEEE Press, pp 388-398.

Shamma, J. & Athans, M. (1991). Guaranteed properties of gain scheduled control for linear parameter-varying plants, *Automatica* 27(3), 559–564.

Simoes, M., Bose, B. & Spiegel, R. (1997). Fuzzy logic based intelligent con-rol of a variable speed cage machine wind generation system, *IEEE Trans. Power Electron.*, vol. 12, no. 1, pp. 87–95.

Song Y., Dhinakaran B. & Bao X. (2000). Variable speed control of wind turbines using nonlinear and adaptive algorithms, *J. Wind Eng. Ind. Aerodyn.*, vol. 85, no. 3, pp. 293–308.

Thiringer, T. & Petersson, A. (2005). *Control of a variable-speed pitch-regulated wind turbine*, Technical report, Chalmers University of Technology, Göteborg, Sweden.

Vihriälä, H., Perälä, R., Mäkilä, P. & Söderlund, L. (July 2001). A Gearless Wind Power Drive: Part 2: Performance of Control System, *Proceedings of the European Wind Energy Conference*, Copenhagen, Denmark.

Vivas, V.C., Castano, F., Khamlichi, A., Bezzazi, M. & Rubio F. (July 2008). *On the robust adaptive control of wind turbines without wind speed measurements*, Controlo'2008. 8th Portuguese Conference On Automatic Control. UTAD, Vila Real.

Wilkie, J., Leithead, W.E. & Anderson, C. (1990). Modelling of wind turbines by simple models, *Wind Engineering* 4:247-274.

Wright, A. (2003) *Modern Control Design for Flexible Wind Turbines*, Ph.D. thesis. Boulder, CO: University of Colorado, USA.

14

A Complete Control Scheme for Variable Speed Stall Regulated Wind Turbines

Dimitris Bourlis
University of Leicester
United Kingdom

1. Introduction

Wind turbine generators comprise the most efficient renewable energy source. Nowadays, in order to meet the increasing demand for electrical power produced by the wind, wind turbines with gradually increasing power rating are preferred.

The variable speed pitch regulated wind turbine is the most dominant wind turbine technology so far, since it achieves high aerodynamic efficiency for a wide range of wind speeds and at the same time good power control to meet the variable utility grid power requirements. In particular, the power control is performed by altering the pitch angle of the rotor blades and consequently the aerodynamic efficiency of the rotor, through closed loop control, in order to keep the power at the specified level.

Although the above technology has been proved to be quite effective, limitations and challenges appear in the construction of Mega Watt scale wind turbines where larger rotor diameters are required. Specifically, as the rotor diameter increases, the challenges and the cost associated with the pitch mechanism increase too, since this mechanism now has to cope with very large and heavy rotor blades. In addition, due to the increasing height of the tower and the associated increase of the cost, lighter constructions are preferred, which are also more flexible and entail lightly damped tower vibration modes. These vibration modes can be easily excited by the action of the pitch controller (Bossanyi, 2003). Consequently, the stable operation of the whole system poses additional challenges on the design of effective pitch controllers and actuators, while at the same time the cost should be kept as low as possible.

The variable speed stall regulated wind turbine comprises a technology that has several advantages over pitch regulated wind turbines and has been of particular interest in the literature (Biachi et al., 2007). In particular, this type of wind turbine uses a rotor of fixed blade angle and therefore has a simpler and more robust construction and can have lower requirements for maintenance than the existing pitch regulated wind turbines. Due to these features, these wind turbines can have reduced cost, which is a crucial parameter especially for large scale wind turbines. In addition, they can be more economically efficient for offshore applications, where the maintenance is a major consideration. However, this type of wind turbine is not yet commercially available due to existing challenges in its control. Specifically, a variable speed stall regulated wind turbine is not an unconditionally stable system and has a dynamic behaviour which depends on the operating conditions (Biachi et al., 2007). Due to this feature, the control and the consequent construction of variable speed stall regulated wind turbines has not been feasible so far, since more sophisticated control methods than the existing ones are required.

In this chapter a novel control system for variable speed stall regulated wind turbines is presented. The presentation starts with background issues in wind turbines and control, including a brief review of existing attempts to solve the aforementioned control problem and continues with the detailed description of the design and operation of the proposed system. Next, simulation results obtained using a hardware-in-loop simulator are presented and analyzed and useful conclusions are drawn. Finally, recommendations and future work are presented in the last section.

2. General background in the control of variable speed stall regulated wind turbines

The main components of a variable speed wind turbine are the turbine rotor, usually three bladed, the drivetrain, the generator and the power electronics. Fig. 1(a) gives a simple schematic of a wind turbine.

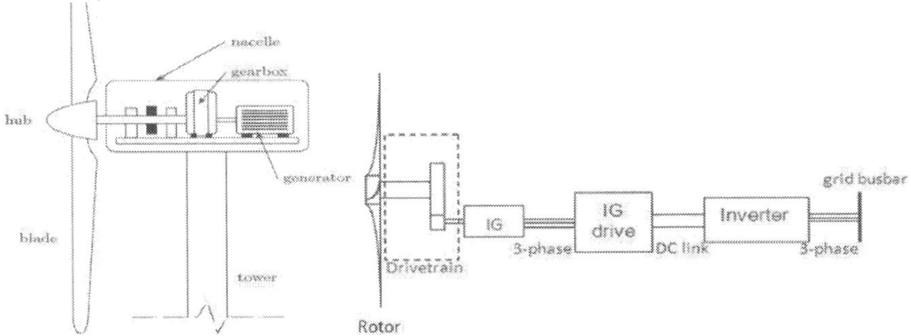

Fig. 1. (a) Wind turbine schematic mechanical (Biachi et al., 2007) and (b) with electrical subsystem simple schematics, (utilizing an Induction Generator-IG).

The rotor blades can either be rigidly mounted on the rotor hub at a fixed "pitch angle" or through a variable pitch mechanism, for power limitation purposes. The interaction of the rotor blades with the oncoming wind results in the development of an aerodynamic torque T_a which rotates the rotor. For the transmission of this torque from the rotor to the generator, either a direct coupling or a step-up gearbox may be used, depending on the type and the number of pole pairs of the generator (induction, synchronous, synchronous with permanent magnets). In the case of a gearbox, the drivetrain also contains a Low Speed Shaft and a High Speed Shaft, at the rotor and generator side respectively.

The power electronics of a variable speed wind turbine are comprised of a generator-side converter and a grid-side converter, both connected back-to-back via a DC link, as can be seen by the diagram in Fig. 1(b). The first converter, which can also work as a variable speed drive for the generator, acts as a rectifier, converting the variable frequency/variable amplitude AC voltage of the generator to DC voltage of variable level, while the second acts as an inverter, converting the DC voltage into AC of a frequency and amplitude, matching that of the grid.

2.1 Stall regulation

Stall regulation refers to the controlled intentional enforcement of the rotor blades to stall and it can be achieved at constant speed, constant torque or constant power (Connor & Leithead, 1994; Goodfellow et al., 1988; Leithead & Connor, 2000).

Fig. 2 gives a schematic of a wind turbine rotor blade element, which helps to understand how stall works. In the figure θ is the angle between the plane of rotation and the blade chord (pitch angle), where the chord is the line connecting the two ends of the blade. If the undisturbed wind velocity towards the blade is $\vec{V}_w$ and the blade tip velocity is $\vec{V}_b$, then the wind velocity seen by the rotating blade is $\vec{W}=\vec{V}_w - \vec{V}_b$, which creates with the blade chord an angle a, the "angle of attack". Due to the impinging wind $\vec{W}$, two forces are developed on the blade element, one perpendicular and one parallel to it, the Lift force, L and the Drag force D respectively.

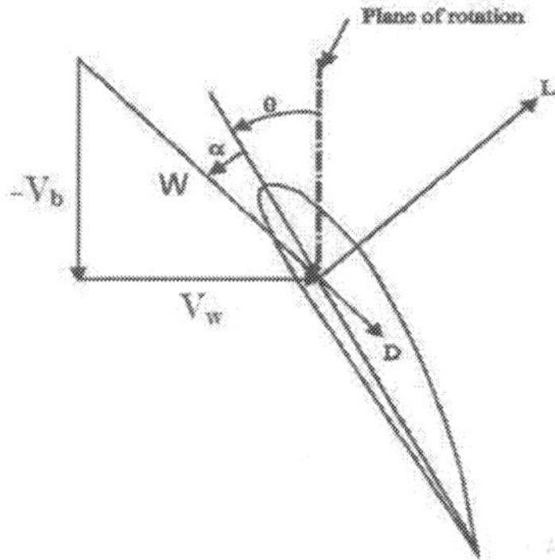

Fig. 2. Wind and rotor blade element velocities and forces acting on it (Kurtulmus et al., 2007).

In a wind turbine when the wind speed $|\vec{V}_w|$ increases relative to the blade tip speed $|\vec{V}_b|$, the angle a increases too, which results in an increase of L and consequently an increase of T_a. However, if the wind further increases and a exceeds a certain value, the air flow detaches from the upper side of the blade and turbulence is created. This results to a drop of the lift force and in turn to drops of T_a and the aerodynamic power P_a, while at the same time, the drag force increases.

In a variable speed stall regulated wind turbine the objective is to keep the power P equal to the rated P_N, so $P=P_N$ for every wind speed $V>V_N$, where V_N is the rated wind speed and this can be theoretically achieved by reducing the speed of the rotor via control of the reaction torque of the generator. This method comprises an implementation of stall regulation at constant power and it is still an open research area due to the nonlinear dynamics involved.

2.2 Aerodynamic torque

When the rotor of the wind turbine is subjected to an oncoming flow of wind, an aerodynamic torque T_a is developed as a result of the interaction between the wind and the rotor blades, which rotate with angular speed ω. Using simplified aerodynamics, an expression of the aerodynamic power P_a has been derived (Biachi et al., 2007; Manwell, 2002). This is given in Eqn. 1:

$$P_a = \frac{1}{2}\pi\rho R^2 C_p V^3 \tag{1}$$

where ρ is the air density, R the radius of the rotor, C_p is the power coefficient of the rotor and V the effective wind speed seen by the rotor, which is a result of a number of phenomena due to the interaction of the rotor and the oncoming wind (Biachi et al., 2007). In particular, V is a quantity used in the equations that attempts to represent the effect on the produced torque of a 2-dimensional wind field with a 1-dimensional quantity. That way, harmonic components of the aerodynamic torque caused by the rotational sampling of the rotor (due to the wind shear or the small spatial correlation of the wind turbulence as well as to the tower shadow) are assumed to be present in the V timeseries. From the above it is obvious that V is a non-measurable quantity, since an anemometer gives only a point wind speed far from the turbine rotor (Leithead & Connor, 2000).

C_p is defined as ratio of the power extracted from the wind to the power available in the wind (Manwell, 2002; Parker, 2000) and it is a measure of the aerodynamic efficiency of the rotor, which indicates the ability of the rotor to extract power from the wind. It is also a nonlinear function of the tip-speed ratio $\lambda = \omega R/V$ and the pitch angle θ and it is particular for each rotor, with its shape depending on the rotor blade profile. C_p has a theoretical maximum of $C_{p_{max}}$=0.593, known as the Betz limit, which indicates that the maximum ability to extract power from the wind is less than 60% (Biachi et al., 2007; Parker, 2000). In practice this value is lower, usually $C_{p_{max}}$=0.45. In general, for a wind turbine it is desirable to operate at $C_{p_{max}}$ for every V and so to have maximum aerodynamic efficiency for every V, unless the rated power of the wind turbine P_N is reached. Also, the torque coefficient of the rotor is defined as $C_q=C_p/\lambda$ and expresses the ability of the wind turbine rotor to produce torque. Typical C_p and C_q curves of a rotor with blades at a fixed pitch angle θ are given in Fig. 3.

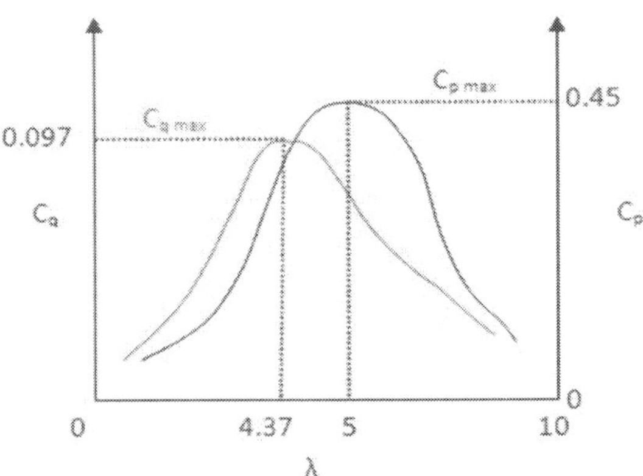

Fig. 3. Typical C_p (black) and C_q (red) curves of a stall regulated wind turbine.

In Fig. 3 it can be observed that the maximum of the torque coefficient ($C_{q_{max}}$) is obtained at a lower tip speed ratio than the maximum power point ($C_{p_{max}}$), which is the case in general. The value of λ that corresponds to $C_{p_{max}}$ is the optimum tip speed ratio, λ_o:

In order for the Newton-Raphson method to be applied, the C_p-λ characteristic of the rotor is analytically expressed using a polynomial. Fig. 9 shows the C_p curve of a Windharvester wind turbine rotor and its approximation by a 5th order polynomial.

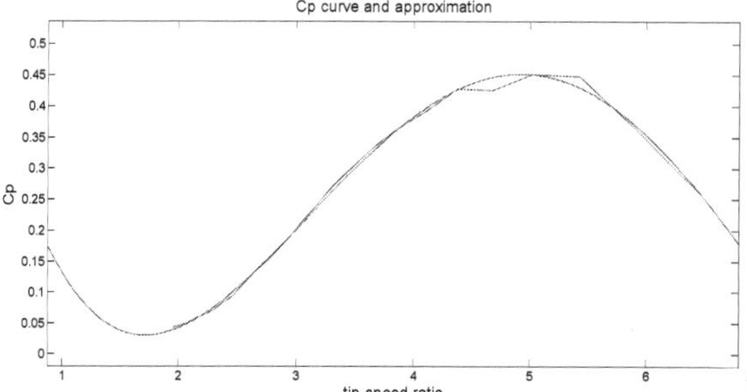

Fig. 9. Actual C_p curve (red) and approximation using a 5th order polynomial (blue).

Fig. 10 shows T_a versus V for a fixed value of ω, for a stall regulated wind turbine. As can be seen, T_a after exhibiting a peak, drops and then starts rising again towards higher wind speeds (Biachi et al., 2007). Fig. 10 also displays three possible V solutions V_1, V_2 and V_3 corresponding to an arbitrary aerodynamic torque level $T_a{=}T_{aM}$, given the fixed ω.

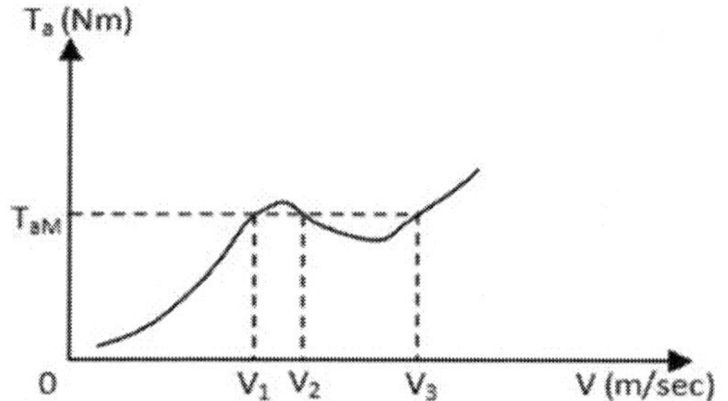

Fig. 10. T_a versus V for fixed ω.

Also, Fig. 11 shows a graph similar to that of Fig. 10 for $\omega{=}\omega_N$, where T_{ω_N} and P_N/ω_N are the aerodynamic torque levels corresponding to the points B and C of Fig. 5 respectively.

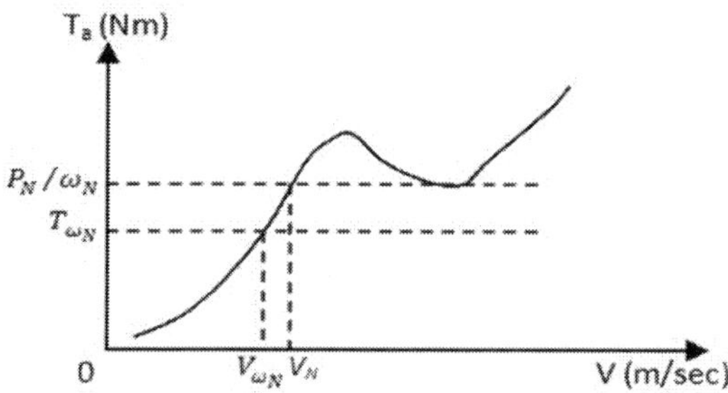

Fig. 11. T_a versus V for $\omega{=}\omega_N$.

For the part AB of Fig. 5, the optimum speed reference is: $\omega_{ref} = \frac{\lambda_0 V_1}{R}$, where V_1 is the lowest V solution seen in Fig. 10. Also, for the part BC the speed reference is: $\omega_{ref} = \omega_N$. In addition, from Fig. 11 it can be seen that for $\omega = \omega_N$ when $V_1 > V_{\omega_N}$, the aerodynamic torque is always $T_a > T_{\omega_N}$, so there is a monotonic relation between V_1 and T_a. Therefore, V_1 can be effectively used in order to switch between the parts AB and BC. So, ω_{ref} for the part ABC can be expressed as:

$$\omega_{ref} = \begin{cases} \dfrac{\lambda_0 V_1}{R}, & V_1 < V_{\omega_N}, \\ \omega_N, & V_1 > V_{\omega_N} \end{cases} \tag{18}$$

Regarding V_1, it can be easily obtained with a Newton-Raphson if this is initialized at an appropriate point, as seen in Fig. 12, where the expression $T_{a_M} - T_a$ versus V is shown.

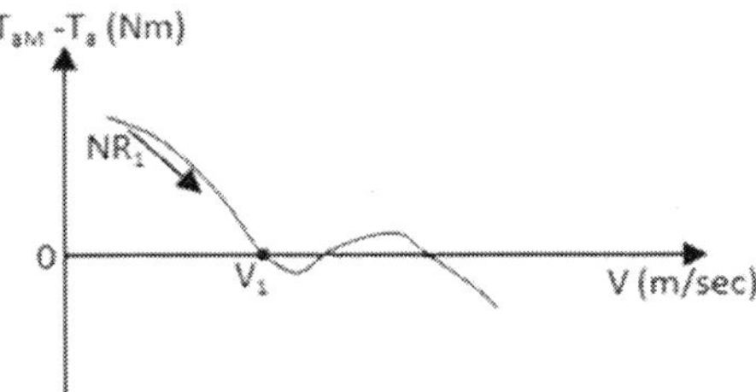

Fig. 12. Newton-Raphson routine NR$_1$ used for V solution extraction of Eqn. 3.

Fig. 13 shows the actual V and its estimate, $\hat{V}$ obtained in Simulink using the Newton-Raphson routine for the model of the aforementioned Windharvester wind turbine.

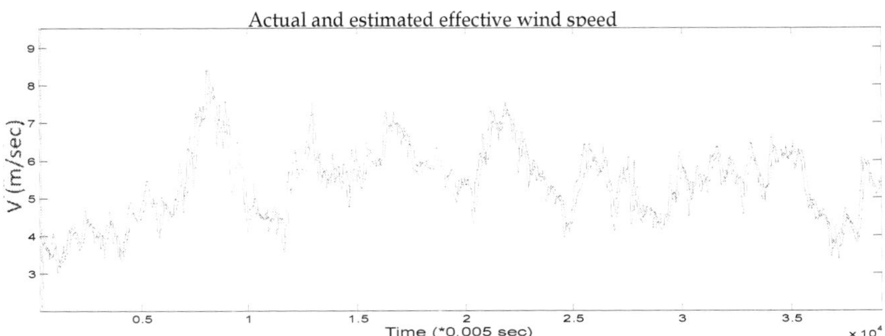

Fig. 13. Actual V (blue) and estimated $\hat{V}$ (red) using NR$_1$.

As can be seen, the wind speed estimation is very accurate. In the next section, the speed control design is described.

7. Gain scheduled proportional-integral speed controller

The speed controller should satisfy conflicting requirements, such as accurate speed reference tracking and effective disturbance rejection due to high frequency components of the aerodynamic torque, but at the same time should not induce high cyclical torque loads to the drivetrain, via excessive control action. In addition, the controller should limit the torque of the generator to its rated torque, T_N and also not impose motoring torque.

Although all the above objectives can be satisfied by a single PI controller, as shown in (Bourlis & Bleijs, 2010a), this cannot be the case in general, due to the highly nonlinear behavior of the wind turbine, due to the rotor aerodynamics. Specifically, the nonlinear dependence of T_a to ω through Eqn. 3, establishes a nonlinear feedback from ω to T_a and due to this feedback, the wind turbine is not unconditionally stable. The dynamics are stable for below rated operation, close to the C_{pmax}, where the slope of the C_q curve is negative (see Fig. 3) and therefore causes a negative feedback, but unstable for stall operation (operation on the left hand side of the C_q curve, where its slope is positive), (Biachi et al., 2007; Novak et al., 1995).

A single PI controller may marginally satisfy stability and performance requirements, but in general it cannot be used when high control performance is required. High performance requires very effective maximum power point tracking and at the same time very effective power regulation for above rated conditions and for Mega Watt scale wind turbines, which are now under demand, trading off between these two objectives is not acceptable, due to economic reasons.

Specifically, for below rated operation and until ω_N is reached, the speed reference for the controller follows the wind variations. For this operating region moderate values of the control bandwidth are required for acceptable reference tracking. Although tracking of higher frequency components of the wind would increase the energy yield, it would simultaneously increase the torque demand variations, which would induce higher cyclical loads to the drivetrain.

For constant speed operation (part BC in Fig. 5) the requirements are a bit different. At this region, the wind acts as a disturbance that tries to alter the fixed rotational speed of the wind turbine. Considering that at this region the aerodynamic torque increases considerably, before it reaches its peak (see Fig. 11), where stall starts occurring, the controller should be able to withstand to potential rotational speed increases, as this could lead to catastrophic wind up of the rotor. For this reason, at this operating region a higher control bandwidth is required.

Further, in the stall region, it is known from (Biachi et al., 2007) that the wind turbine has unstable dynamics, with Right Half Plane zeros and poles. Therefore, different bandwidth requirements exist for this region too.

A type of speed controller that can effectively overcome the above challenges, while at the same time is easy to implement and tune in actual systems, is the gain scheduled PI controller. This type of controller consists of several PI controllers, each one tuned for a particular part of the operating region. Depending on the operating conditions, the appropriate controller is selected each time by the system, satisfying that way the local performance requirements.

In order to avoid bumps of the torque demand that can occur during the switching from one controller to another, the controller is equipped with a bumpless transfer controller, which guarantees a smooth transition between them. The bumpless transfer controller in principle ensures that all the neighbouring controllers have exactly the same output with the active one, so no transient will happen during the transition. For this reason for every PI controller there is a bumpless transfer controller, which measures the difference of its output with the active one and drives it appropriately through its input. Fig. 14 shows a schematic of a gain scheduled controller consisting of two PI controllers.

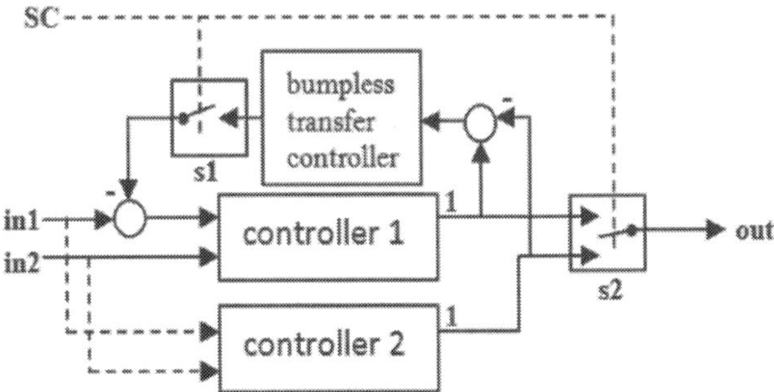

Fig. 14. Gain scheduled controller with bumpless transfer circuit.

As can be seen in Fig. 14, there is a Switch Command (SC) signal that selects the control output via switch "s2". The same signal is responsible for the activation of the bumpless transfer controller. Specifically, when "controller 2" is activated, the bumpless transfer controller for "controller 1" is activated too. The bumpless transfer controller receives as input the difference of the outputs of the two controllers and drives "controller 1" through one of its inputs such that this difference becomes zero. It is mentioned the same bumpless transfer controller exists for "controller 2", but if the dynamic characteristics of the two controllers are not very different, a single bumpless transfer controller can be used for both of them, when only two of them are used.

Regarding the PI controllers used, they have the proportional term applied only to the feedback signal, (known as I-P controller (Johnson & Moradi, 2005; Wilkie et al, 2002)). The I-P controller exhibits a reduced proportional kick and smoother control action under abrupt changes of the reference. The structure of this controller is shown in Fig. 15(a). In Fig. 15(b) the discrete time implementation of the controller with Matlab/Simulink blocks is shown. The implementation also includes a saturation block, which limits the output torque demand to the specified levels (generating demands up to T_N) and an anti-windup circuit, which stops the integrating action during saturation.

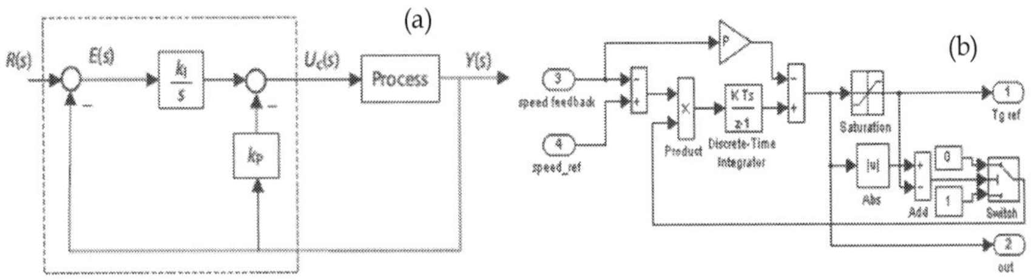

Fig. 15. (a) I-P controller diagram (Johnson & Moradi, 2005) and (b) Simulink implementation.

In the following section, a case design study for the Windharvester wind turbine is presented.

8. Case design study

The analysis that follows is based on data from a 25kW Windharvester constant speed stall regulated wind turbine that has been installed at the Rutherford Appleton Laboratory in Oxfordshire of England. The control system that has been described in the previous sections has been designed for this wind turbine and the complete system has been simulated in a hardware-in-loop wind turbine simulator.

8.1 Description and parameters of the Windharvester wind turbine

This wind turbine has a 3-bladed rotor and its drivetrain consists of a low speed shaft, a step-up gearbox and a high speed shaft. In fact, the gear arrangement consists of a fixed-ratio gearbox, followed by a belt drive. This was originally intended to accommodate different rotor speeds during the low wind and high wind seasons. The drivetrain can be seen in Fig. 16, where the belt drive is obvious. The generator is a 4-pole induction generator.

Fig. 16. Drivetrain of the Windharvester wind turbine.

The data for this wind turbine are given in Table 1.

Rotor inertia, I_1	14145 Kgm^2
Gearbox inertia, I_g	34.2 Kgm^2
Generator inertia, I_2	0.3897 Kgm^2
LSS stiffness, K_1	$3.36 \bullet 10^6$ Nm/rad
HSS stifness, K_2	$2.13 \bullet 10^3$ Nm/rad
Rotor radius, R	8.45 m
Gearbox ratio, N	1:39.16
LSS rated rotational frequency, ω_1	4.01 rad/sec

Table 1. Wind turbine data.

The C_p and C_q curves of the rotor of the wind turbines are shown in Fig. 17 (a) and (b) respectively (in blue). In addition, the data have been slightly modified in order to obtain the steeper C_p and C_q curves, shown in red colour. As mentioned before, the steeper C_p curve requires less speed reduction during stall regulation at constant power and therefore it can be preferred for a variable speed stall regulated wind turbine. However, such a C_p curve requires more accurate control in below rated operation. Thus, the modified curves are also used to assess the performance of the proposed control methods for below rated operation.

The maximum power coefficient C_{pmax}=0.45 is obtained for a tip speed ratio λ_{Cpmax}=5.02, while the maximum torque coefficient is C_{qmax}=0.098 for a tip speed ratio λ_{Cqmax}=4.37.

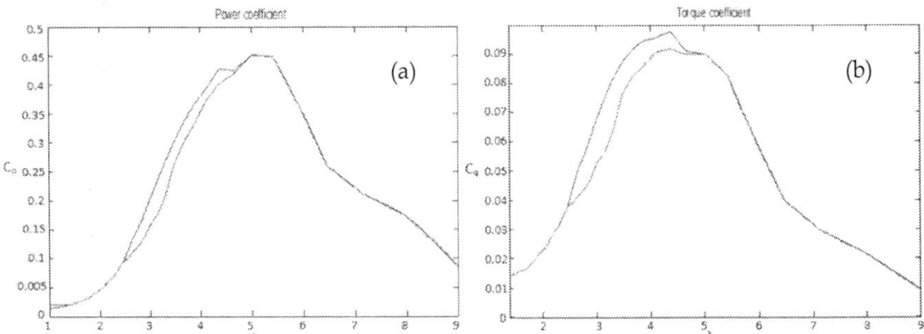

Fig. 17. (a) Power and (b) torque coefficient curve of the rotor of the Windharvester wind turbine.

8.2 Dynamic analysis of the wind turbine

The dynamics of the wind turbine are mainly represented by Eqns. 19-23 after the drivetrain has been modeled as a system with three masses and two stifnesses as shown in Fig. 18.

$$I_1\dot{\omega}_1 = \frac{1}{2}\rho\pi R^3 C_q(\omega_1, V)V^2 - \gamma_1\omega_1 - K_1\theta_1 \tag{19}$$

$$I_2\dot{\omega}_2 = K_2\theta_2 - \gamma_2\omega_2 - T_g \tag{20}$$

$$I_g\dot{\omega}_L = K_1\theta_1 - NK_2\theta_2 \tag{21}$$

$$\dot{\theta}_1 = \omega_1 - \omega_L \tag{22}$$

$$\dot{\theta}_2 = N\omega_L - \omega_2 \tag{23}$$

As can be seen, the dynamic model of Eqns. 19-23 is nonlinear with two inputs V and T_g (generator torque). Output of the model is the generator speed ω_2, which is the only speed measurement available in commercial wind turbines. In order for the model to be analyzed, the term C_q of Eqn. 19, shown in Fig. 17(b), is approximated with a polynomial and the whole model is linearized (Biachi et al., 2007). Then, the transfer functions from its inputs to

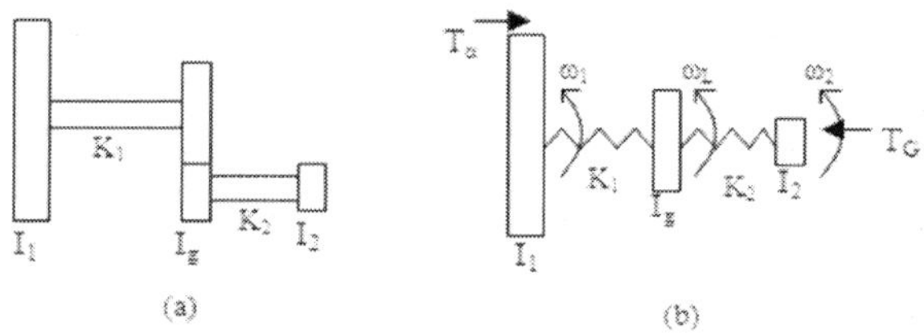

Fig. 18. Wind turbine drivetrain: (a) schematic, (b) dynamic model.

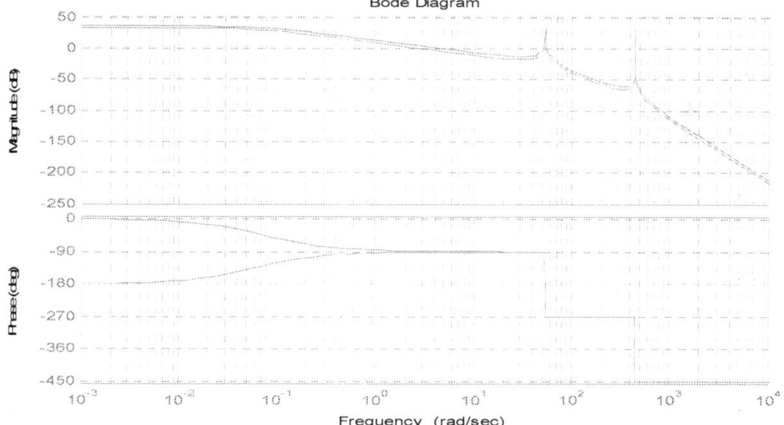

Fig. 19. Bode plots of $G_{V\omega_2}$ for below rated (blue) and above rated (stall) operation (red).

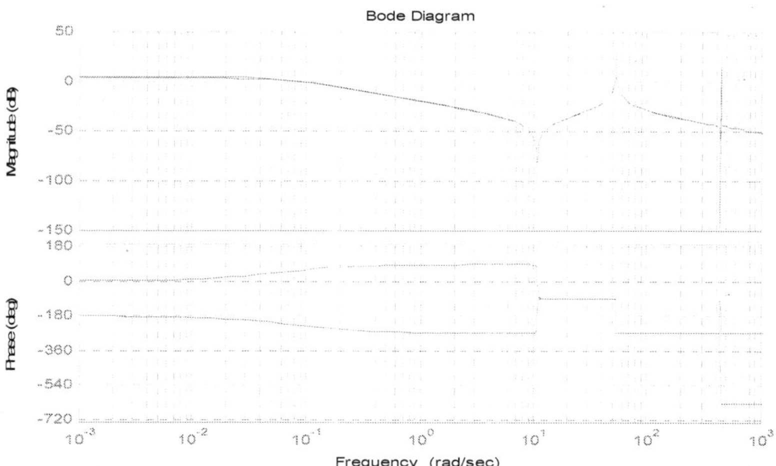

Fig. 20. Bode plots of $G_{T_g\omega_2}$ for below rated (blue) and above rated (stall) operation (red).

its output, $G_{V\omega_2}$ and $G_{T_g\omega_2}$ are examined for different operating conditions. The Bode plots of $G_{V\omega_2}$ and $G_{T_g\omega_2}$ are shown in Figs. 19 and 20 respectively, for two operating points, namely one for below rated operation (ω_1, V)=(4rad/sec, 6.76m/sec) and one for above rated operation, (4 rad/sec, 8.76m/sec).

As can be seen from the above plots, a phase change of 180° occurs, for frequencies less than 0.1rad/sec as the operating point of the wind turbine moves from below rated to stall operation, for both transfer functions. In addition, the first drivetrain mode can be observed at 53rad/sec.

8.3 Control design

In this section the design of the speed controller for the Windharvester wind turbine is presented. In Fig. 21 the actual T_a-ω plot for the simulated wind turbine including the operating point locus (black), is shown. In the plot T_a-ω characteristics are shown in blue colour and the characteristics for wind speeds above 20m/sec are shown with bold line. The brown curve corresponds to operation for $V_{\omega_N} = 6.76$m/sec where operation at constant speed $\omega=\omega_N$ starts. The green curve corresponds to V_N=8.3m/sec, where P_N=25kW. Also the hyperbolic curve of constant power P_N=25kW is shown in red.

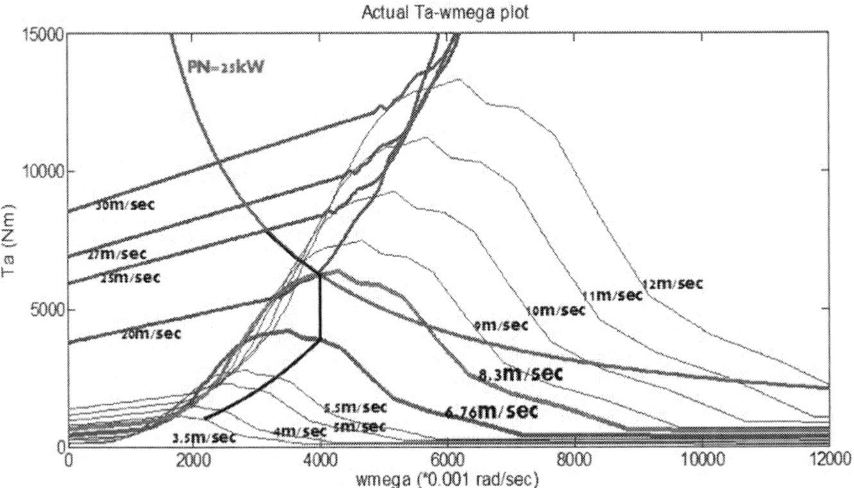

Fig. 21. Actual T_a-ω plot of the simulated wind turbine.
The operating point locus is shown in black and starts at ω_A=2.1rad/sec for V_{cut-in}=3.5m/sec.

Regarding the gain scheduled controller, two PI controllers are used, with PI gains of 20 and 10 Nm/rad/sec for operation below ω_N and 30 and 50Nm/rad/sec for operation above ω_N. Fig. 22 shows the Bode plots of the closed loop transfer function from the reference rotational speed ω_{ref} (see Fig. 7) to the generator speed ω_2, $G_{\omega_{ref}\omega_2}$ for the two controllers used. Fig. 23 shows the corresponding Bode plots for the disturbance transfer function from the wind speed V to ω_2, $G_{V\omega_2}$. These Bode plots have been obtained for operating conditions (V,ω)=(6.76m/sec, 4rad/sec).

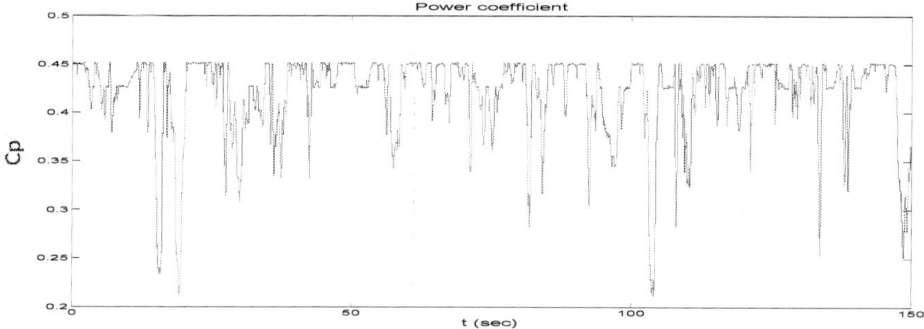

Fig. 31. Power coefficient in time.

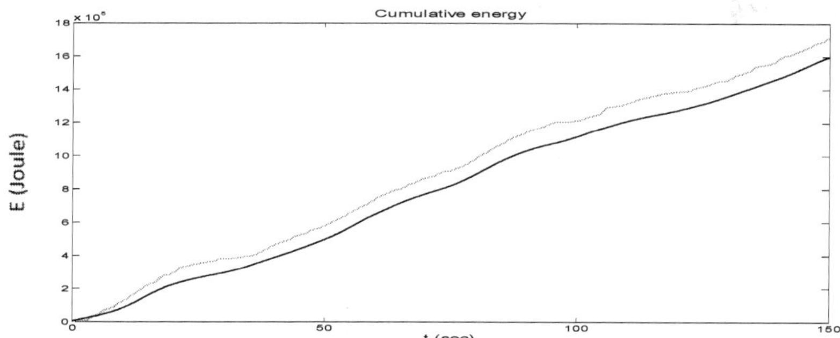

Fig. 32. Cumulative energy with the conventional control (Eqn. 6), (black) and with the proposed control, (green).

As can be seen from Fig. 27, the wind speed estimation is very accurate and the resulting speed reference is quite close to the ideal one (Fig. 28). The speed reference for the generator is low-pass filtered before it is used by the speed controller, in order to smooth out high frequency variations. Furthermore, from Figs. 29-30 it can be seen that the speed of the generator ($N^*\omega$) closely follows its reference and this is achieved without excessive control action. Fig. 31 shows very effective maximum power point operation (close to $C_{p\ max}$=0.45). Finally, Fig. 32 shows a remarkable gain of 6.5% in the cumulative produced energy using the proposed control method, compared to the conventional control method. This is a very important result, which shows that it is possible to very effectively control a variable speed stall regulated wind turbine for maximum power point operation, using the proposed method.

Furthermore, the performance of the control system has been tested at constant speed operation, at $\omega=\omega_N$=4rad/sec, using the original scale of the wind speed series of Fig. 25. At this operating region, the PI speed controller with higher gains is switched on (see Section 8.2). The performance of this controller in terms of speed reference tracking is compared with the performance that is achieved when only the controller of lower gains is used, according to (Bourlis & Bleijs, 2010a). Fig. 33 shows these results, while Fig. 34 shows the control torque and the IG torque using the PI controller with higher gains, when the original C_p-λ curve is used.

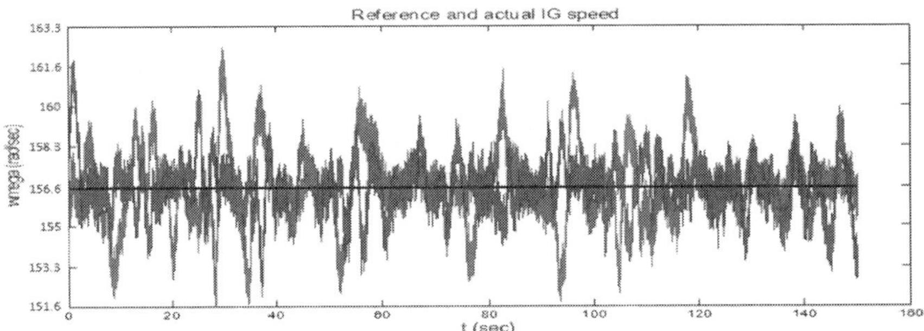

Fig. 33. Reference speed (black), generator speed response with (a) PI controller with low gains (red) and (b) with dedicated PI controller with higher gains (blue).

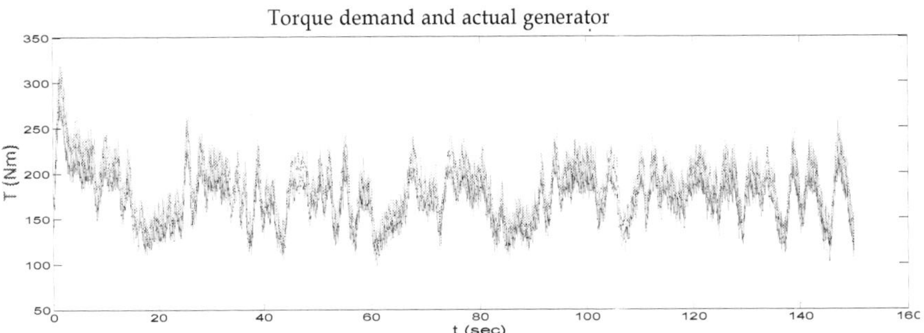

Fig. 34. Torque demand ($T_{g_{ref}}$) (black) and actual (blue) IG torque for the controller with higher gains.

As can be observed, the speed reference tracking is considerably improved using a controller with higher gains. Specifically, the generator speed very rarely diverges further than 2% of its reference, while with the controller with the lower gains, the speed error very often reaches 3.1% and higher. Furthermore, from Fig. 34 it can be seen that the control torque variations are limited to less than 40% around its average value (180Nm), which is absolutely acceptable. This performance can be even better when the steep C_p-λ curve is used.

Using the proposed control method, very accurate reference tracking can be achieved during stall regulation at constant power too. Fig. 35 shows power regulation at 25 and 20kW at above rated wind speeds, when the steeper C_p-λ curve is simulated and using a PI controller with PI gains of 30 and 30 Nm/rad/sec, respectively. Fig. 36 shows the generator torque during this experiment.

As can be seen, Figs. 35 and 36 exhibit very effective power control. This is a result of accurate speed reference tracking and very smooth control action by the control system (examination of the speed reference determination algorithm for stall regulation is outside of the scope of this paper).

(c)

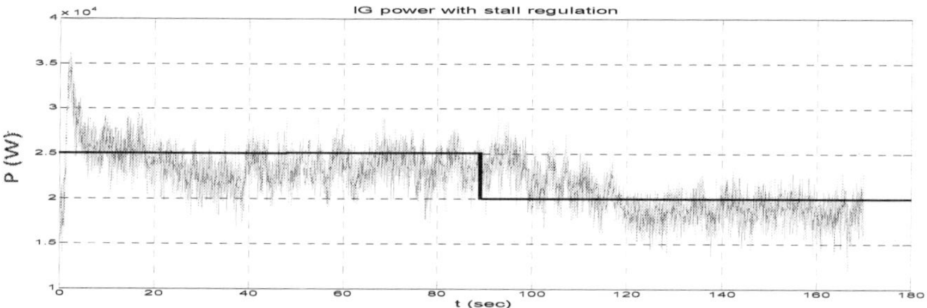

Fig. 35. Reference (black) and actual (blue) induction generator power.

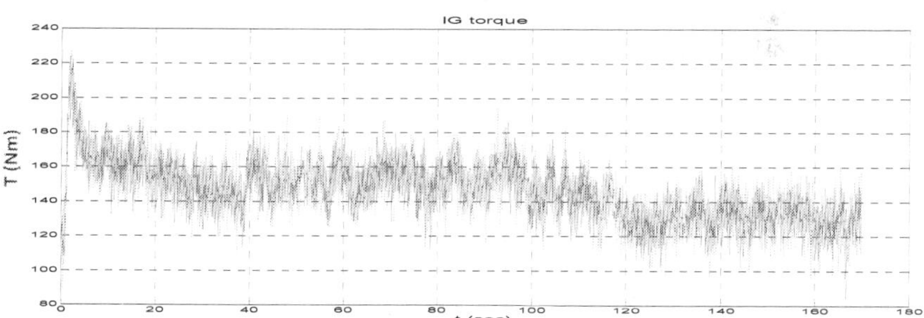

Fig. 36. IG torque.

9. Conclusions

In this paper a control scheme for variable speed stall regulated wind turbines was presented. The control system aims to continuously provide the optimum rotational speed for the wind turbine in order to achieve maximum power production until the rated rotational speed is reached and effective power limitation when the wind turbine operates at wind speeds higher than the rated. The proposed control system consists of an aerodynamic torque estimation stage, a speed reference determination stage and a gain scheduled proportional-integral speed controller. The first two stages are used to produce the optimum speed reference for the speed controller. The speed controller is responsible for the system to closely follow the speed reference and at the same time for the torque loading of the drivetrain as well as the effect of external disturbances to be kept up to specified levels.

In this paper emphasis is put on the examination of the performance of the control system in below rated operation, while the ability of the proposed gain scheduled speed controller to effectively achieve power limitation for above rated wind speeds is also exhibited.

The whole control scheme has been implemented in a high performance hardware-in-loop simulator, which is driven by real wind site data. The hardware-in-loop simulator has been developed using industrial machines and drives and is controlled by an accurate dynamic model of an actual wind turbine, such that it closely approximates the dynamics of the wind turbine.

The hardware simulation results exhibited a very good performance of the proposed control scheme in below rated operation. Specifically, the aerodynamic torque and effective wind speed were accurately estimated, which in turn resulted in very accurate speed reference extraction by the control system. Furthermore, the proposed gain scheduled speed controller very effectively satisfied different bandwidth requirements for different operating regions and at the same time provided adequate damping to the drivetrain oscillation modes and eliminated the effects of external disturbances.

Through simulations using a steep power coefficient curve for the wind turbine rotor, which is a requirement for a variable speed stall regulated wind turbine, the control system achieved accurate reference tracking, which resulted in effective maximum power point operation, as this was observed by the high values of the power coefficient achieved during the operation. As a result, the produced cumulative energy for maximum power point operation using the proposed control system was increased by 6.5%, when compared with the one achieved using conventional control methods that are used in commercial wind turbines. It is also notable that this performance was achieved without excessive control torque action by the generator and this is possible to be achieved in general by appropriately adjusting the bandwidth of the PI controller used, as well as the bandwidth of the low-pass filter at the speed reference.

Furthermore, the hardware simulation results for operation at constant speed for above rated wind speeds exhibited a very good performance of the proposed gain scheduled PI speed controller when compared with previous implementations using a single PI controller for the whole operating region. The proposed controller can be effectively used for speed control during stall regulation at constant power, as this was also shown through hardware simulations. So, this type of controller provides a suitable solution for high performance control of stall regulated wind turbines. In addition, this controller is easy to implement and its tuning only requires basic knowledge of control systems so it can be performed by any experienced engineers.

A key feature of the proposed control scheme is that it can run on commercial digital signal processor boards. From there it can communicate with the drive of the generator of the wind turbine and the whole scheme requires only a speed measurement of the generator, which is always available in commercial wind turbines. Also, in general for the operation of the proposed control scheme there are no considerable requirements for computing power (a sample time of 5msec was used here).

The proposed control scheme provides a novel and easy to implement solution, which as was shown from the hardware simulation results provided, it can be effectively applied for high performance control of variable speed stall regulated wind turbines, outperforming conventional control methods, which is something that is presented for the first time.

To sum up, the control scheme for variable speed stall regulated wind turbines that is proposed here and the simulation results that are presented are very important, because they show that it is possible to effectively control this type of wind turbine using existing technology. That way, the proposed control scheme gives confidence for the development of variable speed stall regulated wind turbines in the near future and this is very important due to the economic advantages that these wind turbines can have.

For the above reasons, future work should be directed on developing this control system in an actual wind turbine. Challenges that have to overcome then are the uncertainty in the knowledge of the exact parameters of the actual wind turbine as well as stochastic changes of the dynamics of the wind turbine, due to the aerodynamic phenomena. The effects of these uncertainties in the operation of the control system and in particular in the wind speed estimation as well as in the performance of the speed controller need therefore to be examined experimentally. That way the robustness of the proposed control system to this kind of uncertainties can be increased appropriately, if required. Therefore, industrial funded research can further contribute to the development of variable speed stall regulated wind turbines.

10. Acknowledgment

I would like to thank Dr. J.A.M. Bleijs from the Electrical Power and Power Electronics Group of the University of Leicester for his help and the Engineering and Physical Sciences Research Council of United Kingdom for providing the funding for this study.

11. References

Anderson B.D.O. & Moore J.B. (1979). *Optimal Filtering*, Prentice - Hall Information and System Sciences Series, Englewood Cliffs, N.J.

Biachi, F. D., et al. (2007). *Wind Turbine Control Systems. Principles Modelling and Gain Scheduling Design* (1st ed.), Springer, ISBN 9871846284922, London UK

Bossanyi, E. A. (2003). The Design Of Closed Loop Controllers For Wind Turbines. *Wind Energy*, Vol. 3, No. 3, pp. (149-163)

Bossanyi E.A. (2003). Wind Turbine Control for Load Reduction. *Wind Energy*, Vol. 6, No. 3, (3 Jun 2003), pp. (229-244)

Boukhezzar B. & Siguerdidjane H. (2005). Nonlinear control of variable speed wind turbines without wind speed measurement, *Proceedings of the 44th IEEE Conference on Decision and Control, and the European Control Conference*, Seville, Spain, (December 12-15, 2005), pp. (3456-3461)

Bourlis D. & Bleijs J.A.M. (2010a). Control of stall regulated variable speed wind turbine based on wind speed estimation using an adaptive Kalman filter, *Proceedings of the European Wind Energy Conference*, Warsaw, Poland, (20-23 April 2010), pp. (242-246)

Bourlis D. & Bleijs J.A.M. (2010b). A wind estimation method using adaptive Kalman filtering for a variable speed stall regulated wind turbine, *Proceedings of the 11th IEEE International Conference on Probabilistic Methods Applied to Power Systems*, Singapore, (14-17 July 2010), pp. (89-94)

Chui C.K. & Chen G. (1999). *Kalman filtering. With real time applications* (3rd ed.) Springer

Connor B. & Leithead W.E. (1994). Control strategies for variable speed stall regulated wind turbines, *Proceedings of the 5th European Wind Energy Association Conference and Exhibition*, Thessaloniki, Greece, (10-14 Oct 1994), Vol. 1, pp. (420-424)

Goodfellow D., Smith G.A. & Gardner G. (1988). Control strategies for variable-speed wind energy recovery, *Proceedings of the BWECS Conference*

Johnson M.A. & Moradi M.H. (2005). *PID Control. New Identification and Design Methods*, Springer Kurtulmus F., Vardar A. & Izli N. (2007). Aerodynamic Analyses of Different Wind Blade Profiles. *Journal of Applied Sciences*, pp. (663-670)

Leithead, W. E. (1990). Dependence of performance of variable speed wind turbines on the turbulence, dynamics and control, *IEE Proceedings*, Vol. 137, No. 6, (November 1990)

Leithead W.E. & Connor B. (2000). Control of Variable Speed Wind Turbines: Design Task. *International Journal of Control*, Vol. 73, No. 13, pp. (1189-1212)

Manwell, J. (2002). *Wind energy explained: theory, design and application*, Willey

Mercer A.S. & Bossanyi E.A. (1996). Stall regulation of variable speed HAWTS, *Proceedings of the European Wind Energy* Conference, Göteborg, Sweden, (20-24 May 1996), pp. (825-828)

Novak P., et al. (1995). Modelling and Control of Variable-Speed Wind-Turbine Drive-System Dynamics. *IEEE Control Systems Magazine*, Vol. 15, No. 4, (Aug 1995), pp. (28-38)

Østergaard, K.Z., et al. (2007). Estimation of Effective Wind Speed. *Journal of Physics: Conference Series*, Vol. 75, No. 1, pp. (1-9)

Parker D.A. (2000). The design and development of a fully dynamic simulator for renewable energy converters, *Phd Thesis*, University of Leicester

Wilkie J., et al. (2002). *Control Engineering. An introductory course*, Palgrave

15

MPPT Control Methods in Wind Energy Conversion Systems

Jogendra Singh Thongam[1] and Mohand Ouhrouche[2]
[1]Department of Renewable Energy Systems, STAS Inc.
[2]Electric Machines Identification and Control Laboratory, Department of Applied Sciences,
University of Quebec at Chicoutimi
Quebec
Canada

1. Introduction

Wind energy conversion systems have been attracting wide attention as a renewable energy source due to depleting fossil fuel reserves and environmental concerns as a direct consequence of using fossil fuel and nuclear energy sources. Wind energy, even though abundant, varies continually as wind speed changes throughout the day. The amount of power output from a wind energy conversion system (WECS) depends upon the accuracy with which the peak power points are tracked by the maximum power point tracking (MPPT) controller of the WECS control system irrespective of the type of generator used.

This study provides a review of past and present MPPT controllers used for extracting maximum power from the WECS using permanent magnet synchronous generators (PMSG), squirrel cage induction generators (SCIG) and doubly fed induction generator (DFIG). These controllers can be classified into three main control methods, namely tip speed ratio (TSR) control, power signal feedback (PSF) control and hill-climb search (HCS) control. The chapter starts with a brief background of wind energy conversion systems. Then, main MPPT control methods are presented, after which, MPPT controllers used for extracting maximum possible power in WECS are presented.

2. Wind energy background

Power produced by a wind turbine is given by [1]

$$P_m = 0.5\pi\rho C_p(\lambda, \beta)R^2 v_w^3 \tag{1}$$

where R is the turbine radius, v_w is the wind speed, ρ is the air density, C_p is the power coefficient, λ is the tip speed ratio and β is the pitch angle. In this work β is set to zero. The tip speed ratio is given by:

$$\lambda = \omega_r R / v_w \tag{2}$$

where ω_r is the turbine angular speed. The dynamic equation of the wind turbine is given as

$$d\omega_r / dt = (1 / J)\left[T_m - T_L - F\omega_r\right] \tag{3}$$

where J is the system inertia, F is the viscous friction coefficient, T_m is the torque developed by the turbine, T_L is the torque due to load which in this case is the generator torque. The target optimum power from a wind turbine can be written as

$$P_{\max} = K_{opt}\omega^3_{r\_opt} \tag{4}$$

where

$$K_{opt} = \frac{0.5\pi\rho C_{p\max}R^5}{\lambda^3_{opt}} \tag{5}$$

$$\omega_{opt} = \frac{\lambda_{opt}v_w}{R} \tag{6}$$

Fig.1 shows turbine mechanical power as a function of rotor speed at various wind speeds. The power for a certain wind speed is maximum at a certain value of rotor speed called optimum rotor speed ω_{opt}. This is the speed which corresponds to optimum tip speed ratio λ_{opt}. In order to have maximum possible power, the turbine should always operate at λ_{opt}. This is possible by controlling the rotational speed of the turbine so that it always rotates at the optimum speed of rotation.

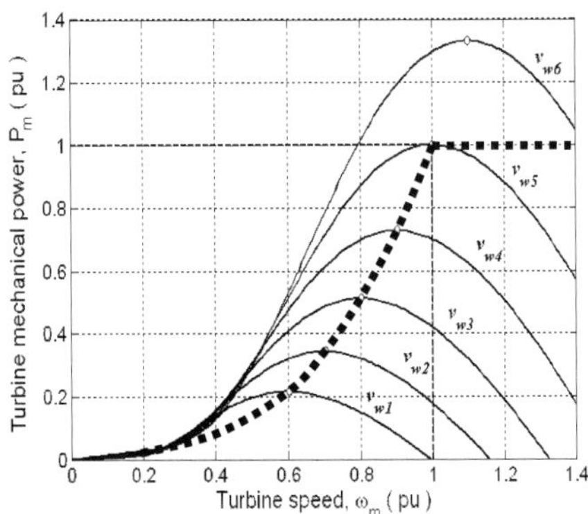

Fig. 1. Turbine mechanical power as a function of rotor speed for various wind speeds.

3. Maximum power point tracking control

Wind generation system has been attracting wide attention as a renewable energy source due to depleting fossil fuel reserves and environmental concerns as a direct consequence of using fossil fuel and nuclear energy sources. Wind energy, even though abundant, varies continually as wind speed changes throughout the day. Amount of power output from a WECS depends upon the accuracy with which the peak power points are tracked by the MPPT controller of the WECS control system irrespective of the type of generator used. The maximum power extraction algorithms researched so far can be classified into three main control methods, namely tip speed ratio (TSR) control, power signal feedback (PSF) control and hill-climb search (HCS) control [2].

The TSR control method regulates the rotational speed of the generator in order to maintain the TSR to an optimum value at which power extracted is maximum. This method requires both the wind speed and the turbine speed to be measured or estimated in addition to requiring the knowledge of optimum TSR of the turbine in order for the system to be able extract maximum possible power. Fig. 2 shows the block diagram of a WECS with TSR control.

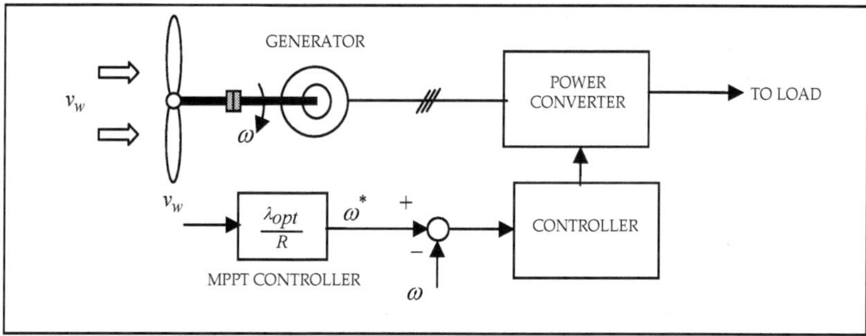

Fig. 2. Tip speed ratio control of WECS.

In PSF control, it is required to have the knowledge of the wind turbine's maximum power curve, and track this curve through its control mechanisms. The maximum power curves need to be obtained via simulations or off-line experiment on individual wind turbines. In this method, reference power is generated either using a recorded maximum power curve or using the mechanical power equation of the wind turbine where wind speed or the rotor speed is used as the input. Fig. 3 shows the block diagram of a WECS with PSF controller for maximum power extraction.

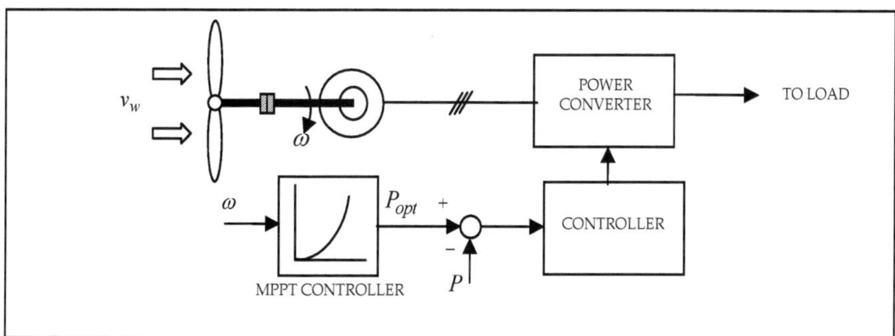

Fig. 3. Power signal feedback control.

The HCS control algorithm continuously searches for the peak power of the wind turbine. It can overcome some of the common problems normally associated with the other two methods. The tracking algorithm, depending upon the location of the operating point and relation between the changes in power and speed, computes the desired optimum signal in order to drive the system to the point of maximum power. Fig. 4 shows the principle of HCS control and Fig. 5 shows a WECS with HCS controller for tracking maximum power points.

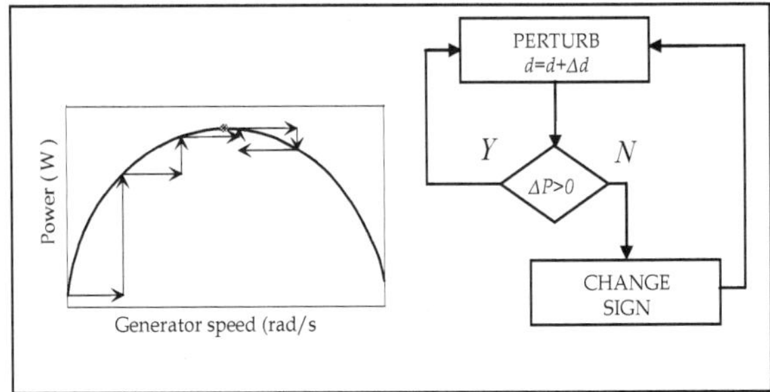

Fig. 4. HCS Control Principle.

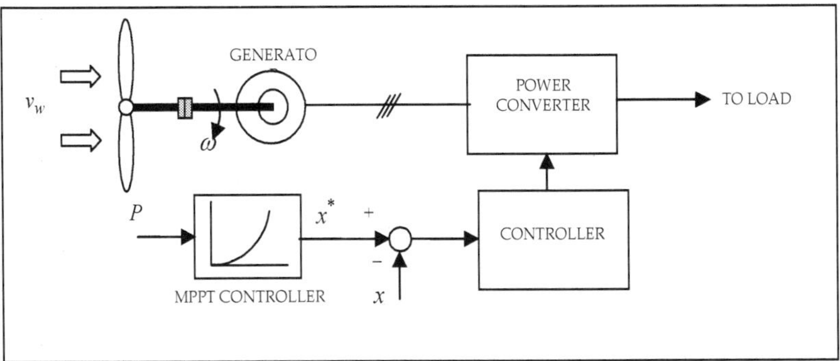

Fig. 5. WECS with hill climb search control.

4. MPPT control methods for PMSG based WECS

Permanent Magnet Synchronous Generator is favoured more and more in developing new designs because of higher efficiency, high power density, availability of high-energy permanent magnet material at reasonable price, and possibility of smaller turbine diameter in direct drive applications. Presently, a lot of research efforts are directed towards designing of WECS which is reliable, having low wear and tear, compact, efficient, having low noise and maintenance cost; such a WECS is realisable in the form of a direct drive PMSG wind energy conversion system.

There are three commonly used configurations for WECS with these machines for converting variable voltage and variable frequency power to a fixed frequency and fixed voltage power. The power electronics converter configurations most commonly used for PMSG WECS are shown in Fig. 6.

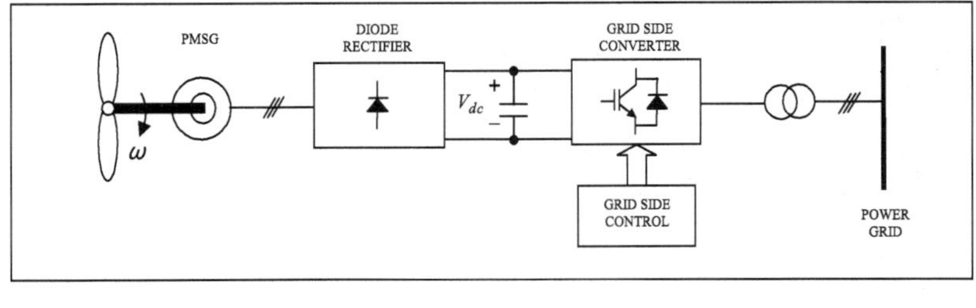

(a)

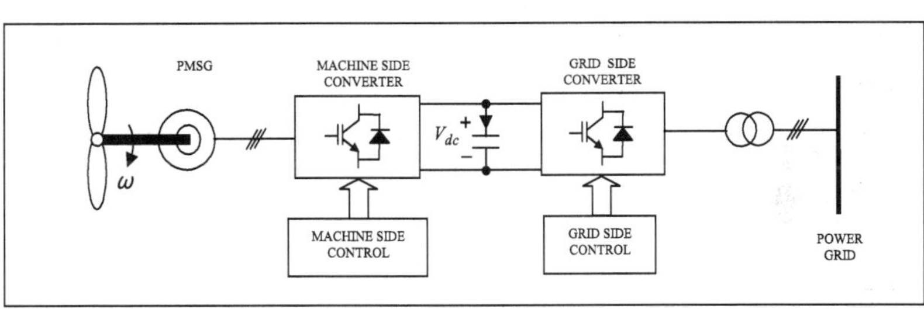

(b)

(c)

Fig. 6. PMSG wind energy conversion systems

Depending upon the power electronics converter configuration used with a particular PMSG WECS a suitable MPPT controller is developed for its control. All the three methods of MPPT control algorithm are found to be in use for the control of PMSG WECS.

4.1 Tip speed ratio control

A wind speed estimation based TSR control is proposed in [3] in order to track the peak power points. The wind speed is estimated using neural networks, and further, using the estimated wind speed and knowledge of optimal TSR, the optimal rotor speed command is computed. The generated optimal speed command is applied to the speed control loop of the WECS control system. The PI controller controls the actual rotor speed to the desired value by varying the switching ratio of the PWM inverter. The control target of the inverter is the output power delivered to the load. This WECS uses the power converter configuration shown in Fig. 6 (a). The block diagram of the ANN-based MPPT controller module is shown in Fig. 7. The inputs to the ANN are the rotor speed ω_r and mechanical power P_m. The P_m is obtained using the relation

$$P_m = \omega_r \left(J \frac{d\omega_r}{dt} \right) + P_e \qquad (7)$$

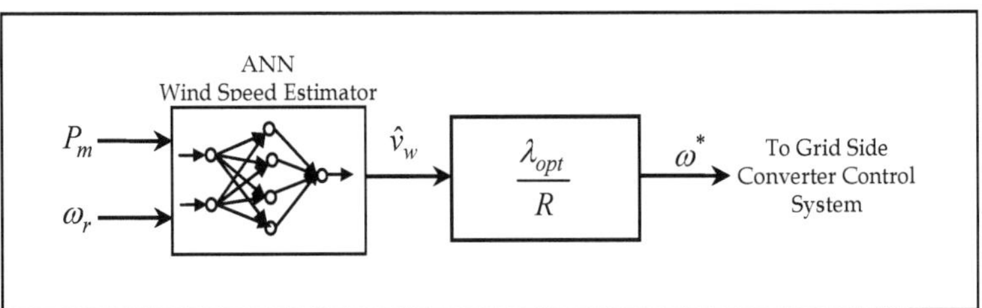

Fig. 7. ANN-based MPPT control module of turbine rotor speed.

4.2 Power signal feedback

In [4], the turbine power equation is used for obtaining reference power for PSF based MPPT control of PMSG WECS. Fig. 8 shows the block diagram for the PSF control signal generation. Using equation (8) we have:

$$P_{opt} = K_{opt} \omega_r^3 \qquad (8)$$

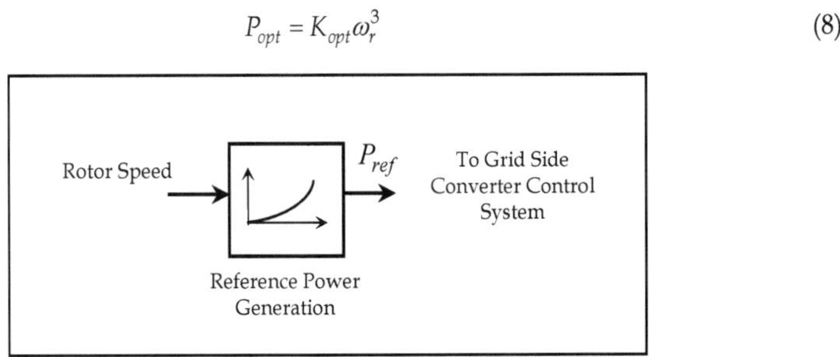

Fig. 8. Reference power generation for PSF control.

The PSF control block generates the reference power command P_{ref} using (8) which is then applied to the grid side converter control system for maximum power extraction.

4.3 Hill climb search control

In [2-17] Hill climb search method of MPPT control for PMSG WECS are proposed. The control algorithm proposed in [2] uses the principle of search-remember-reuse technique. The method uses memory for storing peak power points, obtained during training process, which are used later for tracking maximum power points. The principle behind this algorithm is a search-remember-reuse process. The algorithm will start from an empty intelligent memory with a relatively poor initial performance. During the execution, training

when the operating point is far away from the peak due to the larger magnitude of $P-\omega$ slope and as the peak gets nearer, the step size should automatically approach to zero. The method uses torque current and speed to compute power. The speed step is computed as

$$\Delta\omega = k.\frac{dP}{d\omega} \tag{10}$$

and the speed reference for the machine side converter control system is computed as

$$\omega^*(k+1) = \omega(k) + \Delta\omega(k) \tag{11}$$

In [11, 12] adaptive control algorithm for MPPT control is proposed. The control algorithm allows the generator to track the maximum power points of the wind turbine system under fluctuating wind conditions. The algorithm proposed initiates the TSR control with an approximate optimal TSR value. When the measured wind velocity is found to be stable, the algorithm switches to HCS to search for the true optimal point. When the true peak is reached, a memory table of the optimum generator speed versus the corresponding wind velocity is updated and then, the TSR is corrected. When the wind speed varies, the rotor speed reference is applied from the memory if a recorded data at current wind speed is present in the memory and if not, it is calculated using TSR. The MPPT control signal is given to the boost chopper for tracking the maximum power points. The method requires both wind speed and rotor speed measurement.

In [13] MPPT control of PMSG WECS is implemented via a dc-dc boost converter. The proposed MPPT strategy is based on directly adjusting the dc-dc converter duty cycle according to the result of the comparison of successive WTG output power measurements. The WECS MPPT algorithm operates by constantly perturbing the rectified output voltage V_{dc} of the WECS via the dc-dc boost converter duty cycle and comparing the actual output power with the previous perturbation sample. If the power is increasing, the perturbation will continue in the same direction in the following cycle so that the rotor speed will be increased, otherwise the perturbation direction will be inverted. When the optimal rotational speed of the rotor for a specific wind speed is reached, the HCS algorithm will have tracked the maximum power point and then will settle at or around this point. In [14] a buck-boost converter circuit is used to achieve the maximum power control of wind turbine driven PMSG WECS. The PMSG is suitably controlled according to the generator speed and thus the power from a wind turbine settles down on the maximum power point using the proposed MPPT control method. The method does not require the knowledge of wind turbine's maximum power curve or the information on wind velocity. It uses the dc link power as its input and the output is the chopper duty cycle.

The HCS MPPT control method in [15] uses power as the input and torque as the controller output. The optimum torque output of the controller is applied to the torque control loop of a DTC controlled PMSG. The controller does not require wind speed sensing.

The HCS MPPT control method presented in [16] combines the benefits of two of the commonly used MPPT methods: (i) the tracking method based on the optimum power versus speed characteristic and (ii) the HCS. The algorithm measures generator rotor speed and computes optimum torque T_{opt}, the torque which maximizes power. The actual torque T_t is also calculated. For a small error between the optimal and measured torque ΔT, the system performs a perturb and observe (P&O) process, based on the calculation of actual power, overlooking the use of the optimum $T-\omega$ characteristic. However, if the ΔT exceeds a certain limit the duty cycle is commanded according to the optimum characteristic. In other words, the system tracks the maximum power point through a P&O process under normal circumstances; however, it uses the predefined $T - \omega$ characteristic in case the P&O algorithm is thrown off due to heavy disturbances such as sudden wind speed changes or improper initialization.

Neural networks based MPPT controller is presented in [17]. The method proposed uses Jordan recurrent multilayer ANN with one hidden layer. The weights of the networks are continuously modified by back propagation during the operation of the WECS with online training. The control system continuously searches ways to reach the peak power point. Optimum rotor speed, which is the output of the controller, is used as the reference speed for the vector controlled machine side converter control system.

5. MPPT control methods for SCIG based WECS

The use of induction generators (IG) is advantageous since they are relatively inexpensive, robust, and require low maintenance. The nature of IG is unlike that of PMSG; they need bi-directional power flow in the generator-side converter since they require external reactive power support from the grid. Modern IG WECS are equipped with PWM back-to-back frequency converter which also allows advanced control algorithms to be implemented. However, other converter configurations are possible and can be found in the literature. SCIG WECS with a back-to-back converter configuration is shown in Fig. 10. The MPPT control in such system is realized using the machine side control system. All the existing MPPT control algorithm scan be implemented for the control of IG WECS.

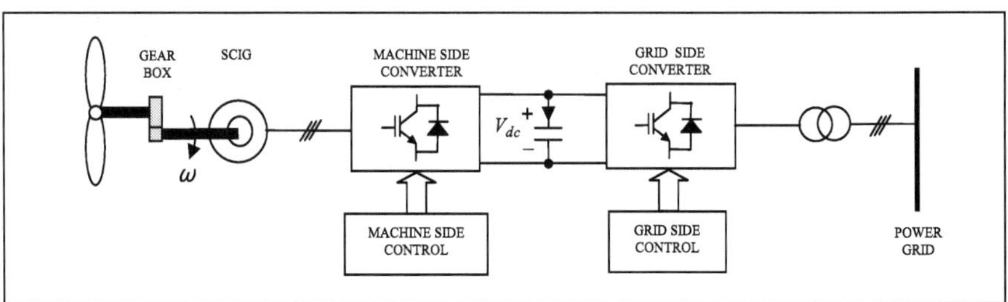

Fig. 10. SCIG WECS

5.1 Tip speed ratio control

In [13, 18, 19], TSR control method of MPPT control for SCIG WECS are presented. In the TSR control method presented in [18] the wind speed is measured for obtaining optimum rotor speed using the value of optimum tip speed ratio. The optimum TSR is obtained from the turbine's C_p-λ curve. The rotor speed required for implementing speed feed-back control is estimated using a speed observer. The speed control is exercised using a fuzzy neural network controller.

Wind-speed estimation based MPPT control are proposed in [3, 19]. In [3], an ANN wind speed estimation based TSR control method was used for implementing MPPT control of SCIG WECS. Here, the optimum speed command was generated by the MPPT controller for speed control loop of machine side converter control system enabling the WECS to extract optimum energy. The method has been presented in section (4.1). The wind speed estimation method in [19] is based on the theory of support-vector regression (SVR). The inputs to the wind-speed estimator are the wind-turbine power and rotational speed. A specified model, which relates the inputs to the output, is obtained by offline training. Then, the wind speed is determined online from the instantaneous inputs. The estimated wind speed is used for MPPT control of SCIG WECS.

5.2 Power signal feedback

In [20], fuzzy logic controller is used to track the maximum power point. The method uses wind speed as the input in order to generate reference power signal. Maximum power output P_{max} of the WECS at different wind velocity v_w is computed and the data obtained is used to relate P_{max} to v_w using polynomial curve fit as given by

$$P_{\max} = -0.3 + 1.08v_w - 0.125v_w^2 + 0.842v_w^3 \qquad (12)$$

The reference power at the rectifier output is computed using the maximum power given by (12) as

$$P_{ref} = \eta_G \eta_R P_{\max} \qquad (13)$$

The actual power output of the rectifier P_o is compared to the reference power P_{ref} and any mismatch is used by the fuzzy logic controller to change the modulation index M for the grid side converter control.

5.3 Hill climb search control

HCS control of SCIG WECS are presented in [21, 22]. In [21], a fuzzy logic based HCS controller for MPPT control is proposed. The block diagram of the fuzzy controller is shown in Fig. 11. In the proposed method, the controller, using Po as input generates at its output the optimum rotor speed. Further, the controller uses rotor speed in order to reduce sensitivity to speed variation. The increments or decrements in output power due to an increment or decrement in speed is estimated. If change in power ΔP_0 is positive with last positive change in speed $\Delta\omega_r$, indicated in Fig. 11 by $L\Delta\omega_r\left(pu\right)$, the search is continued in the same direction. If, on the other hand, $+\Delta\omega_r$ causes $-\Delta P_0$, the direction of search is reversed. The variables ΔP_0 , $\Delta\omega_r$ and $L\Delta\omega_r$ are described by membership functions and rule table. In order to avoid getting trapped in local minima, the output $\Delta\omega_r$ is added to some amount of $L\Delta\omega_r$ in order to give some momentum and continue the search. The scale factors KPO and KWR are generated as a function of generator speed so that the control becomes somewhat insensitive to speed variation. For details please refer to [21].

In [22], a fuzzy logic control is applied to generate the generator reference speed, which tracks the maximum power point at varying wind speeds. The principle of the FLC is to perturb the generator reference speed and to estimate the corresponding change of output power P_0. If the output power increases with the last increment, the searching process continues in the same direction. On the other hand, if the speed increment reduces the output power, the direction of the searching is reversed. The block diagram of the proposed controller is shown in Fig. 12. The fuzzy logic controller is efficient to track the maximum power point, especially in case of frequently changing wind conditions. The controller tracks the maximum power point and extracts the maximum output power under varying wind

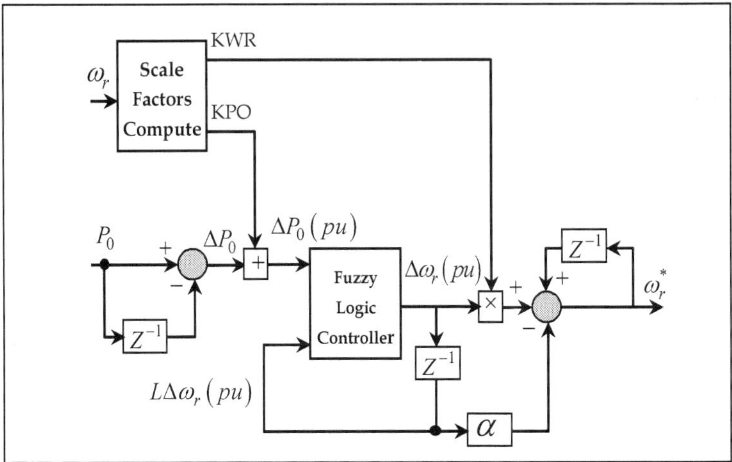

Fig. 11. Block diagram of fuzzy logic MPPT controller

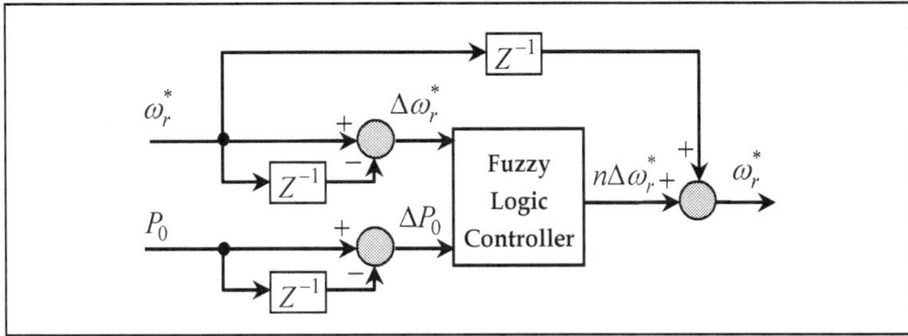

Fig. 12. Fuzzy MPPT controller

conditions. Two inputs $\Delta\omega_r^*$ and ΔP_0 are used as the control input signals and the output of the controller is the new speed reference speed which, after adding with previous speed command, forms the present reference speed. For more details, please refer to [22].

6. MPPT control methods for DFIG based WECS

The PMSG WECS and SCIG WECS have the disadvantages of having power converter rated at 1 $p.u.$ of total system power making them more expensive. Inverter output filters and EMI filters are rated for 1 $p.u.$ output power, making the filter design difficult and costly. Moreover, converter efficiency plays an important role in total system efficiency over the entire operating range. WECS with DFIG uses back to back converter configuration as is shown in Fig. 13. The power rating of such converter is lower than the machine total rating as the converter does not have to transfer the complete power developed by the DFIG. Such WECS has reduced inverter cost, as the inverter rating is typically 25% of total system power, while the speed range of variable speed WECS is 33% around the synchronous speed. It also has reduced cost of the inverter filters and EMI filters, because filters are rated for 0.25 *pu* total system power, and inverter harmonics present a smaller fraction of total system harmonics. In this system power factor control can be implemented at lower cost, because the DFIG system basically operates similar to a synchronous generator. The converter has to provide only the excitation energy. The higher cost of the wound rotor induction machine over SCIG is compensated by the reduction in the sizing of the power converters and the increase in energy output. The DFIG is superior to the caged induction machine, due to its ability to produce above rated power. The MPPT control in such system is realized using the machine side control system.

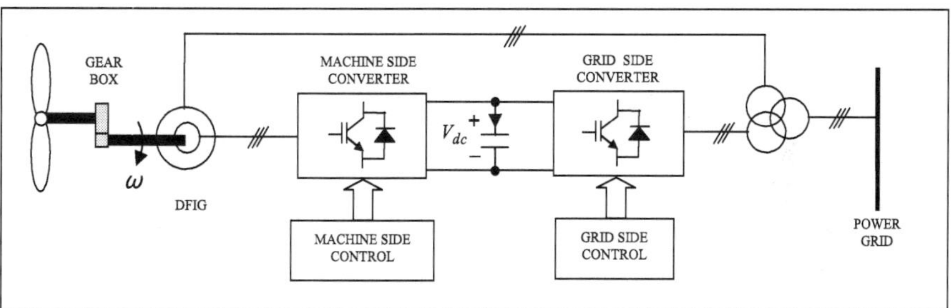

Fig. 13. DFIG WECS

6.1 Tip speed ratio control

TSR control is possible with wind speed measurement or estimation. In [23], a wind speed estimation based MPPT controller is proposed for controlling a brushless doubly fed induction generator WECS. The block diagram of the TSR controller is shown in Fig. 14. The optimum rotor speed ω_{opt}, which is the output of the controller, is used as the reference signal for the speed control loop of the machine side converter control system.

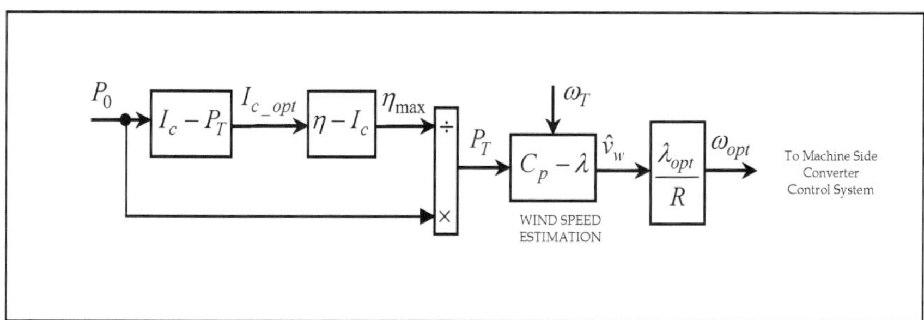

Fig. 14. Generation of optimum speed command

The method requires the total output power P_0 of the WECS and rotor speed as input to the MPPT controller. Using P_0 as the input to a look-up table of $I_c - P_0$ profile, optimum winding current I_{c_opt} is obtained. The maximum generator efficiency η_{max} is estimated at a particular control current optimized operating point using a stored efficiency versus optimum current characteristic of the generator. In the algorithm presented the relations I_c

versus P_T and I_c versus η were implemented using RBF neural networks. Then, generator input power P_T is calculated from the maximum efficiency η_{max} and the measured output power P_0. The next step involves wind speed estimation which is achieved using Newton-Raphson or bisection method. The estimated wind speed information is used to generate command optimum generator speed for optimum power extraction from WECS. For details of the proposed method please refer to [23]. The method is not new; similar work was earlier implemented for controlling a Brushless Doubly Fed Generator by Bhowmik et al [24]. In this method the Brushless Doubly Fed Generator was operated in synchronous mode and input to the controller was only the output power of the WECS.

6.2 Power signal feedback control

PSF control along with feedback linearization is used by [25] for tracking maximum power point. The input-output feedback linearization is done using active-reactive powers, d-q rotor voltages, and active-reactive powers as the state, input and output vectors respectively. The references to the feedback linearization controller are the command active and reactive powers. The reference active power is obtained by subtracting the inertia power from the mechanical power which is obtained by multiplying speed with torque. A disturbance torque observer is designed in order to obtain the torque.

A fuzzy logic based PSF controller is presented in [26]. Here, a data driven design methodology capable of generating a Takagi-Sugeno-Kang (TSK) fuzzy model for maximum power extraction is proposed. The controller has two inputs and one output. The rotor speed and generator output power are the inputs, while the output is the estimated maximum power that can be generated. The TSK fuzzy system, by acquiring and processing the inputs at each sampling instant, calculates the maximum power that may be generated by the wind generator, as shown in Fig. 15.

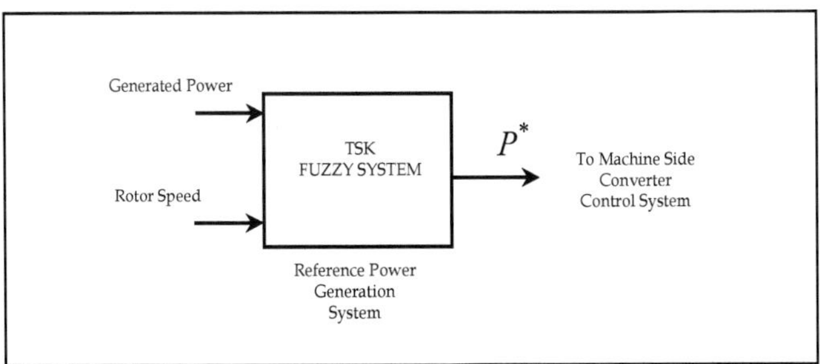

Fig. 15. TSK fuzzy MPPT controller

The approach is explained by considering the turbine power curves, as shown in Fig. 16. If the wind turbine initially operates at point A, the control system, using rotor speed and turbine power information, is able to derive the corresponding optimum operating point B, giving the desired rotor speed reference ω_B. The generator speed will therefore be controlled in order to reach the speed ω_B allowing the extraction of the maximum power P_B from the turbine.

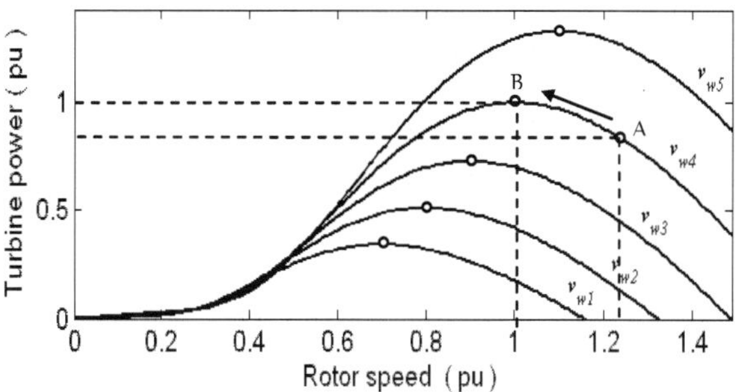

Fig. 16. Turbine power curves

6.3 Hill climb search control

HCS control method of MPPT control are presented in [27-29]. In [27], a simple HCS method is proposed wherein output power information required by the MPPT control algorithm is obtained using the dc link current and generator speed information. These two signals are the inputs to the MPPT controller whose output is the command speed signal required for maximum power extraction. The optimum speed command is applied to the speed control loop of the grid side converter control system. In this method, the signals proportional to the P_m is computed and compared with the previous value. When the result is positive, the process is repeated for a lower speed. The outcome of this next calculation then decides whether the generator speed is again to be increased or decreased by decrease or increase of the dc link current through setting the reference value of the current loop of the grid side converter control system. Once started, the controller continues to perturb itself by running through the loop, tracking to a new maximum once the operating point changes slightly. The output power increases until a maximum value is attained thus extracting maximum possible power.

The HCS control method presented in [28] operates the generator in speed control mode with the speed reference dynamically modified in accordance with the magnitude and direction of change of active power. Optimum power search algorithm proposed here uses the fact that $dP_o/d\omega=0$ at peak power point. The algorithm dynamically modifies the speed command in accordance with the magnitude and direction of change of active power in order to reach the peak power point.

In [29], the proposed MPPT method combines the ideas of sliding mode (SM) control and extremum seeking control (ESC). In this method only the active power of the generator is required as the input. The method does not require wind velocity measurement, wind-turbine parameters or rotor speed etc. The block diagram of the control system is shown in Fig. 17. In the figure ρ is the acceleration of P_{opt}. When the sign of derivative of ε changes, a sliding mode motion occurs and ω^* is steered towards the optimum value while P_o tracks P_{opt}. The speed reference for the vector control system is the optimal value resulting from the MPPT based on sliding mode ESC.

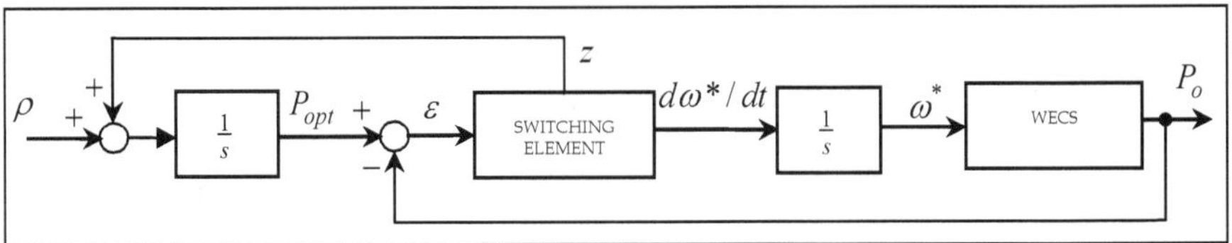

Fig. 17. Sliding mode extremum seeking MPPT control

7. Case study

An MPPT controller for variable speed WECS proposed in [30] is presented in this work as a case study. The method proposed in [30], does not require the knowledge of wind speed, air density or turbine parameters. The MPPT controller generates at its output the optimum speed command for speed control loop of rotor flux oriented vector controlled machine side converter control system using only the instantaneous active power as its input. The optimum speed commands, which enable the WECS to track peak power points, are generated in accordance with the variation of the active power output due to the change in the command speed generated by the controller. The proposed concept was analyzed in a direct drive variable speed PMSG WECS with back-to-back IGBT frequency converter. Vector control of the grid side converter was realized in the grid voltage vector reference frame. The complete WECS control system is shown in Fig. 18.

The MPPT controller computes the optimum speed for maximum power point using information on magnitude and direction of change in power output due to the change in command speed. The flow chart in Fig. 19 shows how the proposed MPPT controller is executed. The operation of the controller is explained below.

The active power $P_o(k)$ is measured, and if the difference between its values at present and previous sampling instants $\Delta P_o(k)$ is within a specified lower and upper power limits P_L and P_M respectively then, no action is taken; however, if the difference is outside this range, then certain necessary control action is taken. The control action taken depends upon the magnitude and direction of change in the active power due to the change in command speed.

- If the power in the present sampling instant is found to be increased i.e. $\Delta P_o(k) > 0$ either due to an increase in command speed or command speed remaining unchanged in the previous sampling instant i.e. $\Delta\omega^*(k-1) \geq 0$, then the command speed is incremented.

- If the power in present sampling instant is found to be increased i.e. $\Delta P_o(k) > 0$ due to reduction in command speed in the previous sampling instant i.e. $\Delta\omega^*(k-1) < 0$, then the command speed is decremented.

- Further, if the power in the present sampling instant is found to be decreased i.e.either due to a constant or increased command speed in the previous sampling instant i.e. $\Delta\omega^*(k-1) \geq 0$, then the command speed is decremented.

- Finally, if the power in the present sampling instant is found to be decreased i.e. $\Delta P_o(k) < 0$ due to a decrease in command speed in the previous sampling instant i.e. $\Delta\omega^*(k-1) < 0$, then the command speed is incremente

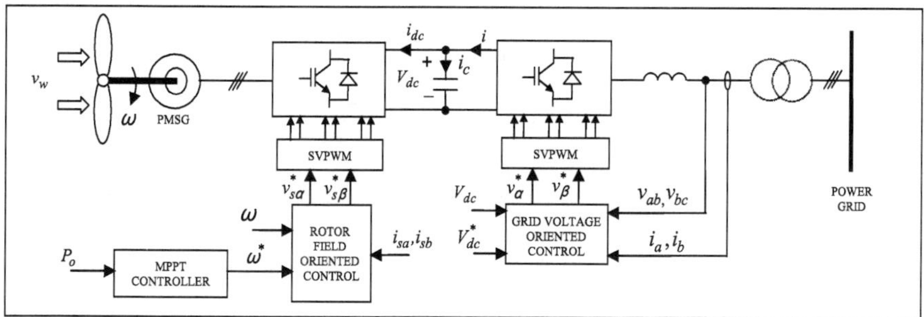

Fig. 18. PMSG wind energy conversion system.

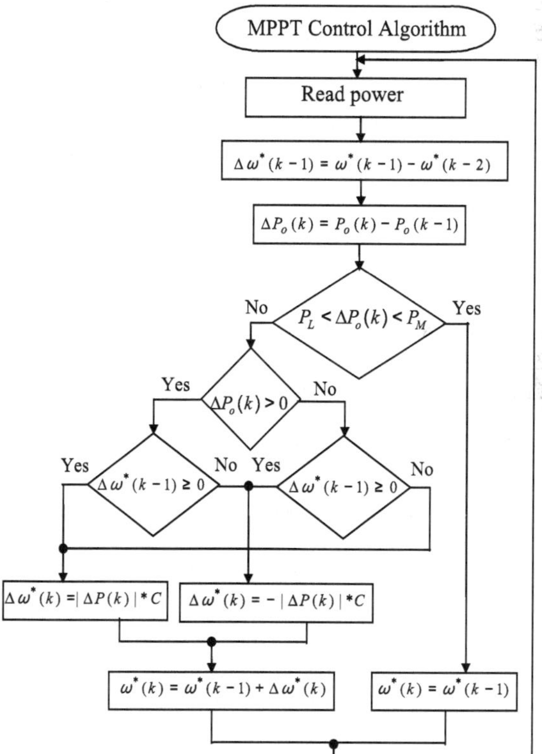

Fig. 19. Flow chart of MPPT controller.

The magnitude of change, if any, in the command speed in a control cycle is decided by the product of magnitude of power error $\Delta P_o(k)$ and C. The values C are decided by the speed of the wind. During the maximum power point tracking control process the product mentioned above decreases slowly and finally equals to zero at the peak power point.

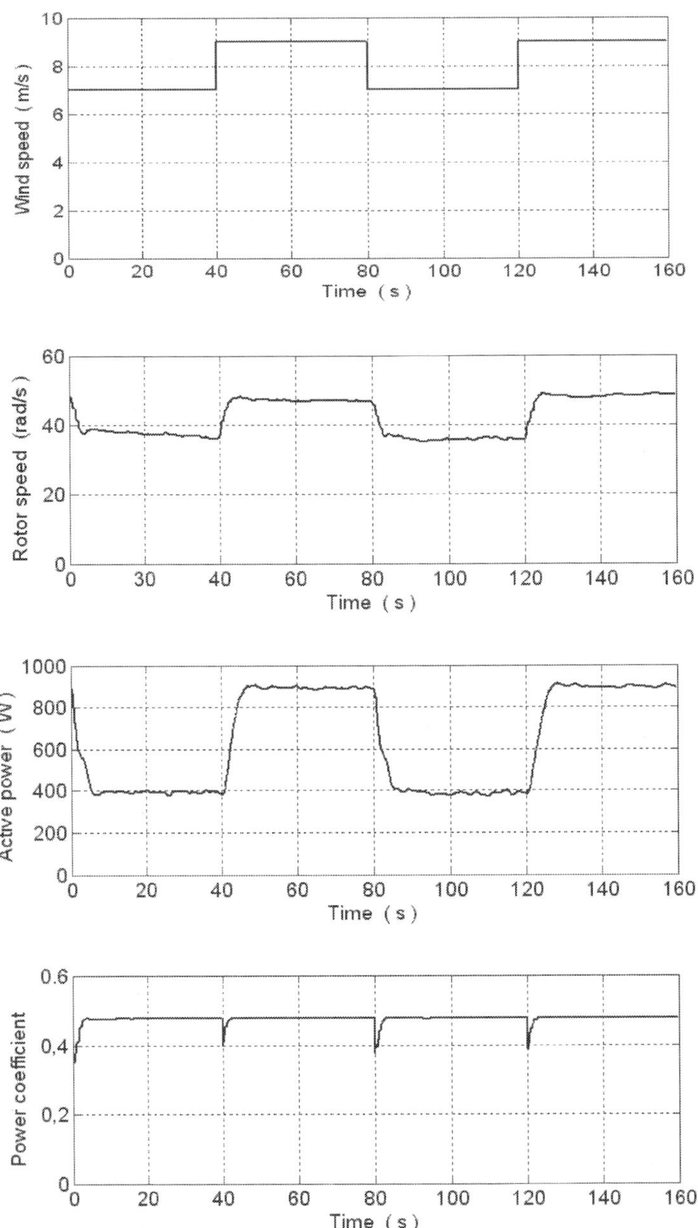

Fig. 20. Operation of the WECS under step wind speed profile.

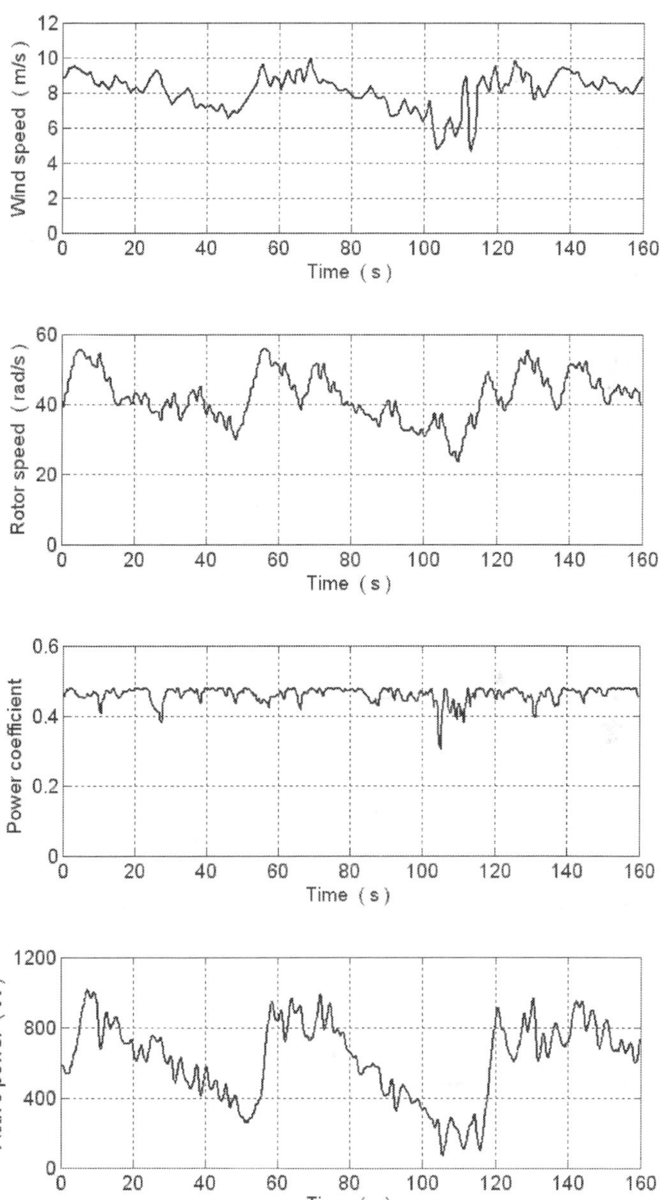

Fig. 21. Operation of the WECS under real wind speed profile.

In order to have good tracking capability at both high and low wind speeds, the value of C should change with the change in the speed of wind. The value of C should vary with variation in wind speed; however, as the wind speed is not measured, the value of command rotor speed is used to set its value. As the change in power with the variation in speed is lower at low speed, the value of C used at low speed is larger and its value decreases as speed increases. In this work, its values are determined by running several simulations with different values and choosing the ones which show best results.

The values of C, used in implementing the control algorithm, are computed by performing linear interpolation of 1.1 at 0 rad/s, 0.9 at 10 rad/s, 0.6 at 20 rad/s, 0.32 at 30 rad/s 0.26 at 40 rad/s, 0.25 at 50 rad/s and 0.24 at 55 rad/s.

During the simulation, the d axis command current of the machine side converter control system is set to zero; whereas, for the grid side converter control system the q axis command current is set to zero. Simulation was carried out for two speed profiles applied to the WECS, incorporating the proposed MPPT controller.

Initially, a rectangular speed profile with a maximum of 9 m/s and a minimum of 7 m/s was applied to the PMSG WECS in order to see the performance of the proposed controller. The wind speed, rotor speed, power coefficient and active power output for this case are shown in Fig. 20. Good tracking capability was observed. Then, a real wind speed profile was applied to the PMSG wind generator system. Fig. 21 shows for this case, the wind speed, rotor speed, power coefficient and active power. The maximum value of C_P of the turbine considered was 0.48, and it was found that in worst case, the value of C_P was 0.33 which shows good performance of the proposed controller. It can therefore be concluded from the results of simulation that the   proposed control algorithm has good capability of tracking peak power points. The method also has good application potential in other types of WECS.

8. Conclusions

Wind energy conversion system has been receiving widest attention among the various renewable energy systems. Extraction of maximum possible power from the available wind power has been an important research area among which wind speed sensorless MPPT control has been a very active area of research. In this chapter, a concise review of MPPT control methods proposed in various literatures for controlling WECS with various generators have been presented. There is a continuing effort to make converter and control schemes more efficient and cost effective in hopes of developing an economically viable solution to increasing environmental issues. Wind power generation has grown at an alarming rate in the past decade and will continue to do so as power electronic technology continues to advance.

9. References

[1] M. Pucci and M. Cirrincione, "Neural MPPT control of wind generators with induction machines without speed sensors," *IEEE Trans. Ind. Elec.*, vol. 58, no. 1, Jan. 2011, pp. 37-47.

[2] Q. Wang and L. Chang, "An intelligent maximum power extraction algorithm for inverter-based variable speed wind turbine systems," *IEEE Trans. Power Electron.*, vol. 19, no. 5, pp. 1242-1249, Sept. 2004.

[3] H. Li, K. L. Shi and P. G. McLaren, "Neural-network-based sensorless maximum wind energy capture with compensated power coefficient," *IEEE Trans. Ind. Appl.*, vol. 41, no. 6, pp. 1548-1556, Nov./Dec. 2005.

[4] A. B. Raju, B. G. Fernandes, and K. Chatterjee, "A UPF power conditioner with maximum power point tracker for grid connected variable speed wind energy conversion system," proc. of *1st International Conf. on Power Electronics Systems and Applications (PESA 2004)*, Bombay, India, 9-11 Nov., 2004, pp. 107-112.

[5] E. Koutroulis and K. Kalaitzakis, "Design of a maximum power tracking system for wind-energy-conversion applications," *IEEE Transactions on Industrial Electronics*, vol. 53, no. 2, April 2006, pp. 486-494.

[6] M. Matsui, D. Xu, L. Kang, and Z. Yang, "Limit Cycle Based Simple MPPT Control Scheme for a Small Sized Wind Turbine Generator System," *Proc. of 4th International Power Electronics and Motion Control Conference*, Xi'an, Aug., 14-16, 2004, vol. 3, pp. 1746-1750.

[7] Y. Higuchi, N. Yamamura, and M Ishida, "An improvement of performance for small-scaled wind power generating system with permanent magnet type synchronous generator," in *Proc. IECON*, 2000.

[8] S. Wang, Z. Qi, and T. Undeland, "State space averaging modeling and analysis of disturbance injection method of MPPT for small wind turbine generating systems," in *Proc. APPEEC*, 2009.

[9] R. J. Wai, C.Y. Lin, and Y.R. Chang, "Novel maximum-power extraction algorithm for PMSG wind generation system," *IET Electric Power Applications*, vol. 1, no. 2, March 2007, pp. 275-283.

[10] J. Yaoqin, Y. Zhongqing, and C. Binggang, "A new maximum power point tracking control scheme for wind generation," in *Proc. International Conference on Power System Technology 2002 (PowerCon 2002)*. 13-17 Oct., 2002. pp.144-148.

[11] J. Hui and A. Bakhshai, "A new adaptive control algorithm for maximum power point tracking for wind energy conversion systems," in *Proc. IEEE PESC 2008*, Rhodes, 15-19 June, 2008. pp. 4003-4007.

[12] J. Hui and A. Bakhshai, "Adaptive algorithm for fast maximum power point tracking in wind energy systems," in *Proc. IEEE IECON 2008*, Orlando, USA, 10-13 No. 2008, pp. 2119-2124.

[13] M. G. Molina and P. E. Mercado, "A new control strategy of variable speed wind turbine generator for three-phase grid-connected applications," in *Proc. IEEE/PES Transmission and Distribution Conference and Exposition: Latin America, 2008,* Bogota, 13-15 Aug., 2008, pp. 1-8.

[14] T. Tafticht, K. Agbossou and A. Chériti, "DC Bus Control of Variable Speed Wind Turbine Using a Buck-Boost Converter," in *Proc. IEEE Power Eng. Society General Meeting*, Montreal, 18-22 June, 2006.

[15] J. M. Kwon, J. H. Kim, S. H. Kwak, and H. H. Lee, "Optimal power extraction algorithm for DTC in wind power generation systems," in *Proc. IEEE International Conf. on Sustainable Energy Technology, (ICEST 2008)*, Singapore, 24-27 Nov., 2008 pp. 639 - 643.

[16] C. Patsios, A. Chaniotis, and A. Kladas, "A Hybrid Maximum Power Point Tracking System for Grid-Connected Variable Speed Wind-Generators," in *Proc. IEEE PESC 2008,* Rhodes, 15-19 June, 2008, pp.1749-1754.

[17] J. S. Thongam, P. Bouchard, H. Ezzaidi, and M. Ouhrouche, "ANN-Based Maximum Power Point Tracking Control of Variable Speed Wind Energy Conversion Systems," Proc. of the 18th *IEEE International Conference on Control Applications 2009*, July 8-10, 2009, Saint Petersburg, Russia.

[18] W. M. Lin, C. M. Hong, and F. S. Cheng, "Fuzzy neural network output maximization control for sensorless wind energy conversion system," *Energy*, vol. 35, no. 2, February 2010, pp. 592-601.

[19] A. G. Abo-Khalil and D. C. Lee, "MPPT control of wind generation systems based on estimated wind speed using SVR," IEEE Trans. Ind. Appl., vol. 55, no. 3, March 2008, pp. 1489-1490.

[20] R. Hilloowala and A. M. Sharaf, "A rule-based fuzzy logic controller for a PWM inverter in a stand alone wind energy conversion scheme," *IEEE Trans. Ind. Applicat.*, vol. IA-32, pp. 57-65, Jan. 1996.

[21] M. G. Simoes, B. K. Bose, and R. J. Spiegal, "Fuzzy logic-based intelligent control of a variable speed cage machine wind generation system," *IEEE Trans. Power Electron.*, vol. 12, no.1, pp. 87-95, Jan. 1997.

[22] A. G. Abo-Khalil, D. C. Lee, and J. K. Seok, "Variable speed wind power generation system based on fuzzy logic control for maximum power output tracking", *in Proc. 35th Annual IEEE-PESC'04*, vol. 3, pp. 2039-2043, 2004.

[23] C. Shao, X. Chen and Z. Liang, "Application research of maximum wind-energy tracing controller based adaptive control Strategy in WECS," Proc. of the CES/IEEE 5th *International Power Electronics and Motion Control Conference (IPEMC 2006)*, Shanghai, Aug. 14-16, 2006.

[24] S. Bhowmik, R. Spee, and J. H. R. Enslin, "Performance optimization for doubly fed wind power generation systems", *IEEE Trans. Ind. Appl.*, Vol. 35, No. 4, pp. 949-958, July/Aug. 1999.

[25] G. Hua and Y. Geng, "A novel control strategy of MPPT taking dynamics of the wind turbine into account," Proc. of 37th IEEE Power Electronics specialist Conf., PESC'06, 18-22 June, 2006, pp. 1-6.

[26] V. Galdi, A. Piccolo, and P. Siano, "Designing an adaptive fuzzy controller for maximum wind energy extraction," *IEEE Trans. Energy Conversion*, vol. 23, no. 2, June 2008, pp. 559-569.

[27] J. H. R. Enslin and J. D. Van Wyk, "A study of a wind power converter with micro-computer based maximal power control utilizing an over-synchronous electronic Scherbius cascade," *Renewable Energy*, vol. 2, no. 6, pp. 551-562, 1992.

[28] R. Datta and V. T. Ranganathan, "A method of tracking the peak power points for a variable speed wind energy conversion system", *IEEE Trans. on Energy Conversion*, vol. 18, no. 1, pp. 163-168, March 2003.

[29] T. Pan, Z. Ji, and Z. Jiang, "Maximum power point tracking of wind energy conversion systems based on sliding mode extremum seeking control," Proc. of the *IEEE Energy 2030*, Atlanta, GA, USA, 17-18 Nov., 2008, pp. 1-5.

[30] J. S. Thongam, P. Bouchard, H. Ezzaidi, and M. Ouhrouche, "Wind speed sensorless maximum power point tracking control of variable speed wind energy conversion systems," Proc. of the *IEEE International Electric Machines and Drives Conference IEMDC 2009*, May 3-6, 2009, Florida, USA.

16

Modelling and Environmental/Economic Power Dispatch of MicroGrid Using MultiObjective Genetic Algorithm Optimization

Faisal A. Mohamed[1] and Heikki N. Koivo[2]
[1]*Department of Electrical Engineering, Omar Al-Mukhtar Universityhe University,*
[2]*Department of Automation and Systems Technology, Aalto University,*
[1]*Libya*
[2]*Finland*

1. Introduction

MultiObjective (MO) optimization has a very wide range of successful applications in engineering and economics. Such applications can be found in optimal control systems (Liu et al., 2007), and communication (Elmusrati et al., 2007). The MO optimization can be applied to find the optimal solution which is a compromise between multiple and contradicting objectives.

In MO optimization we are more interested in the Pareto optimal set which contains all non-inferior solutions. The decision maker can then select the most preferred solution out of the Pareto optimal set.

The weighted sum method to handle MO optimization applied in this paper. Furthermore, the weighted sum is simple and straightforward method to handle MO optimization problems.

The need for more flexible electric systems to cope with changing regulatory and economic scenarios, energy savings and environmental impact is providing impetus to the development of MicroGrids (MG), which are predicted to play an increasing role of the future power systems (Hernandez-Aramburo et al., 2005). One of the important applications of the MG units is the utilization of small-modular residential or commercial units for onsite service. The MG units can be chosen so that they satisfy the customer load demand at compromise cost and emissions all the time. Solving the environmental economic problem in the power generation has received considerable attention. An excellent overview on commonly used environmental economic algorithms can be found in (Talaq et al., 1994). The environmental economic problems have been effectively solved by multiobjective evolutionary in (Abido, 2003) and fuzzy satisfaction-maximizing approach (Huang et al., 1997).

Several strategies have been reported in the literature related to the operation costs as well as minimizing emissions of MG. In (Hernandez-Aramburo et al., 2005) the optimization is aimed at reducing the fuel consumption rate of the system while constraining it to fulfil the local energy demand (both electrical and thermal) and provide a certain minimum reserve power. In (Hernandez-Aramburo et al., 2005) and (Mohamed & Koivo, 2010), the problem is treated as a single objective problem. This formulation, however, has a severe difficulty in finding the best trade-off relations between cost and emission. In (Mohamed & Koivo, 2007) the problem is handled as a multiobjective optimization problem without considering the sold and purchased power.

The algorithm in (Mohamed & Koivo, 2008) is modified in this chapter; the modification is to optimize the MG choices to minimize the total operating cost. Based on sold power produced by Wind turbine and Photovoltaic Cell, then the algorithm determines the optimal selection of power required to meet the electrical load demand in the most economical and environmental fashion.

Furthermore, the algorithm consists of determining at each iteration the optimal use of the natural resources available, such as wind speed, temperature, and irradiation as they are the inputs to windturbine, and photovoltaic cell, respectively. If the produced power from the wind turbine and the photovoltaic cell is less than the load demand then the algorithm goes to the next stage which is the use of the other alternative sources according to the load and the objective function of each one.

This chapter assumes the MG is seeking to minimize total operating costs. MicroGrids could operate independently of the uppergrid, but they are usually assumed to be connected, through power electronics, to the uppergrid. The MG in this paper is assumed to be interconnected to the uppergird, and can purchase some power from utility providers when the production of the MG is insufficient to meet the load demand. There is a daily income to the MG when the generated power exceeds the load demand.

The second objective of this chapter deals with solving an optimization problem using several scenarios to explore the benefits of having optimal management of the MG. The exploration is based on the minimization of running costs and is extended to cover a load demand scenario in the MG. Furthermore, income also considered from sold power of WT and PV. Switching one load is considered in this paper. It will be shown that by developing a good system model, we can use optimization to solve the cost optimization problem accurately and efficiently. The result obtained is compared with the results obtained from (Mohamed & Koivo, 2009).

2. System description

The MG architecture studied is shown in Fig 1. It consists of a group of radial feeders, which could be part of a distribution system. There is a single point of connection to the utility called point of common coupling (PCC). The feeders 1 and 2 have sensitive loads which should be supplied during the events. The feeders also have the microsources consisting of a photovoltaic (PV), a wind turbine (WT), a fuel cell (FC), a microturbine (MT), and a diesel generator (DG). The third feeder has only traditional loads. A static switch (SD) is used to island the feeders 1 and 2 from the utility when events requiring it happen. The fuel input is needed only for the DG, FC, and MT as the fuel for the WT and PV comes from nature. To serve the load demand, electrical power can be produced either directly by PV, WT, DG, MT, or FC. The diesel oil is a fuel input to the DG, whereas natural gas is a fuel input to a fuel processor to produce hydrogen for the FC. The gas is also the input to the MT. Each component of the MG system is modeled separately based on its characteristics and constraints. The characteristics of some equipment like wind turbines and diesel generators are available from manufacturers.

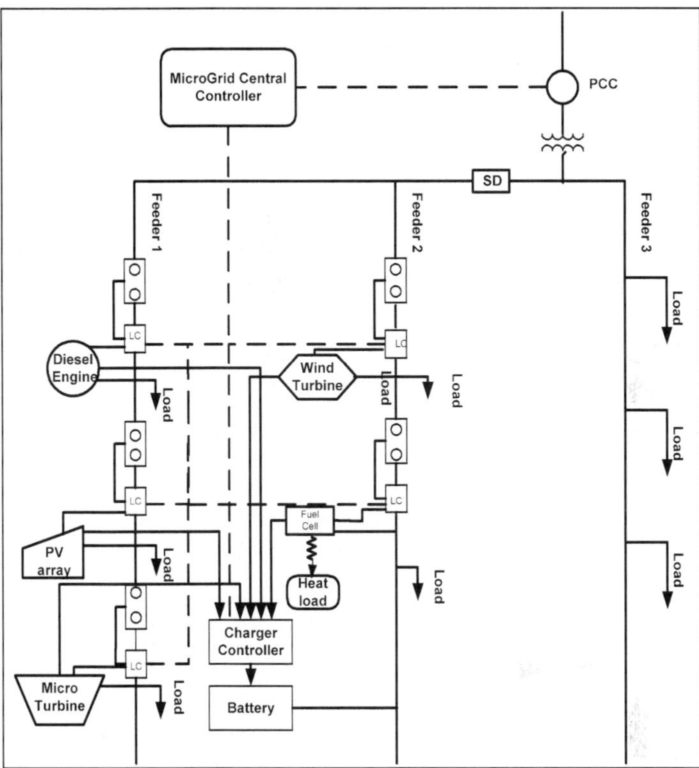

Fig. 1. MicroGrid Architecture.

3. Optimization model

The power optimization model is formulated as follows. The output of this model is the optimal configuration of a MG taking into account the technical performance of supply options, locally available energy resources, load demand characteristics, environmental costs, start up cost, daily purchased-sold power tariffs, and operating and maintenance costs.

Figure 2. illustrates the optimization model, when its inputs are:

- Power demand by the load.
- Data about locally available energy resources: These include wind speed (m/s) Figure 3, temperature (C°) Figure 4 , solar irradiation data (W/m²) Figure 5 , as well as cost of fuels ($/liter) for the DG and natural gas price for supplying the FC and MT ($/kW).
- Daily purchased and sold power tariffs in (kWh).
- Start up costs in ($/h).
- Technical and economic performance of supply options: These characteristics include, for example, rated power for PV, power curve for WT, fuel consumption characteristics DG and FC.

- Operating and maintenance costs and emission factors: Operating and maintenance costs must be given (\$/h) for all generators; emission factors must be given in kg/h for DG, FC, and MT.

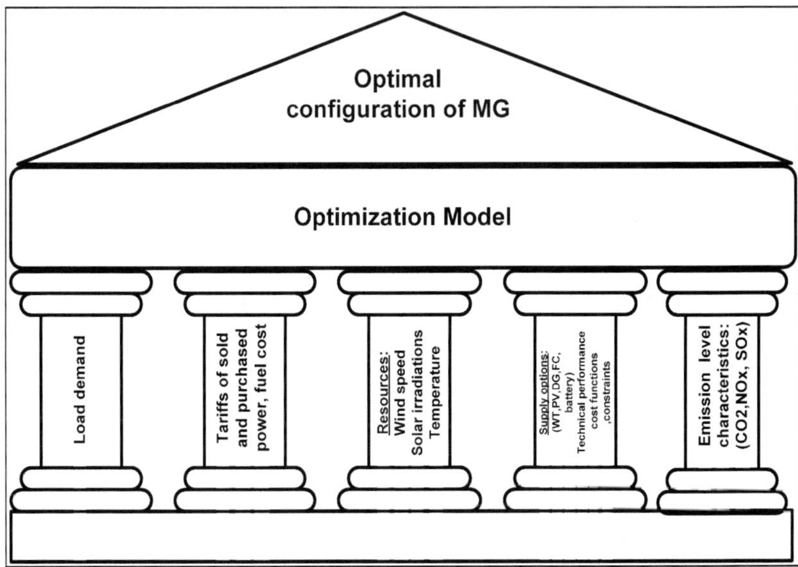

Fig. 2. The Optimization Model.

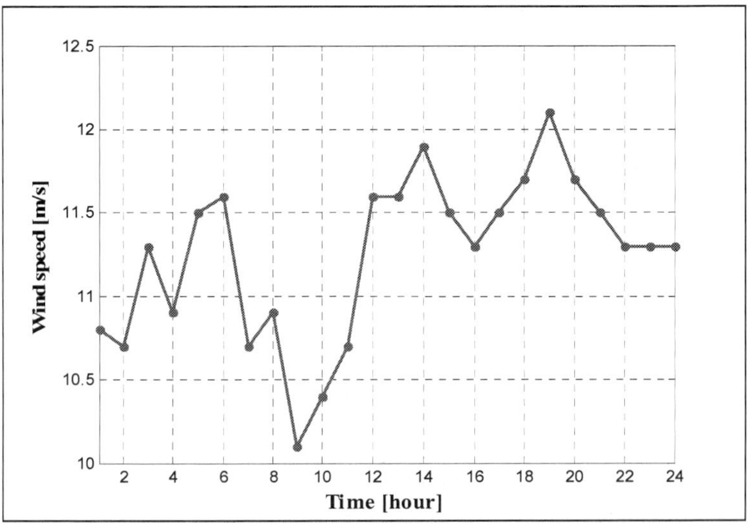

Fig. 3. The input wind speed as used in the model.

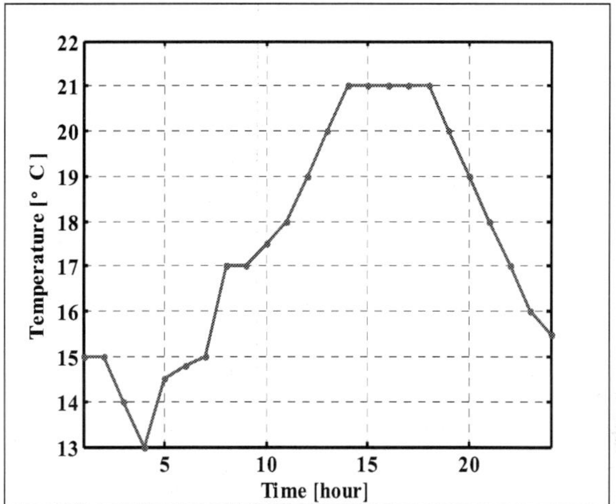

Fig. 4. The input temperature data as used in the model.

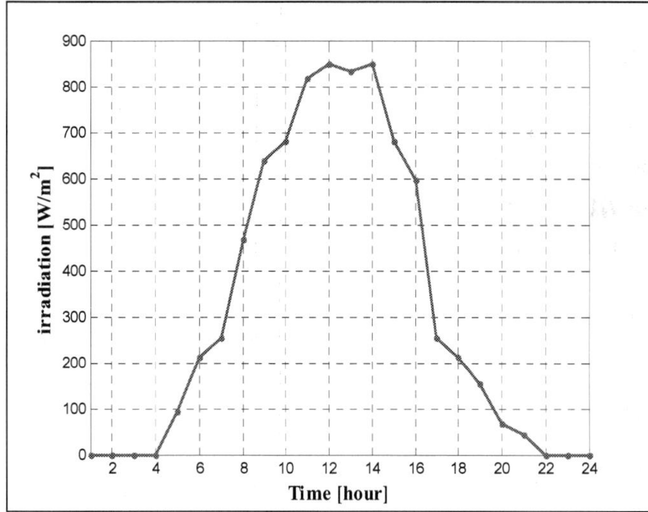

Fig. 5. The input irradiation data as used in the model.

4. Optimization model

4.1 Wind turbine

To model the wind turbine, several important factors should be known. They are the availability of the wind and the wind turbine power curve. The following is the model used to calculate the output power generated by the wind turbine generator as a function of the wind velocity (Chedid et al., 1998):

$$\begin{cases} P_{WT} = 0, & V_{ac} < V_{ci} \\ P_{WT} = aV_{ac}^2 + bV_{ac} + c, & V_{ci} \le V_{ac} < V_r \\ P_{WT,r} = 130, & V_r \le V_{ac} > V_{co} \end{cases} \qquad (1)$$

where $P_{WT,r}$, V_{ci}, and V_{co} are the rated power, cut-in and cut-out wind speed respectively. Furthermore V_r, and V are the rated, and actual wind speed. Constants **a**, **b**, and **c** depend on the type of the WT.

We assume AIR403 wind turbine model in this paper. According to the data from the manufacturer, the turbine output $\mathbf{P_{WT,r}}$ is roughly 130 W if the wind speed is greater than approximately 18 m/s.

In Fig 6. we model the wind turbine power curve according to equation 1, with the actual power curve obtained from the owner's manual. The parameters used to model the power curve are as follows:

a = 3.4; b = -12; c = 9.2; $\mathbf{P_{WT,r}}$= 130 watt; $\mathbf{V_{ci}}$ = 3.5 m/s; $\mathbf{V_{co}}$= 18 m/s; $\mathbf{V_r}$ =17.5 m/s.

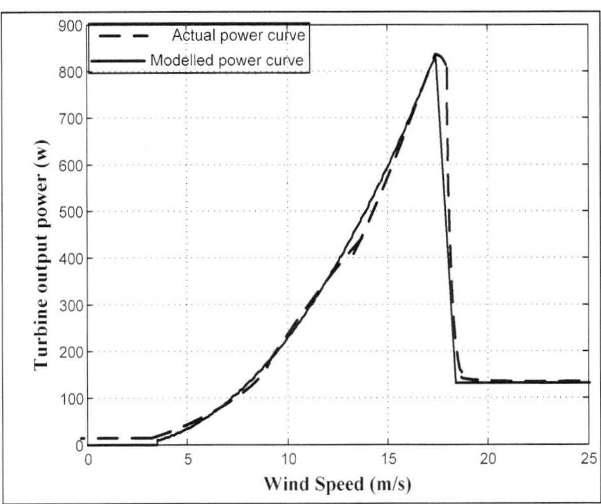

Fig. 6. The actual and modeled power curve of AIR403.

4.2 Photovoltaic

Photovoltaic generations are systems which convert the sunlight directly to electricity. The characteristics of the PV in operating conditions that differ from the standard condition (1000 W/m², 25C° cell temperature), the effect of solar irradiation and ambient temperature on PV characteristics are modeled. The influence of solar intensity is modeled by considering the power output of the module to be proportional to the irradiance (Gavanidou & Bakirtzis 1992), (Lasnier & Ang 1990). The PV Modules are treated at Standard Test Condition (STC). The output power of the module can be calculated using:

$$P_{PV} = P_{STC}\frac{G_{ING}}{G_{STC}}\left[1 + k(T_c - T_r)\right] \qquad (2)$$

where:

P_{PV} The output power of the module at Irradiance G_{ING},

P_{STC} The Module maximum power at Standard Test Condition (STC),

G_{ING} Incident Irradiance,

G_{STC} Irradiance at STC 1000 (W/m²,

k   Temperature coefficient of power,

T_c The cell temperature,

T_r The reference temperature.

We assume that SOLAREX MSX-83 modules are used in this paper. Their output characteristics are: peak power = 83W, voltage at peak power = 17.1V, current at peak power = 4.84A, short circuit current = 5.27A, and open circuit voltage =21.2V at STC.

4.3 Diesel generator costs

Diesel engines are the most common type of MG technology in use today. The traditional roles of diesel generation have been the provision of stand-by power and peak shaving. The fuel cost of a power system can be expressed mainly as a function of its real power output and can be modeled by a quadratic polynomial (Wood & Wollenberg 1996. The total $/h DG fuel cost $F_{DG,i}$ can be expressed as:

$$F_{DG,i} = \sum_{i=1}^{N}(d_i + e_i P_{DG,i} + f_i P_{DG,i}^2) \tag{3}$$

where N is the number of generators, d_i, e_i, and f_i are the coefficients of the generator, $P_{DG,i}$, $i = 1,2,...,N$ is the diesel generator i output power (kW), assumed to be numerically known. Typically, the constants d_i, e_i, and f_i are given by the manufacturer. For example, diesel fuel consumption data of a 6-kW diesel generator set (Cummins Power) model DNAC 50 Hz is available in L/h at 1/4, 1/2, 3/4 and full loads. From the data sheet the parameters in eq (3) are: d_i=0.4333, e_i=0.2333, and f_i=0.0074. Figure 7 shows the fuel consumption as function of power of the DNAC 50 Hz diesel engine.

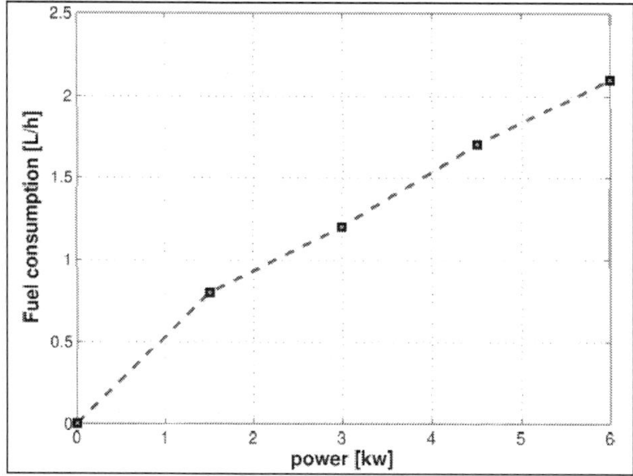

Fig. 7. Fuel consumption of DNAC 50 Hz diesel engine.

4.4 Fuel cell cost

Fuel cells work by combining hydrogen with oxygen to produce electricity, heat, and water. DC current and heat are produced by a chemical reaction rather than by a mechanical process driven by combustion. Fuel cells can operate as long as fuel is being supplied, as opposed to the fixed supply of chemical energy in a battery.

The efficiency of the FC depends on the operating point, and it refers to the ratio of the stack output power to the input energy content in the natural gas. It is normally calculated as the ratio of the actual operating voltage of a single cell to the reversible potential (1.482V) (Barbir & Gomez, 1996). The overall unit efficiency is the efficiency of the entire system including auxiliary devices.

We assume the typical efficiency curves of the Protone Exchange Membrane (PEM) fuel cell including the cell and the overall efficiencies (Barbir & Gomez, 1996). The efficiency of any fuel cell is the ratio between the electrical power output and the fuel input, both of which must be in the same units (W), (Azmy & Erlic, 2005):

The fuel cost for the fuel cell is calculated as follows:

$$F_{FC,i} = C_{nl} \sum_J \frac{P_J}{\eta_J} \tag{4}$$

where

C_{nl} the natural gas price to supply the fuel cell (\$/kWh),

P_J the net electrical power produced at interval \$J\$,

η_J the cell efficiency at interval J.

To model the technical performance of a PEM fuel cell (Barbir & Gomez, 1996), a typical efficiency curve is used to develop the cell efficiency as a function of the electrical power and used in equation (4).

4.5 Microturbine cost

Microturbines use a simple design with few moving parts to improve reliability and reduce maintenance costs. Microturbine models are similar to those of fuel cells (Campanari & Macchi,2004). However, the parameters and curves are modified to properly describe the performance of the MT unit. From a typical electrical and thermal efficiency curves of a 25 kW Capstone- C30 microturbine (Yinger, 2001) we obtain the efficiency as function of the generated power. The total efficiency of a microturbine can be written as:

$$\eta_J = \frac{P_{el} + P_{th,rec}}{m_f LHV_f} \tag{5}$$

where

P_{el} the net electrical output power (kW),

$P_{th,rec}$ the thermal power recovered (kW),

LHV_f the fuel lower heating rate (kJ/kgf),

m_f the mass flow rate of the fuel (kg/s).

Unlike the fuel cell, the efficiency of the MT increases with the increase of the supplied power. The MT fuel cost is as follows:

$$C_{MT} = C_{nl} \sum_{J} \frac{P_J}{\eta_J} \tag{6}$$

where

C_{nl} is the natural gas price to supply the MT,

P_J is the net electrical power produced at interval J,

η_J is the cell efficiency at interval J.

Since power required by our assumed MG is much lower than the level in (Yinger, 2001), the curves of this MT are rescaled to be suitable for a unit with a 4kW rating. These curves are used to derive the electrical efficiency and power as functions of the electrical power to be used in the economic model of the MT.

4.6 Battery storage

Battery banks are electrochemical devices that store energy from other AC or DC sources for later use. The power from the battery is needed whenever the microsources are insufficient to supply the load, or when both the microsources and the main grid fail to meet the total load demand. On the other hand, the energy is stored whenever the supply from the microsources exceeds the load demand.

The following assumptions are used to model the battery bank: the charge and discharge current is limited to 10% of battery AH capacity (the storage capacity of a battery is measured in terms of its ampere-hour (AH) capacity) (Chedid & Rahman, 1997) and (Manisaet, 2004); the round- trip efficiency is 95 %.

When determining the state of charge for an energy storage device, two constraint equations must be satisfied at all times: First, because it is impossible for an energy storage device to contain negative energy, the maximum state of charge (SOC_{max}) and the minimum state of charge (SOC_{min}) of the battery are 100 % and 20 % of its AH capacity, respectively.

The constraints that represent the maximum allowable charge and discharge current to be less than 10% of battery AH capacity are shown in the following equations, respectively.

$$P_+ \leq (0.1 \times V_{sys} \times U_{batt}) / \Delta_t \tag{7}$$

$$P_- \leq (0.1 \times V_{sys} \times U_{batt}) / \Delta_t \tag{8}$$

where the parameter V_{sys} is the system voltage at the DC bus, Δ_t is the time in hours, and the parameter U_{batt} is the battery capacity in AH. The state of charge (SOC) of the battery can be obtained by monitoring the charge/discharge power of the battery, as shown in eq.(9).

$$SOC = SOC_{max} - P_- + P_+ \tag{9}$$

It is important that the SOC of the battery prevents the battery from overcharging or undercharging. The associated constraints can be formulated by comparing the battery SOC in any hour i with the battery SOC_{min} and the battery SOC_{max}, as shown in eq(10).

This research assumes that SOC_{min} and SOC_{max} equal 20% and 100% of the battery AH capacity, respectively. It is also assumed that the initial SOC of the battery is 100% at the beginning of the simulation.

The constraints on battery SOC are:

$$SOC_{min} \leq SOC \leq SOC_{max} \tag{10}$$

Finally, in order for the system with battery to be sustained over a long period of time, the battery SOC at the end must be greater than a given percentage of its $\mathbf{SOC_{max}}$. This study assumes 90%.

5. Proposed objective function

The major concern in the design of an electrical system that utilizes MG sources is the accurate selection of output power that can economically satisfy the load demand, while minimizing the emission. Hence the system components are found subject to:
1.   Minimize the operation cost ($/h).
2.   Minimize the emissions (kg/h).
3.   Ensure that the load is served according to the constraints.

5.1 Operating cost
As shown in Fig.1, the main utility balances the difference between the load demand and the generated output power from microsources. Therefore there is a cost to be paid for the purchased power whenever the generated power is insufficient to cover the load demand. On the other hand, there is income because of sold power when the power generated is higher than the load demand but the price of the sold power is lower than the purchased power tariff. It is possible that there will be no sold power at all. To model the purchased and sold power, the cost function takes the form

$$CF(\boldsymbol{P}) = \sum_{i=1}^{N} (C_i F_i(P_i) + OM_i(P_i) + STC_i + DCPE_i - IPSE_i) \tag{11}$$

$CF(\boldsymbol{P})$ represents the operating costs in $/h, C_i fuel costs of the generating unit i in $/1 for the DG, natural gas price for supplying the FC and MT ($/kWh), $F_i(P_i)$ fuel consumption rate of a generator unit i, $OM_i(P_i)$ Operation and maintenance cost of the generating unit i in $/h, P_i decision variables, representing the real power output from generating unit i in kW and defined as: $P_i = P_{FC}$ or P_{MT} or P_{DG}, $\boldsymbol{P}$ is the vector of the generators active power and is defined as: $\boldsymbol{P} = [P_1, P_2, ..., P_N]$, N is the total number of generating units.
where STC_i is the start-up costs of the unit generator i $/h. The start-up cost in any given time interval can be represented by an exponential cost curve:

$$STC_i = \sigma_i + \delta_i \left[1 - \exp\left(\frac{-T_{off,i}}{\tau_i} \right) \right] \tag{12}$$

The start-up cost depends on the time the unit has been off prior to a start up.
where, σ_i is the hot start-up cost, δ_i the cold start-up cost, τ_i the unit cooling time constant and $T_{off,i}$ is the time a unit has been off.
$DCPE_i$ is the daily purchased electricity of unit i if the load demand exceeds the generated power in $/h. $IPSE_i$ is the daily income for sold electricity of unit i if the output generated

power exceeds the load demand in \$/h. the expression for $DCPE_i$ and $IPSE_i$ can be written as

$$DCPE_i = C_p \times \max(P_L - P_i, 0)$$
$$IPSE_i = C_s \times \max(P_i - P_L, 0) \tag{13}$$

where, C_p and C_s are the tariffs of the purchased and sold power respectively in (\$/kWh).
System Constraints:
Power balance constraints: To meet the active power balance, an equality constraint is imposed

$$\sum_{i=1}^{N} P_i = P_L - P_{PV} - P_{WT} - P_{batt} \tag{14}$$

where P_L is the total power demanded in kW, P_{PV} the output power of the photovoltaic cell in kW, P_{WT} the output power of the wind turbine in kW,
P_{batt} the output power of the battery in kW.
Generation capacity constraints: For stable operation, real power output of each generator is restricted by lower and upper limits as follows:

$$P_i^{\min} \le P_i \le P_i^{\max} \quad \forall_i = 1, ..., N \tag{15}$$

where, $P_i^{\min}$ is the minimum operating power of unit i and $P_i^{\max}$ the maximum operating power of unit i.
Each generating unit has a minimum up/down time limit (MUT/MDT). Once the generating unit is switched on, it has to operate continuously for a certain minimum time before switching it off again. On the other hand, a certain stop time has to be terminated before starting the unit. The violation of such constraints can cause shortness in the life time of the unit. These constraints are formulated as continuous run/stop time constraints as follows (Abido, 2003).

$$(T_{t-1,i}^{on} - MUT_i)(u_{t-1,i} - u_{t,i}) \ge 0$$
$$(T_{t-1,i}^{off} - MDT_i)(u_{t,i} - u_{t-1,i}) \ge 0 \tag{16}$$

$T_{t-1,i}^{off} / T_{t-1,i}^{on}$ represent the unit i off/on time, at time $t-1$, while $u_{t-1,i}$ denotes the unit off/on [0,1] status.
Finally the number of starts and stops ($\varepsilon_{start-stop}$) should not exceed a certain number ($N_{\max}$).

$$\varepsilon_{start-stop} \le N_{\max} \tag{17}$$

The operating and maintenance costs OM are assumed to be proportional with the produced energy, where the proportionally constant is K_{OM_i} for unit i .

$$OM = \sum_{i=1}^{N} K_{OM_i} P_i \tag{18}$$

The values of K_{OM_i} for different generation units are as follows :
where,

$$K_{OM_1} = K_{OM}(DG) = 0.01258 \ \$/kWh.$$

$$K_{OM_2} = K_{OM}(FC) = 0.00419 \ \$/kWh.$$

$$K_{OM_2} = K_{OM}(MT) = 0.00587 \ \$/kWh.$$

5.2 Emission level

The atmospheric pollutants such as sulphur oxides SO_2, carbon oxides CO_2, and nitrogen oxides NOx caused by fossil-fueled thermal units can be modeled separately. The total kg/h emission of these pollutants can be expressed as (Morgantown, 2001):

$$E(\boldsymbol{P}) = \sum_{i=1}^{N} 10^{-2}(\alpha_i + \beta_i P_i + \gamma_i P_i^2) + \zeta_i \exp(\lambda_i P_i) \tag{19}$$

where α_i, β_i, γ_i, ζ_i, and λ_i are nonnegative coefficients of the ith generator emission characteristics.

For the emission model introduced in (Talaq et al., 1994) and (Morgantown, 2001), we propose to evaluate the parameters α_i, β_i, γ_i, ζ_i, and λ_i using the data available in (Orero & Irving, 1997) Thus, the emission per day for the DG, FC, and MT is estimated, and the characteristics of each generator will be detached accordingly.

6. Implementation of the algorithm

The following items summarize the key characteristics of the proposed algorithm:

- Power output of WT is calculated according to power the relation between the wind speed and the output power.
- Power output of PV is calculated according to the effect of the temperature and the solar radiation that are different from the standard test condition.
- We assume that the WT and PV deliver free cost power in terms of running as well being emission free. Furthermore, their output power is treated as a negative load, determine the different between the actual load and WT and PV output power. If the output from PV and WT is greater than the load, the excess power is directed to charge the battery.
- The power from the battery is needed whenever the PV and WT are insufficient to serve the load. Meanwhile the charge and discharge of the battery is monitored.
- The net load is calculated if the output from PV and WT is smaller than the total load demand.
- Choose serving the load by other sources (FC or MT or DG) according to the objective functions.
- If the output power is not sufficient then purchase power from the main gird, and if the output power is more than the load demand, sell the exceed power to the main grid.

7. Multiobjective genetic algorithm

There are two common goals in all multiobjective GA implementations. First, to move the population toward the Pareto optimal front; and second, to maintain diversity (either in parameter space or objective space) in the population so that multiple solutions can be developed. GA approaches to multiobjective optimization can be grouped into three categories: approaches that use aggregating functions, non-Pareto based approaches, and Pareto based approaches. The simplest and most obvious approach to multiobjective optimization is to combine the objectives into one aggregating function, and to treat the problem like a single objective optimization problem. Therefore, it is commonly used because of its simplicity and computational efficiency. The weighted sum approach combines objectives using weights.

The weighted sum approach combines k objectives f_i using weights, $w_i, i = 1,...,k$

$$fitness = w_1 f_2(\boldsymbol{P}) + w_2 f_2(\boldsymbol{P}) + ... + w_k f_k(\boldsymbol{P})$$

(20)

The weights are real numbers $w_i \geq 0$ and $\boldsymbol{P}$ as in (11)

8. Results and discussion

At first, the optimization model is applied to the load. The load demand varies from 4 kW to 14 kW. The available power from the PV and the WT are used first. The best results of the cost and emission functions, when optimized individually, are given in Table 1. Convergence of operation cost and emission objectives for both approaches, when the purchased tariff is 0.12 $/kWh and the sold tariff 0.07 $/kWh, is as shown in Figure 8, where faster convergence is achieved.

Figure 9 illustrates the hourly operating costs and emissions. However, the costs and emissions are high when the generators are on and the load is high. Operational cost and emission objectives are optimized individually in order to explore the extreme points of the trade-off surface. The first case is when the cost objective function is optimized and the second when the emission objective function is optimized.

The set of power curve found by the optimization algorithms is shown in Figure 10. The figure confirms that when the load demand is low, the best choice in terms of cost is to use the output power from MT. The second best choice is the use of the fuel cell.

When the load is high at the peak time, all the generators are used to serve the load.

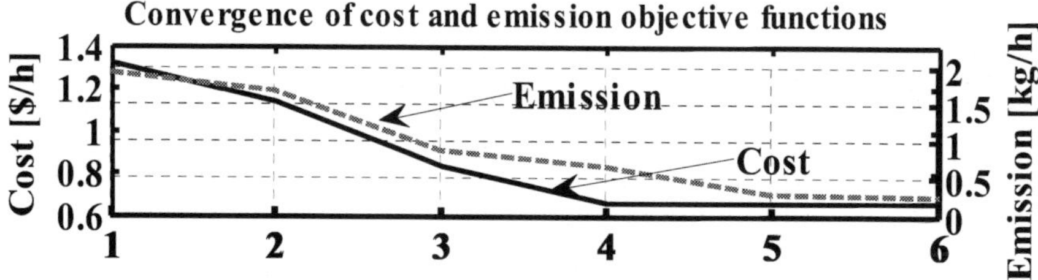

Fig. 8. Convergence of cost and emission objective functions.

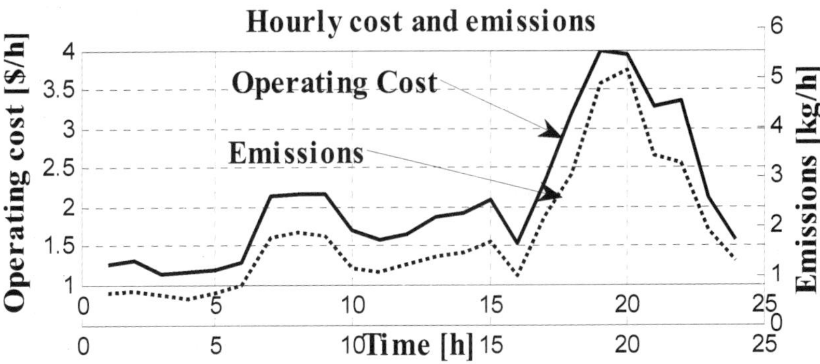

Fig. 9. Hourly operating cost and emission

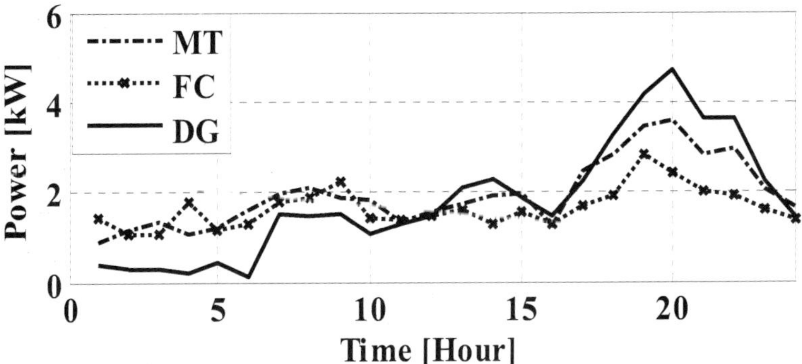

Fig. 10. Power Generation distribution.

	Total Load (kW/Day)	Total Emissions (kg/Day)	Total Costs ($/Day)	Optimal Generation (kW/Day)	Sold Power (kW7Day)	Purchased Power (kW/Day)
Case 1	171.4009	127.250	86.3466	352.8773	70.5188	14.8635
Case 2	171.4009	9.6045	224.4015	31.8217	00.0000	135.3149

Table 1. The Objective Function when Optimized Individually.

Figure 11 shows the relationship (trade-off curve) of the operating cost and emission objectives of the non-dominated solutions obtained for different purchased and sold tariffs. The operating costs of the non-dominated solutions thus appear to be inversely proportional to their emissions.

Table 2. shows the effect of changing the purchased and sold tariffs on the optimal setting of the MG. There are all together four cases.

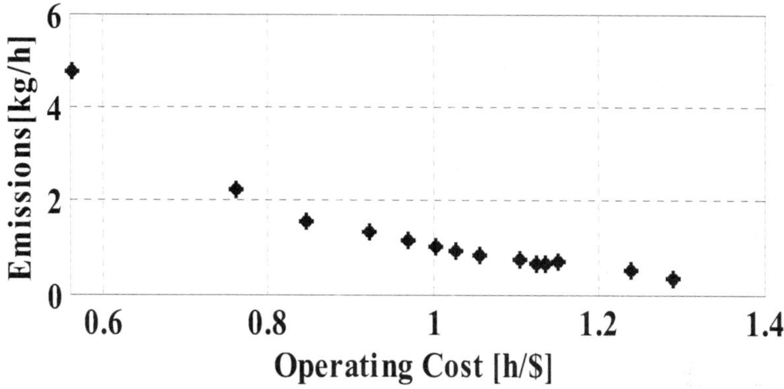

Fig. 11. Trade- off in cost and emission.

	TOTAL LOAD (KW/DAY)	Optimal Generation (KW/DAY)	Total Cost ($/Day)	TOTAL EMISSION (kg/Day)
Case 1	171.4009	88.3430	38.8254	25.9907
Case 2	171.4009	117.2192	45.1067	37.2475
Case 3	171.4009	110.5423	43.8823	34.3452
Case 4	171.4009	110.3820	43.9964	34.5317

Table 2. The effect of the purchased and sold tariffs on the optimal generation using MOGA.

In case 1, the effect of the changing the purchased ariffs is studied, when the sold power was 0.04 \$/kWh and the purchased tariffs were 0.1 \$/kWh, while in Case 2 the value of the purchased tariffs has increased to 0.16 \$/kWh and the sold was the same as in Case 1. During changing the purchased tariffs values, it was noticed that when the tariffs were low, it was preferable to buy as much power from the main grid as possible. However, when the tariffs were higher, it was more economic to generate the required power from the MG.

In Cases 3 and 4, the purchase power tariff is kept constant at 0.12 \$/kWh, while the sold tariffs was 0.0 \$/kWh in case 3 and 0.04 \$/kWh in case 4. It is noticeable that, the changing of the sold tariffs has no effect for such a small change. It only reacts if the change is much larger.

Table 3. illustrates the cost savings and emission reductions of the MG using different cases and compares them with the proposed technique.

The results obtained using the proposed technique to minimize the total cost and total emissions are compared with some conventional strategies of settings. The first case is when the DG, FC, and MT operate at their rated power for the whole day (Case A). The second is to optimize the cost individually (Case B). The third scenario is to optimize the emissions objective function individually (Case C).

Case A gives higher operating cost and higher emissions which indicates that it is not relevant. The larger generating power, the larger costs and emissions are attained. In the

Case B, the cost is relatively reduced, while the emissions were increased. In the third case, the cost increased while the emissions decreased and the optimal choice was to purchase more power from the main grid.

For achieving the completeness and checking the effectiveness of the proposed cost function and proposed solution, the problem was treated as single objective optimization problem by a linear combination of the cost and emission objectives as follows (Abido, 2003a) (case D):

$$\underset{P}{Minimize} \quad \omega CF(\boldsymbol{P}) + (1-\omega)\rho E(\boldsymbol{P}) \tag{21}$$

where ρ is the scaling factor and ω is the weighting factor.

With the proposed free power sale the total operating cost was reduced to 42.3054 \$/day and 33.9223 kg/day for the emissions after increasing the number of WT and PV, also switching off one of the load comparing to other settings.

Table 3. confirms that the MO optimization technique has made reasonable selections. The selections were not so straightforward; because of the existing the start-stop time limit constraints which have a big effect on the performance of the algorithm. It can be seen that the load was served perfectly.

	AVERAGE COST & EMISSION			AVERAGE DIFFERENCE WITH RESPECT TO THE OPTIMAL CASE		
	COST ($/DAY)	EMISSIONS (KG/DAY)	COST ($/DAY)	EMISSIONS (KG/DAY)	COST %	Emission %
Case A	99.9283	229.4892	57.6229	195.5672	136.2%	576.5151%
Case B	92.6276	224.6428	25.2862	27.3513	59.77%	80.6292%
Case C	119.3319	8.1371	70.3539	-25.7852	166.3%	-76.01%
Case D	54.1359	57.7628	11.8205	24.9295	27.94%	70.279%
Ref[10]	43.3754	34.0102	1.07	0.0879	2.5292%	0.2637%
Optimal Setting	42.3054	33.9223	00.0000	00.0000	00.000%	00.0000%

Table 3. Saving and emissions reductions of the MG using MOGA.

9. Conclusion

This chapter has presented modling and the GA approach to solve the multiobjective problem. From the results obtained, optimization of the above-formulated objective functions using MOGA yields not only a single optimal solution, but a set of Pareto optimal solutions, in which one objective cannot be improved without sacrificing other objectives. For practical applications, however, one solution is needed to be selected, which will satisfy the different goals to some extent. Such a solution is called the best compromise solution. One of the challenging factors of the trade-off decision is the imprecise nature of the decision maker's judgment.

Initially in all the three scenarios, minimum and maximum values of each original objective function are computed in order to obtain the last compromise solution. Minimum values of the objectives are obtained by giving full consideration to one of the objectives and neglecting the others. In this study, two objective functions are considered. Operating costs and emission level are optimized individually to obtain minimum values of the objectives. Owing to the conflicting nature of the objectives, emission level has to have maximum values when operating cost is minimum. The GA transforms the original multiobjective optimization problem into a single-objective problem and, thus, the set of noninferior solutions can be easily obtained. Compared with the other strategies of settings, the proposed approach significantly reduces the operating cost and emission level, while satisfying the load demand required by the multiobjective MG problem.

10. References

Abido M. A, (2003) A niched Pareto genetic algorithm for Multiobjective environmental/economic dispatch", *Electr Power Energy Syst.*, Vol. 25(2), ., p. 97–105.

Abido M. A, (Nov. 2003a) Enverionmental/Economic Power Disparch Using Multiobjective Evolutionary Algorithms, *IEEE Trans. on Power Syst*, Vol. 18, No. 4 p. 1529 – 1537.

Azmy, A.M., and Erlich, I., Online Optimal Management of PEM Fuel Cells Using Neural Networks, *IEEE Transactions on Power Delivery*. Vol. 29, No. 2, p. 1051–1058, Appril 2005.

Barbir, F., and Gomez, T., (Oct. 1996) Efficiency and Economics of Proton Exchange Membrane {PEM} Fuel Cell *Int. Journal of Hydrogen Energy*, Vol. 21, No. 10 p. 891–901.

Campanari, S., and Macchi, E, (July,2004) Technical and Tariff Scenarios effect on Microturbine Trigenerative Applications *Journal of Engineering for Gas turbines and Power*, Vol.126, p. 581–589.

Chedid, R., and Rahman, S., Unit sizing and control of hybrid wind-solar power systems, *IEEE Transactions on Energy Conversion*, (1997),Vol. 12, No. 1, pp. 79-85,March 1997.

Chedid, R., and Akiki, H., and Rahman, S.,(March ,1998) A Decision Support Technique For The Design Of Hybrid Solar- Wind Power Systems, *IEEE Transaction on Energy Conversion.*, Vol. 13, No. 1, p.76–83, 1.

Elmusrati, M., Riku, J., and Koivo, N. H, Mutliobjective Distributed Power control Algorithm for CDMA Wireless Commuincation Systems, *IEEE Transactions on Vehicular Technology*, Vol. 56, NO. 2, , p. 779-788.

Gavanidou,E. S and Bakirtzis, A. G, (March 1992) Design of a Stand Alone System with Renewable Energy Sources using Trade off Methods *IEEE Transaction on Energy Conversion*, Vol. 7, No. 1 p. 42–48.

Hernandez-Aramburo, C. A., and Green, T. C., and Mugniot, N, (May/June. 2005)"Fuel Consumption Minimization of a Microgrid", *IEEE Transactions on Industry Applications*, Vol. 41, NO. 3, , p. 673-681.

Huang, C. M., Yang, H. T., and Huang, C. L., (Nov. 1997)Bi-objective power dispatch using fuzzy satisfaction-maximizing decision approach, *IEEE Trans. Power Syst.*, Vol. 12, ., pp. 1715–1721.

Lasnier, F. and Ang, T. G.,(1990) *Photovoltaic Engineering Handbook* IOP Publishing Ltd.,1990, ISBN0-85274-311-4},1990.

Liu, G., Yang, J., and Whidborne J, (March. 2007) *Multiobjective optimisation and control, Research Studies Press LTD.*

Manisa Pipattanasomporn., (2004) *A Study of Remote Area Internet Access with Embedded Power Generation,* Faculty of the Virginia Polytechnic Institute and State University, PhD thesis,} December 2004.

Mohamed Faisal A, Koivo Heikki, (April 2007) Online Management of MicroGrid with Battery Storage Using Multiobjective Optimization, *in Proc. the first International Conference on Power Engineering, Energy and Electrical Drives (POWERENG07),* Setubal, Portugal, 12--14.

Mohamed Faisal A, Koivo Heikki, (April, 2009) Environmental/economic power of microgrid using multiobjective optimization", *in Proc. The International Conference on Renewable Energies and Power Quality (ICREPQ'09),* Valencia, Spain, 15 - 17.

Mohamed Faisal A, Koivo Heikki. (2010), System modelling and online optimal management of MicroGrid using Mesh Adaptive Direct Search, *International Journal of Electrical Power & Energy Systems.,* Vol. 32,no 5 . , pp. 398–407.

Mohamed Faisal A, Koivo Heikki.,(2008) Multiobjective Genetic Algorithms for Online Management Problem of Microgrid, *Journal of International Review of Electrical Engineering (IREE)* Vol. 3,no 1 . , pp. 46-54.

Morgantown, W, Emission rates for new DG technologies, *the Regulatory Assistance Project.,* Available,http://www.raponline.org ProjDocs/DREmsRul/Collfile/DGEmissionsMay2001.pdf.

Orero S. O, and Irving M. R, (1997). *Large scale unit commitment using a hybrid genetic algorithm,* International Journal of Electrical Power & Energy systems, Vol. 19, No. 1, pp. 45- 55.

Talaq,J. H. El-Hawary, F., and El-Hawary, M. E, (Aug. 1994)A summary of environmental/ economic dispatch algorithms, *IEEE Transactions. Power   Syst.,* Vol. 9, , p. 1508– 1516.

Wood, A. J. and Wollenberg, B. F, (1996) *Power Generation, Operation and Control,* Book., John Wiley $\&$ Sons, Ltd ,New York.

Yinger, R. J. (July 2001) Behaviour of capstone and Honeywell microturbine generators during load changes," Lawrence Berkeley National Laboratory (LBNL-49095) for CERTS,[Online]. Available http://certs.lbl.gov/pdf/LBNL$_$49095.pdf.

17

Size Optimization of a Solar-wind Hybrid Energy System Using Two Simulation Based Optimization Techniques

Orhan Ekren[1] and Banu Yetkin Ekren[2]
[1]Department of HVAC, Ege Vocational Training School, Ege University, Bornova
[2] Industrial Engineering Department, Pamukkale University, Kinikli
Turkey

1. Introduction[1]

Energy is an important fact for a country for both its socio-economic development and economic growth. Main energy source on earth is the fossil fuels. However, the usage of fossil fuels causes global warming whose negative effects have recently been felt by all over the world. Also, because it is limited on earth, increased energy demand and high energy prices increase concerns on fossil fuels. For instance, petroleum reservoirs are present only in few countries therefore the other countries mostly purchase petroleum from these countries more than their production amount. Hence, decrease of fossil fuel reservoirs may create a new 'energy crisis' and/or "energy wars" as in 1970s in near future.

For a sustainable world the usage of fossil fuels must be decreased, in fact ended. Instead, the usage of renewable energy sources must be increased. As it is known, the interest in renewable energy sources has increased because it does not cause greenhouse effect in contrary to the fossil fuels. These energy sources are indigenous, environmental friendly and, they help to reduce the usage of fossil fuels. Solar, wind, wave, biomass and geothermal energies are renewable energy sources. Sun is the source of all energies. Solar energy is usually used in two aims: for thermal applications and for electricity production. Wind is the indirect form of solar energy and is always replenished. The past and the predicted amount of global renewable energy source status by 2040 are presented in Table 1. As seen in this table the renewable energy consumption is predicted to increase in the future.

There are several studies on renewable energy sources and hybrid combination of these sources for electricity production. Electricity has high cost mostly due to centralized energy systems which operate mostly on fossil fuels and require large investments for establishing transmission and distribution of grids that can penetrate remote regions (Deepak, 2009). Unlike the centralized energy systems, decentralized energy systems are mostly based on renewable energy sources. They operate at lower scales (a few kWh scale) both in the presence and absence of grid, and easily accessible to remote locations due to generation of

---

[1] Many of the parts in this chapter have been reproduced from Ekren and Ekren, 2008; Ekren and Ekren, 2009; Ekren et al., 2009 with permission.

	2001	2010	2020	2030	2040
Total energy consumption (million tons oil equivalent)	*10,038*	*10,549*	*11,425*	*12,352*	*13,310*
Biomass	1080	1313	1791	2483	3271
Large hydro	22.7	266	309	341	358
Geothermal	43.2	86	186	333	493
Small hydro	9.5	19	49	106	189
Wind	4.7	44	266	542	688
Solar thermal	4.1	15	66	244	480
Photovoltaic	0.1	2	24	221	784
Solar thermal electricity	0.1	0.4	3	16	68
Marine(tidal/wave/ocean)	0.05	0.1	0.4	3	20
Total renewable energy consumption	*1,365.5*	*1,745.5*	*2,964.4*	*4,289*	*6,351*
Renewable energy source consumption (%)	*13.6*	*16.6*	*23.6*	*34.7*	*47.7*

Table 1. World global renewable energy sources scenario by 2040 (adapted from Kralova and Sjöblom, 2010)

power in the propinquity of demand site. They are also called "stand-alone energy systems" and produce power independently from the utility grid. Because these systems are not connected to the utility grid, they usually need batteries for storage of electricity produced during off-peak demand periods.

Solar and wind energies are usually available for most of the remote areas as renewable sources. However, it is prudent that neither a standalone solar energy nor a wind energy system can provide a continuous supply of energy due to seasonal and periodical variations. Simultaneous utilization of multiple energy resources greatly enhances the certainty of meeting demands. These systems are called hybrid energy systems (see Figure 1). Because they are good complementary energy sources of each other, solar and wind energies have been widely used as hybrid combination for electricity supply in isolated locations far from the distribution network. However, they suffer from the fluctuating characteristics of available solar and wind energy sources. Therefore, properly sized wind turbine, photovoltaic panel and storage unit provides high reliability and low initial investment cost. In this chapter, we aim to show two simulation based size optimization procedures for a solar-wind hybrid energy system providing minimum cost. The case study is completed to meet the electricity demand of a GSM base station located near a seaside region in western of Turkey. As in Figure 1, the studied hybrid system's electricity is also produced via photovoltaic array and wind turbine which are regulated by voltage regulator components and, the excess electricity produced is stored by the battery banks to be used for later lacking loads. Here, the amount of the electricity produced via the solar energy and the wind depends on the total solar radiation on horizontal surface and the wind speed respectively.

2. Background and motivation

Several researchers have studied hybrid renewable energy sources. Panwar et al. (2011) reviewed the renewable energy sources to define the role of the renewable energy sources for environmental protection. In their study, it is emphasized that renewable technologies are clean energy sources and optimal use of these resources minimize negative environmental

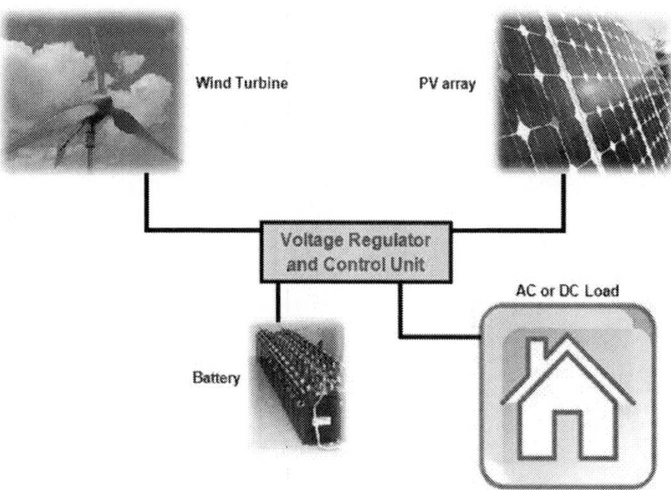

Fig. 1. A solar-wind hybrid energy system

impacts, produce minimum secondary wastes and provide a sustainable world. In another study, Angelis-Dimakis et al. (2011) evaluated the availability of renewable energy sources such as solar, wind, wave, biomass and geothermal energy. In their research, a detailed survey including existing methods and tools to determine the potential energy in renewable resources is presented. Also, tendency of using the renewable energy by the most developed countries in order to reduce the concentration of carbon dioxide in the atmosphere is emphasized. Their study also mentions the usability of hybrid energy system by mixing different renewable sources.

A great deal of research has been carried out on hybrid energy systems with respect to performance and optimization of these systems (Kellogg et al., 1996; Protegeropoulos et al., 1997; Seeling-Hochmuth, 1997; Markvart, 1996; Bagul et al., 1996; Borowy and Salameh, 1996; Morgan et al., 1997; Celik, 2002; Yang et al., 2003; Ashok, 2007; Bernal-Agustin and Dufo-Lo´ pez, 2009; Yang et al., 2009; Bilal et al., 2010; Zhou et al., 2010). Hybrid energy system studies in the past mostly based upon a particular design scenario with a certain set of design values yielding the near optimum design solution only (Kellogg et al., 1996; Protegeropoulos et al., 1997; Seeling-Hochmuth, 1997; Bagul et al., 1996; Morgan et al., 1997; Celik, 2002; Yang et al., 2003; Ashok, 2007). Such an approach, although providing the near optimum solution, unfortunately lacks the ability to provide a general understanding about how the total system cost changes with the size of the design parameters. A graphical optimization technique to optimize the size of the solar-wind hybrid energy system is studied by Markvart (1996). He considered monthly average solar and wind energy data in optimization. On the other hand, unlike the methods based on hourly, daily, and monthly average basis, a statistical approach for optimizing the size of PV arrays and the number of batteries for a standalone solar-wind hybrid system is presented by Bagul et al. (1996). They proposed a three-event probabilistic approach to overcome the limitations of the conventional two-event approach in matching the actual distribution of the energy generated by hybrid systems. Borowy and Salameh (1996) developed an algorithm to optimize a photovoltaic-array with battery bank for a standalone solar-wind hybrid energy system. Their model is based on a long-term hourly solar radiation and peak load demand data from the site. In this study, direct cost of the solar-wind hybrid energy system is considered. Different from these studies, Morgan et al. (1997) studied performance of battery units in a standalone hybrid energy system at various temperatures by taking into account the state of voltage (SOV) instead of the state of charge (SOC). Their algorithm is able to predict the performance of a hybrid energy system at various battery temperatures. This study is important for efficiency of a hybrid energy system because temperature affects the performance of a PV array and battery unit. Celik (2002) carried out techno-economic analysis and optimization of a solar-wind hybrid energy system. Yang et al. (2003) proposed an optimization technique by considering the loss of power supply probability (LPSP) model of a solar-wind hybrid system. They demonstrated the utility of the model for a hybrid energy system for a telecommunication system. Ashok (2007) presented a model based on different components of a hybrid energy system and developed a general model to define the optimal combination of renewable energy source components for a typical rural community. Recently, Bernal-Agustin and Dufo-Lo´ pez (2009) have analyzed usability of renewable energy systems. They use simulation to optimize the generation of electricity by hybrid energy systems. According to the authors stand-alone hybrid renewable energy systems are usually more suitable than only photovoltaic (PV) or wind systems in terms of lower cost and higher reliability. On the other hand, the design, control, and optimization of the hybrid systems are usually very complex tasks because of the high number of variables and the non-linearity. They come up with that the most reliable system is composed of PV–Wind–Battery. Although PV–Diesel–Battery is also reliable, this system uses fossil fuel. Optimal design and techno-economic analysis of a hybrid solar–wind power generation system is studied by Yang et al. (2009). They show sizing, optimizing and selecting the most suitable renewable source couples which are crucial for a hybrid system. They also come up with that solar and wind energies are the most suitable renewable energy resources. The authors also propose an optimal design model for solar–wind hybrid energy system. Obtaining the optimum configurations ensures decreased annual cost and, the loss of power supply probability (LPSP) is satisfied. Bilal et al. (2010) studied size optimization of a solar-wind-battery hybrid system for Potou which is an isolated site located in the northern coast of Senegal. This area is far away from the electricity supply. The methodology used in their study consists of sizing and optimization of a hybrid energy system by multi-objective genetic algorithm and the influence of the load profiles on the optimal configuration. Optimal configurations are examined for three profiles. Profile 1 is for the load for operation of refrigerators, domestic mill, welding machines, and other equipment in the village. Profile 2 is for the load for operation of a desalination and water pumping system, commercial refrigerators and domestic equipment, etc. The third load illustrates low consumption during the day (population working in the fields in the morning), and high power demand at night. Zhou et al. (2010) presented current status of researches on optimum sizing of stand-alone solar–wind hybrid power generation systems. The authors use two renewable energy sources because of the fact that their availability and topological advantages for local power generations. Also, these combinations of hybrid energy systems allow improving the system efficiency and power reliability and reduce the energy storage requirements for stand-alone applications. It is concluded that continued research and development in this area are still needed to improve these systems' performance.

Most of the existing studies use historical data and/or intervals for the input variables – solar-wind energies and electricity demand - of the system. Different from the existing studies, we utilize probabilistic distributions in order to carry out random input simulation. A detailed studied carried out to fit the input variables to probabilistic distributions. We fit the probabilistic distributions based on hours for each month for the solar radiation and the wind speed values. The electricity consumption of the GSM base station is also fit hourly basis. Different from the previous studies, we use two different simulation based optimization techniques – RSM and OptQuest - to compare their results. We implement a case study to model a stand-alone solar-wind hybrid energy system at a remote location from the grid system located in western of Turkey (Ekren and Ekren, 2008; Ekren and Ekren, 2009; Ekren et al., 2009).

Section 3 explains the simulation modeling of the hybrid energy system. In this section, the measured values of the system's inputs - solar radiation, wind speed and GSM base station's electricity consumption – are also provided. Besides, hourly fitted distributions for each month of the solar radiation and the wind speed and, hourly fitted distributions of the GSM base station's electricity consumption are also presented. In Section 4, the first optimization methodology - RSM – is introduced and used to optimize the hybrid system. In this section, we also provide the conducted experiments. In Section 5, the second optimization methodology – OptQuest – is explained. Last, we conclude the study.

3. Simulation modeling of the solar-wind hybrid energy system

The hybrid system under study relies on solar and wind energies as the primary power resources, and it is backed up by the batteries (see Figure 1). Batteries are used because of the stochastic characteristics of the system inputs. Namely, it is used to meet the electricity demand while the solar and wind energies are not adequate. The basic input variables of the hybrid model are: solar radiation, wind speed, and the electricity consumption of the GSM base station. Because the characteristics of these variables are non-deterministic, we fit to probability distributions to carry out a Monte Carlo (MC) simulation (see Tables 3-5). The probability distributions are specified in the *input analyzer* tool of the ARENA simulation software. Random data for solar radiation, wind speed, and the electricity consumption are generated using these distributions in ARENA (Kelton et al., 2004). Because in the simulation model hourly data are used, one of the system's assumptions is that the input variables do not change throughout an hour. This means that the solar radiation and the wind speed input values are constant e.g. from 12:00 pm to 1:00 pm. in any month in the model. The length of each simulation run is considered as twenty years of the economical life which consists of 365 days/year, 24 hours/day, in total 175,200 hours. For each run, 5 independent replications are completed. In the simulation model, since it is a popular and useful variance reduction technique to compare two or more alternative configurations, the common random numbers (CRN) variance reduction technique is used. And since a steady state analysis is needed to analyze a long time period non-terminating system, the warm-up period is decided as 12,000 hours (Law, 2007). Hourly mean solar radiation and wind speed data for the period of 2001-2003 (26,280 data = 24hours*365days*3years) are recorded at a meteorological station where the suggested hybrid energy system is to be established. Technical specifications of the meteorological station are given in Table 2 (Ekren, 2003).

Instrument	Specification/Description		
Pyranometer (CM11)	Wieving Angle : 2π Irradiance : 0 - 1400 W / m² Sensitive : 5.11*10⁻⁶ Volts per W/ m² (+/- 0.5 % at 20 ºC and 500 W/m²) Expected Signal Output : 0 – 10 mV Response time for 95 % response : < 15 sec.		
Data Logger	Module capacity:192896 bytes 12 signal inputs		
	Measurement Range	Recording Resolution	Accuracy
Anemometer (for speed)	0.3 to 50 m/s	≤0.1 m/s	± 0.3 m/s
Wind Vane (for direction)	0⁰ - 360⁰	≤1°	± 2⁰
Thermometer	(-30)–(+70) ºC	≤0.1°C	± 0.2 K
Hygrometer	0-100 % RH	1 % RH	±2 % RH
Barometer	800 to 1600 kPa	≤ 1kPa	-

Table 2. Main Characteristics of the Meteorological Station

3.1 Solar radiation

Figure 2 presents average measured hourly total solar radiation on horizontal surface, H, based on months in a year. Hourly total solar radiation on tilted surface, I_T, is calculated using H and optimum tilted angle of the PV panel, β is taken as 38⁰ (Eke et al., 2005).

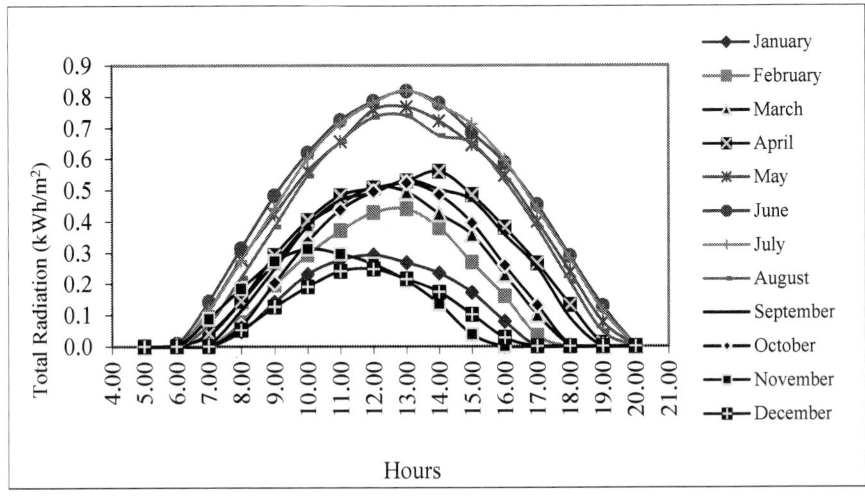

Fig. 2. Average hourly total solar radiation on horizontal surface (Measured)

Table 3 presents the fitted hourly total solar radiation distributions, for example, for three months, in June, July, and August. Each solar radiation distribution is different at each hour of the month. This means that solar radiation may vary in accordance with its distribution at each hour of the months which is not known in advance. Here, the existences of the zero values are due to the sunset.

Hours	Months		
	June	July	August
00:00-01:00	0	0	0
01:00-02:00	0	0	0
02:00-03:00	0	0	0
03:00-04:00	0	0	0
04:00-05:00	0	0	0
05:00-06:00	0	0	0
06:00-07:00	-0.001 + 37 * BETA(0.0269, 0.0857)	0	0
07:00-08:00	48 + 123 * BETA(2.71, 0.787)	NORM(113, 12.9)	18 + 80 * BETA(2.22, 1.22)
08:00-09:00	NORM(313, 16.6)	TRIA(240, 278, 305)	80 + 189 * BETA(3.49, 1.22)
09:00-10:00	NORM(482, 19.9)	270 + 209 * BETA(3.26, 0.724)	183 + 246 * BETA(1.57, 0.542)
10:00-11:00	481 + 173 * BETA(2.39, 0.602)	NORM(602, 17.2)	NORM(545, 55.8)
11:00-12:00	558 + 206 * BETA(2.51, 0.601)	NORM(714, 28.1)	265 + 457 * BETA(2.76, 0.465)
12:00-13:00	538 + 297 * BETA(1.34, 0.431)	517 + 337 * BETA(3.95, 1.11)	221 + 617 * BETA(3.05, 0.62)
13:00-14:00	472 + 399 * BETA(1.5, 0.42)	672 + 187 * BETA(3.27, 0.858)	145 + 704 * BETA(2.09, 0.385)
14:00-15:00	229 + 617 * BETA(1.13, 0.313)	400 + 436 * BETA(1.32, 0.415)	144 + 655 * BETA(0.978, 0.366)
15:00-16:00	359 + 411 * BETA(1.02, 0.402)	453 + 311 * BETA(0.931, 0.405)	334 + 393 * BETA(1.45, 0.507)
16:00-17:00	198 + 450 * BETA(1.09, 0.303)	174 + 481 * BETA(1.34, 0.407)	193 + 464 * BETA(2.78, 1.04)
17:00-18:00	178 + 326 * BETA(1.28, 0.418)	46 + 452 * BETA(1.16, 0.347)	76 + 371 * BETA(1.47, 0.483)
19:00-20:00	103 + 233 * BETA(1.55, 0.387)	69 + 258 * BETA(2.46, 0.527)	TRIA(78, 266, 278)
20:00-21:00	63 + 91 * BETA(1.54, 0.604)	-0.001 + 148 * BETA(3.43, 1.02)	-0.001 + 101 * BETA(0.974, 1.13)
22:00-23:00	0	0	0
23:00-00:00	0	0	0

Table 3. Hourly solar radiation on horizontal surface distributions for months June, July, August (W/m^2)

ARENA simulation software uses nine different theoretical distributions to fit data to a theoretical distribution. These are: Exponential, Gamma, Lognormal, Normal, Triangular, Uniform, Weibull, Erlang, and Beta distributions. Each of the distribution has its own probabilistic characteristics in creating random variables in a stochastic model.

3.2 Wind speed

In order to measure the wind speed and the prevailing wind direction, a three-cup anemometer and a wind vane are used. Hourly average wind speed at 10-meter-height for all months of the year, can be seen in Figure 3. These average hourly measured solar radiation and wind speed data figures are given for a general idea of the energy potential of the area. Otherwise, in the simulation model we use long-term dynamic hourly data. The height used to measure the wind speed is the universally standard meteorological measurement height (AWS Scientific, 1997). Table 4 shows hourly wind speed distributions of three months as an example.

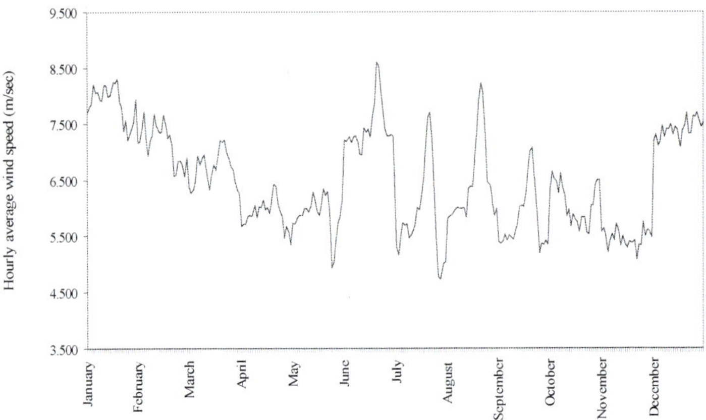

Fig. 3. Average hourly average wind speeds (measured)

	Months		
Hours			
	January	**February**	**March**
00:00-01:00	TRIA(2, 3.08, 18)	2 + 13 * BETA(1.07, 1.57)	1 + 12 * BETA(1.46, 1.79)
01:00-02:00	1 + LOGN(7.02, 5.28)	WEIB(7.75, 1.88)	1 + 11 * BETA(1.54, 1.66)
02:00-03:00	2 + 14 * BETA(0.881, 1.23)	GAMM(2.24, 3.31)	TRIA(2, 3.95, 13)
03:00-04:00	NORM(8.22, 3.96)	1 + LOGN(7.1, 5.85)	TRIA(1, 4.52, 14)
04:00-05:00	2 + WEIB(6.7, 1.52)	1 + LOGN(6.61, 5.58)	NORM(6.93, 2.66)
05:00-06:00	TRIA(2, 4.21, 18)	1 + LOGN(6.3, 5.69)	2 + WEIB(5.3, 1.78)
06:00-07:00	1 + 15 * BETA(1.37, 1.6)	2 + WEIB(5.5, 1.2)	TRIA(1, 7.62, 12)
07:00-08:00	1 + LOGN(7.23, 5.21)	2 + WEIB(5.87, 1.48)	TRIA(1, 7, 13)
08:00-09:00	2 + WEIB(6.84, 1.66)	2 + 17 * BETA(1.14, 2.28)	TRIA(1, 7.69, 10.9)
09:00-10:00	1 + WEIB(8.11, 2.12)	2 + WEIB(5.98, 1.41)	1 + GAMM(1.13, 4.69)
10:00-11:00	2 + WEIB(6.71, 1.8)	1 + WEIB(7.1, 1.61)	2 + WEIB(5.15, 2.25)
11:00-12:00	2 + WEIB(6.78, 1.87)	1 + ERLA(2.11, 3)	NORM(6.77, 2.54)
12:00-13:00	3 + ERLA(2.63, 2)	2 + 22 * BETA(0.955, 2.76)	NORM(6.68, 2.81)
13:00-14:00	2 + GAMM(2.45, 2.54)	1 + WEIB(7.24, 1.56)	1 + ERLA(1.98, 3)
14:00-15:00	2 + WEIB(6.97, 1.58)	1 + WEIB(6.92, 1.57)	1 + WEIB(6.97, 1.76)
15:00-16:00	2 + WEIB(6.54, 1.6)	1 + WEIB(7.06, 1.61)	NORM(7.17, 3.63)
16:00-17:00	2 + WEIB(6.41, 1.63)	1 + WEIB(6.86, 1.6)	2 + 17 * BETA(1.37, 3.1)
17:00-18:00	NORM(7.35, 3.12)	1 + ERLA(2.8, 2)	NORM(6.98, 2.9)
19:00-20:00	TRIA(2, 4.7, 16)	NORM(6.84, 3.82)	1 + LOGN(6.07, 3.79)
20:00-21:00	2 + WEIB(5.76, 1.57)	1 + WEIB(6.46, 1.57)	2 + 12 * BETA(1.06, 1.56)
22:00-23:00	1 + WEIB(7.01, 1.73)	1 + WEIB(6.4, 1.62)	1 + GAMM(1.84, 3.1)
23:00-00:00	NORM(7.41, 4.06)	1 + WEIB(6.4, 1.62)	TRIA(2, 4.39, 13)

Table 4. Hourly average wind speed distributions for months, January, February, March (m/sec)

3.3 Electricity consumption

The third stochastic data is the electricity consumption of the GSM base station. The data are collected from the base station for every hour of the day. In this study, the existence of a seasonal effect on the GSM base station's electricity consumption is ignored. The statistical data are collected in 15 random days in each season, fall, winter, spring, and summer. Hence, totally 15*4 = 60 electricity consumption data are collected for an hour (e.g. for 1.00 pm). Then these data are fit to theoretical distributions without considering seasonal effects. Fig. 4 illustrates the hourly mean electricity consumption values of the GSM base station. And the fitted distributions are given in Table 5.

The output of the wind generator and PV panels are DC power and, inverter converts it to the AC power.

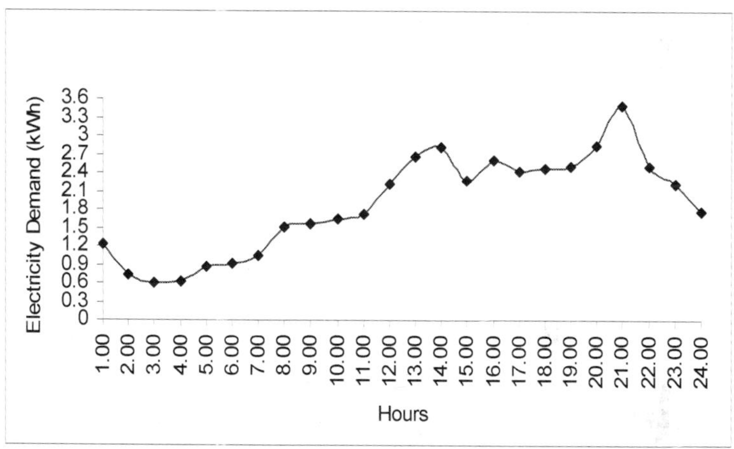

Fig. 4. Hourly average demand of the GSM base station

3.4 Formulations

If the hybrid energy systems are well designed, they provide a reliable service for an extended period of time. In the optimization procedure, the sizes of system components are decision variables, and their costs are objective function. The cost of the system is considered as the total cost of PV, wind turbine rotor, battery, battery charger, installation, maintenance, and engineering. A solar and wind hybrid energy system with the sizes of a_s and a_w, respectively, can be defined by (1)–(2):

$$a_s = \eta . A_s \tag{1}$$

where, η is the PV module efficiency, A_s is the PV array area and:

$$a_w = C_p . (\pi . r^2) \tag{2}$$

where, Cp is the power coefficient, and r is the rotor radius. Here, $\pi . r^2$ represents A_w, rotor swept area. η value is taken as a variable value depending on PV module type and module temperature. In the study, mono-crystal silicon PV module type, rated output of 75 W at

Hours	Demand
00:00-01:00	NORM(1.21, 0.602)
01:00-02:00	1.72 * BETA(1.13, 1.55)
02:00-03:00	WEIB(0.645, 1.28)
03:00-04:00	GAMM(0.565, 1.11)
04:00-05:00	EXPO(0.854)
05:00-06:00	3.31 * BETA(1.28, 3.34)
06:00-07:00	2.86 * BETA(1.15, 1.96)
07:00-08:00	WEIB(1.7, 2.36)
08:00-09:00	TRIA(0, 0.827, 3.84)
09:00-10:00	3.84 * BETA(2.46, 3.25)
10:00-11:00	0.15 + 3.34 * BETA(2.54, 2.87)
11:00-12:00	0.28 + LOGN(1.91, 0.951)
12:00-13:00	1 + ERLA(0.828, 2)
13:00-14:00	1 + LOGN(1.82, 1.43)
14:00-15:00	NORM(2.27, 0.626)
15:00-16:00	1 + GAMM(0.669, 2.4)
16:00-17:00	1 + GAMM(0.417, 3.45)
17:00-18:00	NORM(2.47, 0.685)
19:00-20:00	1 + ERLA(0.377, 4)
20:00-21:00	1.11 + ERLA(0.436, 4)
22:00-23:00	1.29 + LOGN(2.2, 1.59)
23:00-00:00	1 + GAMM(0.435, 3.48)

Table 5. Hourly electricity demands (kW)

1000 W/m² is used. Here, η value changes between 7% and 17% based on module surface temperatures which are between 10ºC and 70ºC (Kemmoku, 2004; Muselli, 1999). In the simulation model, for December, January and February the temperature and the η values are assumed to be 10ºC and 17%, respectively. For March, April and May the temperature and the η values are assumed to be 50ºC and 10%, respectively. For June, July, August the temperature and the η values are assumed to be 70ºC and 7%, respectively. And, for September, October, November the temperature and the η values are assumed to be 30ºC and 13%, respectively. These values are obtained from a manufacturer firm. Cp value is also taken from a manufacturer firm as a graphic value which changes according to the wind speed value. Wind energy density, W, is calculated by (3):

$$W = 1/2 \cdot \rho \cdot V^3 \cdot D \qquad (3)$$

D is length of period. Because of hourly operating state, it is taken as 1 hour. ρ is air density which is considered as 1.225, and V is hourly average wind velocity whose distribution is shown in Table 4.

Solar radiation on tilted plate is calculated for isotropic sky assumption by (4a)-(4b):

$$I_T = I_b \, R_B + I_d \, (1 + \cos \beta) / 2 + (I_b + I_d) \, \rho \, (1 - \cos \beta) / 2 \qquad (4a)$$

$$R_b = \cos \theta \, / \, \cos \theta z \qquad (4b)$$

where, I_T is total solar radiation on tilted surface, I_b is horizontal beam radiation, I_d is horizontal diffuse radiation, R_b is ratio of beam radiation on tilt factor, θ is incidence angle, θ_z is zenith angle, ρ is surface reflectivity, β is tilted angle of the plate.

The supply of hourly solar and wind energies must meet the hourly demand, d. This expression can be formulated as:

$$S.\,a_s + W.\,a_w \geq d \tag{5}$$

where S is the solar energy density on tilted surface, H_T (kWh/m²) and W is the wind energy density (kWh/m²). If (5) is not realized, the stored energy in the battery will be used. The battery's efficiency is assumed as 85%. The inverter's efficiency is considered 90% here. If the total of solar, wind, and battery energies still cannot meet the demand, the energy shortage will be supplied by an auxiliary energy source, whose unit cost, herein, is considered \$0.5 per kWh electricity. In this study, the location of the hybrid system is assumed in such a place where the unit cost of the auxiliary energy is more expensive than the electricity produced by the hybrid system. Therefore, the cost of extra energy here is decided such that it is three times as much of the average unit cost of the electricity produced by the hybrid system. Otherwise, if the unit cost of the auxiliary energy was less than that of the electricity produced by the hybrid system, the shortage could be supplied from the auxiliary energy source all the time without the need for such a hybrid system (Celik, 2002). As seen in (6), the auxiliary energy cost is also a part of the total hybrid system cost. C_s, C_w, C_B, C_{sh} and C_T are the unit cost of photovoltaic, wind energy generator, battery, shortage electricity, and total hybrid energy system, respectively. B_C denotes the battery capacity. E_i is the total amount of the electricity energy shortage because of not meeting the demand during an hour i. C_s are \$5.8/W_p (mono-crystal silicon, rated output of 75 W at 1000 W/m²), \$5.5/W_p (multi-crystal silicon, rated output of 75 W at 1000 W/m²) whereas C_w is US \$3/W (rated output of 5000 W at 10 m/s), and C_B is \$180/kWh (200 Ah 12V lead acid battery) (Eke et al., 2005). C_{sh} is \$0.5 per kWh as explained above, and n is the simulation time period, 175,200 hours. In this study, the inflation rate and time value of money are not considered.

$$C_T = C_s.\,a_s + C_w.\,a_w + C_B.\,B_C + \sum_{i=1}^{n} C_{sh}.E_i \tag{6}$$

In addition, a total of US \$500 battery charger cost, and 5% installation, maintenance and engineering cost of the initial hardware is also added into the total system cost for an assumed 20-year-lifetime.

4. Response surface methodolgy

RSM consists of a group of mathematical and statistical techniques useful for developing, improving and optimizing processes (Myers et al., 2009). It is also useful in the design, development and formulation of new systems as well as improvement of existing systems. In RSM, usually there are several *input variables* that affect some performance measure which is also called *response*. Most real world applications generally involve more than one response. The input variables are sometimes called *independent variables* which can be controlled by the controller for purposes of a test or an experiment (Myers et al., 2009).

Figure 5 shows an example for the relationship between the response variable (bean yield) and the two input variables (PhosAcid and Nitrogen) in a chemical process graphically.

Since there is a response lying above the two input variables' plane the term "response surface" comes from this reason.

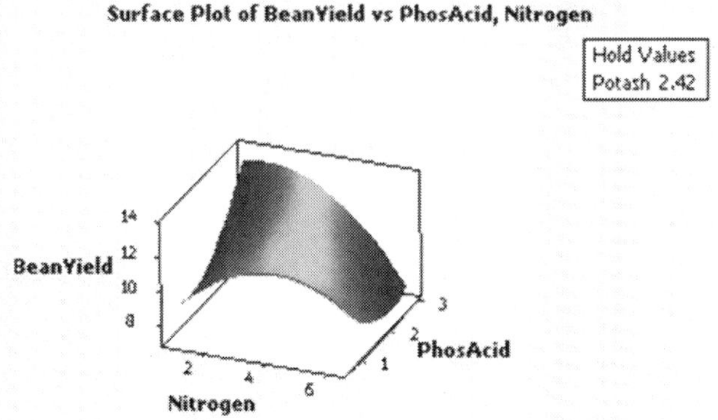

Fig. 5. Response surface of a chemical process

RSM requires developing an approximating model also called metamodel for the true response surface. The metamodel is based on observed data from the system and is an empirical model. Metamodels are developed to obtain a better understanding of the nature of the true relationship between the input variables and the responses of the system. This approximate formula could be used as a proxy for the full-blown simulation itself in order to get at least a rough idea of what would happen for a large number of input-parameter combinations. Multiple regression is a technique useful for building the types of empirical models required in RSM (Kleijnen, 1987; Friedman, 1996; Kleijnen and Sargent, 2000).
A first-order response surface metamodel is given by (7) as an example.

$$y = \beta_0 + \beta_1 x_1 + \beta_2 x_2 + \varepsilon \tag{7}$$

where y represents the response, x_1 and x_2 represent the input variables. β_0 is a fixed variable and β_1 and β_2 are the coefficients of x_1 and x_2, respectively. Hence, this is a multiple linear- regression with two independent (input) variables.
In RSM, first the significant input variables are identified. While identifying the significant input variables, it is also important to determine the current level of these variables. Here, the objective is to select the levels that provide the response close to its optimum. An RSM begins when the process is near optimum. At this phase it is aimed to predict the model that estimates the true response with a small region around the optimum. Because there is usually a curvature near the optimum a second-order model is mostly used. After determining the metamodel of the system, RSM uses steepest ascent method to optimize it (Myers et al., 2009).
The sequential nature of RSM provides the experimenter to have an idea about these five steps: a) the optimum point's location region b) the type of metamodel required c) the proper choice of experimental designs d) how much replication is necessary e) whether or not transformations on responses or any of process variables are required (Myers et al.,

2009). The RSM process typically involves taking observations in a starting region, usually according to an experimental design such as a factorial (2^k) or fractional (2^{k-p}). Here, k is the number of factors (input variables) and $1/2^p$ is called the degree of fractionation, because it represents the fraction of observations from a 2^k design that is required. For example, if a problem having 5 factors is studied, the necessary number of runs in the experiment would be $2^5=32$. Because each run may require time-consuming and costly setting and resetting of machinery, it is often not feasible to require many different production runs for the experiment. In these conditions, fractional factorials (2^{k-p}) are used that "sacrifice" interaction effects so that main effects may still be computed correctly (Law, 2007).

After obtaining the metamodel, it is necessary to examine whether the fitted model provides an adequate estimate on the true system or not. Also, it is necessary to verify whether the pre-defined regression assumptions are satisfied or not. There are several techniques to check the model adequacy (Myers et al., 2009). In this study, since it is easy and popular we use residual analysis to check the model adequacy.

4.1 Residual analysis

Normality assumption should be checked in metamodel fitting. This assumption is satisfied if the normality plot of the residuals is as in Figure 6a and it is not if it is as in Figure 6b. Namely, if the residuals plot approximately along a straight line then the normality assumption is satisfied (Myers et al., 2009).

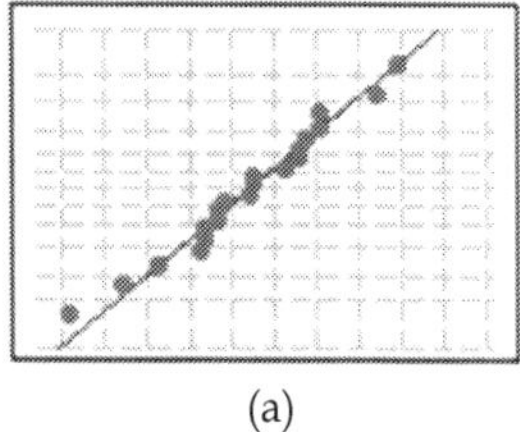
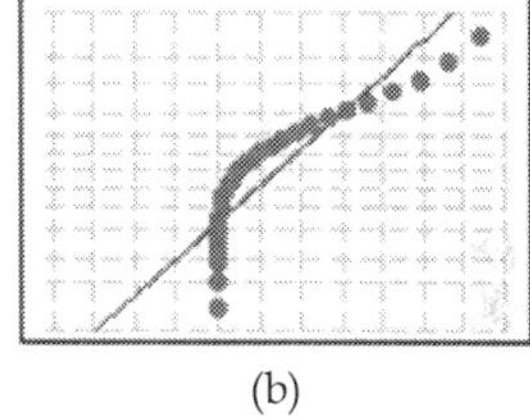

(a)   (b)

Fig. 6. a. Probability plot of normal data   b. Probability plot of nonnormal data

4.2 Lack of fit test

In RSM, it is important to check whether the regression function is a good fit or not. We use the lack of fit test to decide this. This test uses F test to reject or accept the hypothesis "the fitted model adequately describes the data". If the F value is high so, the p value is small (< 0.05) then we reject the hypothesis. Under this condition, the model does not reflect the data adequately (Myers et al., 2009).

4.3 Design of experiment and results

Design of Experiment (DOE) is a design tool that makes changes to the factors (input variables) to determine their effect on the responses. It not only identifies the significant factors that affect the response, but also how these factors affect the response (Montgomery, 2001).

In this study, a three-level, three-factorial Box–Behnken experimental design is used to evaluate the effects of selected independent variables on the responses to characterize the hybrid energy system and to optimize the procedure (Myers et al., 2009). This design is suitable for exploration of second-order (quadratic) response surfaces and for construction of second-order polynomial models, thus helping to optimize them by a small number of experimental runs (Box and Draper, 1987; Myers et al., 2009). Box–Behnken experimental design is an orthogonal design. Therefore, the factor levels are evenly spaced and coded for low, medium, and high settings, as −1, 0 and +1 (Montgomery, 2001; Myers et al., 2009). For the three-level, three-factorial Box–Behnken experimental design, a total of 15 experimental runs, shown in Table 6, are needed. Here, x_1, x_2 and x_3 are the factors that could affect the cost function as given in (6). -1, 0 and +1 show the coded variables of these factor levels. -1 is the low level, 0 is the middle level and +1 is the high level of the factors. In the middle level, three independent replications, namely 13th, 14th, and 15th experiments in Table 6 are needed.

Experiment	Factor and factor level		
	x_1	x_2	x_3
1	−1	−1	0
2	−1	+1	0
3	+1	−1	0
4	+1	+1	0
5	−1	0	−1
6	−1	0	+1
7	+1	0	−1
8	+1	0	+1
9	0	−1	−1
10	0	−1	+1
11	0	+1	−1
12	0	+1	+1
13	0	0	0
14	0	0	0
15	0	0	0

Table 6. Experiments and coded levels for Box–Behnken design

In the design model, three factors are chosen as PV size, A_s, wind turbine rotor swept area, A_w, and the battery capacity, B_C. A list of factors and their levels are provided in Table 7. The levels are 0 and 10, for -1 and +1 levels of PV size; 17 and 37 for -1 and +1 levels of wind turbine rotor swept area; and 10 and 50 for -1 and +1 levels of battery capacity. Since the amount of wind energy in this area is greater than the solar energy, the interval between the low and the high level of the wind turbine rotor swept area is greater than the interval of the low and the high level of the PV size. This also means that at the optimum point wind turbine rotor swept area tends to be greater than the PV size.

Factor	Level		
	−1	0	+1
x_1: PV (m^2)	0	5	10
x_2: wind (m^2)	17	27	37
x_3: B_C (kWh)	10	30	50

Table 7. Factors and factor levels used in Box–Behnken experimental design

4.4 Second-order polynomial (quadratic) model fitting and RSM results

The general representation of second order regression model of the design is shown by (7).

$$Y = b_0 + b_1 x_1 + b_2 x_2 + b_3 x_3 + b_4 x_1 x_2 + b_5 x_2 x_3 + b_6 x_1 x_3 + b_7 x_1^2 + b_8 x_2^2 + b_9 x_3^2 \tag{7}$$

where Y is the selected response; b_0–b_9 are the regression coefficients and x_1-x_3 are the factors. This model is used to estimate the relationship between the cost function, Y, and the three independent factors, PV size, x_1, wind turbine rotor swept area, x_2, and battery capacity, x_3. Here, b_1, b_2, b_3 coefficients denote the main effect of factors x_1, x_2 and x_3, respectively. Besides, b_4 denotes the interaction between factors x_1, x_2; b_5 denotes the interaction between factors x_2, x_3, and b_6 denotes the interaction between factors x_1, x_3. Finally, b_7, b_8, b_9 denote the quadratic effect of factors x_1, x_2 and x_3, respectively.

In order to fit the metamodel given by (7), 15 experiments with three independent replications in the middle are utilized as illustrated in Table 6. The validity of the fitted model is tested by computing a lack-of-fit, F-test and residual analysis. For fitting the metamodel and the optimization procedure, a software called Design Expert 7.1 a popular statistic software package is used (Montgomery, 2001; Myers et al., 2009).

As a result, the fitted standardized metamodel for the hybrid system is given by (8). The model is found to be significant at 95% confidence level by the F-test. In addition, the model does not exhibit lack-of-fit ($p>0.05$). The lack-of-fit test measures the failure of the model to represent data in the experimental domain at points that are not included in the regression. If a model is significant, meaning that the model contains one or more important terms, and the model does not suffer from lack-of-fit, does not necessarily mean that the model is a good one. If the experimental environment is quite noisy or some important variables are left out of the experiment, then it is possible that the portion of the variability in the data not explained by the model, also called the residual, could be large. Thus, a measure of the model's overall performance referred to as the coefficient of determination and denoted by R^2 must be considered. The value R^2 quantifies goodness of fit. It is a fraction between 0 and 1, and has no units. Higher values indicate that the model fits the data better. At the same time, adjusted R^2 allowing for the degrees of freedom associated with the sums of the squares is also considered in the lack-of-fit test, which should be an approximate value of R^2.

$$\hat{Y} = 38812.37 - 5008.24 x_2 + 2816 x_1 x_2 + 10638.18 x_2^2 + 3393.62 x_3^2 \tag{8}$$

R^2 and adjusted R^2 are calculated as 0.984 and 0.956, respectively. If adjusted R^2 is significantly lower than R^2, it normally means that one or more explanatory variables are missing. Here, the two R^2 values are not significantly different, and the normal probability plots of residuals do not show evidence of strong departures from normality as depicted in Figure 7. Therefore, the overall second-order metamodel, as expressed by (8), for the response measure is significant and adequate.

In general, estimated standardized metamodel coefficients provide two types of information. The magnitude of a coefficient indicates how important that particular effect is and its sign indicates whether the factor has a positive or a negative effect on the response.

The estimated cost function, $\hat{Y}$, of the hybrid energy system obtained by RSM in (8) indicates that there is no main effect of PV size and battery capacity on the cost, while wind turbine rotor swept area, x_2, significantly affects the response variable ($p<0.05$). Instead of the main effect of PV size, there is a significant interaction term of PV size and wind turbine

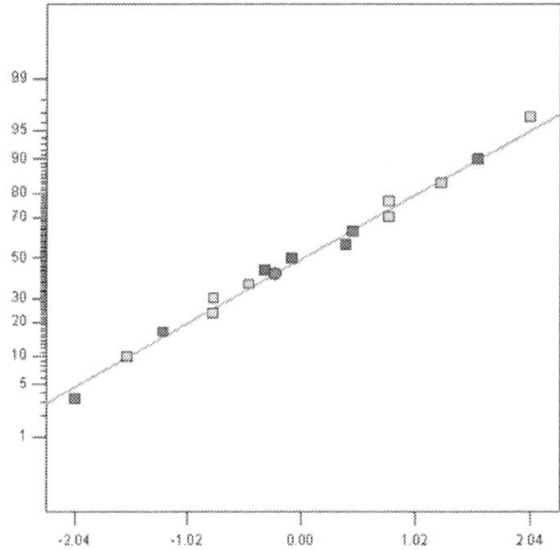

Fig. 7. Normal probability plot of residuals

rotor swept area, $x_1 x_2$. Besides, the quadratic effects of x_2 and x_3 are statistically significant ($p<0.05$). It should be noticed that only the second factor affects the cost negatively while the others affect it positively.

Design Expert 7.1 RSM result indicates that the near optimum value of the fitted response surface, $X_0 = (x_1, x_2, x_3)$, is (-0.21, 0.24, 0.096) as coded variable and (3.95, 29.4, 37.55) as the real value yielding a predicted mean response of $Y_0 = \$37,033.9$ (C_T), a minimum in the experimental region. When we run the simulation model at the optimum point the total energy met by the auxiliary energy source in 175,200 hours is obtained as 2,043.85 kWh, which means a shortage cost of \$1,021.925. This shortage cost can be interpreted as a small portion in the whole hybrid system cost. Figure 8, illustrates the 3D surface graph of cost response at the optimum battery capacity, 31.92 kWh, as a function of two factors, PV size, x_1, and wind turbine rotor swept area, x_2.

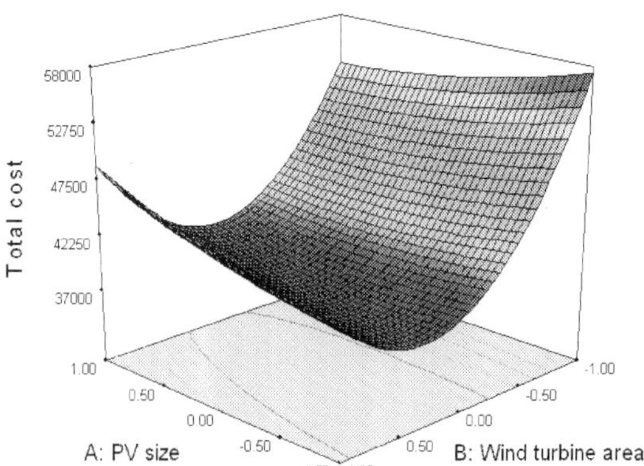

Fig. 8. Response surface of cost at fixed (optimum) battery capacity as a function of PV size, x_1, and wind turbine rotor swept area, x_2.

From this figure, while the battery capacity is fixed at 37.55 kWh, the trade-offs between PV size and wind turbine rotor swept area can be seen easily. And also, by this figure, the coded optimum values can be seen around the minimum point of the convex shape.

5. OptQuest optimization

The OptQuest tool is provided with the student version of the ARENA software with free of charge (Kelton et al., 2004). Like all practical simulation optimization methods, OptQuest is also an iterative heuristic. It treats the simulation model as a black box by observing only the input and the output of the simulation model (Kleijnen and Wan, 2007). It is an optimization tool that combines the meta-heuristics of tabu search, neural networks, and scatter search into a single search heuristic.

OptQuest requires the specification of lower, suggested, and upper values for the variables that are to be optimized. The suggested values determine the starting point of the decision variables. This choice affects the efficiency and effectiveness of the search. In practice, the simulation analyst may run the optimization tool twice so that he/she can use the current solution as a suggested value in the second run. We also run the OptQuest twice in order to determine the region of the initial points. Then, the obtained optimum values by the first run are used as suggested values for the second run of the optimization procedure. We thus narrow the search space based on the suggested values obtained from the first run.

Moreover, OptQuest requires the selection of the (random) simulation output that is the goal or objective variable to be minimized. In our problem, we select the minimization of the hybrid system cost as objective. And, OptQuest allows the user to explicitly define integer and linear constraints on the deterministic simulation inputs. Here, we do not specify any constraints, on the decision variables.

OptQuest allows different precision criteria for both the objective and the constrained simulation outputs. For example, it selects the number of replicates such that the halfwidth of the 95% confidence interval for the average output is within a user-selected percentage of the true mean. Because the pre-defined half-width could be obtained by, we use a fixed number of replications, $m = 5$ (Law, 2007). And, we select 95% confidence interval. Although, OptQuest allows alternative stopping criteria, for example, stop the search either after 300 min (5 h) or after 500 'nonimproving solutions, we use a tolerance value of 0.1 to determine when two solutions are equal and to stop the optimization automatically. Figure 9 shows an option example for OptQuest tool.

We optimize the same system using OptQuest. The optimum total cost of the system is obtained as \$32,962.5 (C_T) which corresponds to (3.04, 32.5, 33.67) for PV size, wind turbine rotor swept area, and battery capacity, respectively.

It should be noted that we obtained better result by OptQuest than RSM. This is probably because OptQuest integrates several heuristics into a single heuristic. Also, running the simulation twice might narrow the search space by getting closer to the optimum point. The only disadvantage using OptQuest is its time consuming property compared to RSM. Also, because the problem is a continuous type optimization in another word, the decision variables - PV size, wind turbine rotor swept area, and battery capacity- can get any numerical values between their levels, the OptQuest tries enormous number of alternative designs while running to find out the near optimum. Therefore, RSM is much faster than the OptQuest.

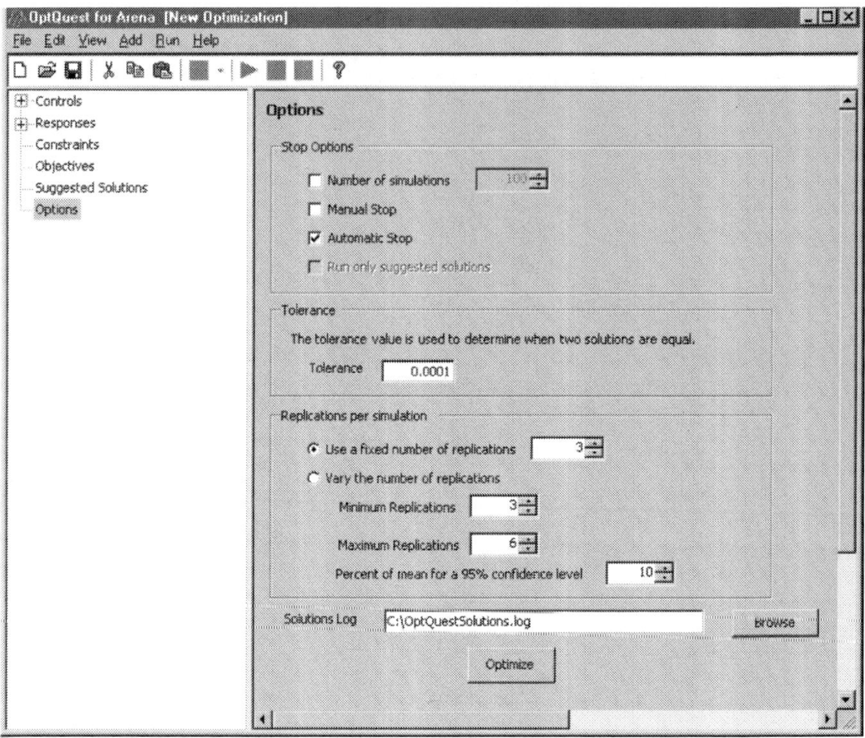

Fig. 9. OptQuest tool options

6. Conclusion

In this study, size of a PV-wind hybrid energy conversion system with battery storage is optimized using two simulation based optimization methods - RSM and OptQuest. The hybrid system is based on an hourly operating cost to meet the electricity demand of a GSM base station located in western Turkey. First, we fit the probabilistic distributions for solar radiation, wind speed and electricity consumption of the GSM base station using the input analyzer in ARENA – a commercial simulation software. Second, we complete a MC simulation and model a detailed hybrid energy system in ARENA. Third, the system's cost is optimized by considering the size of PV, wind turbine rotor swept area and battery capacity as decision variables, using RSM and the OptQuest tool in ARENA. Finally, we noticed that we could obtain better result by OptQuest than RSM corresponding to $32,962.5 cost with (3.04, 32.5, 33.67) PV size (m²), wind turbine rotor swept area (m²), and battery capacity (kWh), respectively.

7. References

Angelis-Dimakis A., Biberacher M., Dominguez J., Fiorese G., Gadocha S., Gnansounou E., Guariso G., Kartalidis A., Panichelli L., Pinedo I., Robba M. Methods and tools to evaluate the availability of renewable energy sources. Renewable and Sustainable Energy Reviews, 15(2):1182–1200, 2011.

Ashok S. Optimised model for community-based hybrid energy system. Renewable Energy, 32(7):1155–1164, 2007.

AWS Scientific. Inc. Wind Resource Assessment Handbook. NREL/SR-440-22223, National Renewable Energy Laboratory, 1997.

Bagul A.D., Salameh Z.M., Borowy B. Sizing of stand-alone hybrid wind–PV system using a three event probability density approximation. Solar Energy, 56(4):323–335, 1996.

Bernal-Agustin J.L., Dufo-Lo´ pez R. Simulation and optimization of stand-alone hybrid renewable energy systems. Renewable and Sustainable Energy Reviews, 13(8):2111–2118, 2009.

Bilal B.O., Sambou V., Ndiaye P.A., Kébé C.M.F., Ndongo M. Optimal design of a hybrid solar wind–battery system using the minimization of the annualized cost system and the minimization of the loss of power supply probability (LPSP). Renewable Energy, 35(10):2388-2390, 2010.

Borowy B.S., Salameh Z.M. Methodology for optimally sizing the combination of a battery bank and PV array in a wind–PV hybrid system. IEEE Transactions Energy Conversion, 11(2):367–375, 1996.

Box G.E.P., Draper N.R. Empirical Model-Building and Response Surfaces. First ed. New York: John Wiley & Sons, 1987.

Celik A.N. Optimisation and techno-economic analysis of autonomous photovoltaic–wind hybrid energy systems in comparison to single photovoltaic and wind systems. Energy Conversion and Management, 43(18):2453–2468, 2002.

Deepak P,K., Balachandra P., Ravindranath N.H. Grid-connected versus stand-alone energy systems for decentralized power—a review of literature. Renewable and Sustainable Energy Reviews, 13(8):2041–2050, 2009.

Eke R., Kara O., Ulgen K. Optimization of a Wind/PV hybrid power generation system. International Journal of Green Energy, 2(1):57-63, 2005.

Ekren B.Y., Ekren O. Simulation based size optimization of a PV/Wind hybrid energy conversion system with battery storage under various load and auxiliary energy conditions. Applied Energy, 86(9):1387-1394, 2009.

Ekren O. Optimization of a hybrid combination of a photovoltaic system and a wind energy conversion system. M.S. Thesis. Department of Mechanical Engineering. Izmir Institute of Technology, 2003.

Ekren O., Ekren B.Y. Size optimization of a PV/Wind hybrid energy conversion system with battery storage using response surface methodology. Applied Energy, 85(11):1086-1101, 2008.

Ekren O., Ekren B.Y., Ozerdem B. Break-even analysis and size optimization of a Pv/wind hybrid energy conversion system with battery storage - a case study. Applied Energy, 86 (7-8):1043-1054, 2009.

Friedman L.W. The Simulation Metamodel. First ed. Netherlands: Kluwer; 1996.

Kellogg W., Nehrir M.H., Venkataramanan G., Gerez V. Optimal unit sizing for a hybrid wind/photovoltaic generating system. Electric Power Systems Research, 39(1):35–38, 1996.

Kelton W.D., Sadowski R.P., Sturrock D.T. Simulation with Arena. 3rd ed. New York: McGrawHill, 2004.

Kleijnen J.P.C. Statistical Tools for Simulation Practitioners. First ed. New York: Marcel Dekker Inc; 1987.

Kleijnen J.P.C., Sargent R.G. A methodology for fitting and validating metamodels in simulation. European Journal of Operational Research, 120(1):14–29, 2000.

Kleijnen J.P.C., Wan J. Optimization of simulated systems: OptQuest and alternatives. Simulation Modelling Practice and Theory, 15(3):354-362, 2007.

Kralova I., Sjöblom J. Biofuels-renewable energy sources: a review. Journal of Dispersion Science and Technology, 31(3):409–425, 2010.

Law A.M. Simulation Modeling and Analysis. 4th ed. McGraw-Hill Higher Education, 2007.

Markvart T, Sizing of hybrid photovoltaic–wind energy systems. Solar Energy, 57(4):277–281, 1996.

Montgomery D.C. Design and Analysis of Experiments, 5th edition, New York: John Wiley & Sons, 2001.

Morgan T.R., Marshall R.H., Brinkworth B.J. ARES – a refined simulation programme for the sizing and optimization of autonomous hybrid energy systems. Solar Energy, 59(4-6):205–215, 1997.

Myers R.H., Montgomery D.C., Anderson-Cook C.M. Response Surface Methodology: Process and Product Optimization Using Designed Experiments, 3rd edition, New York: John Wiley & Sons, 2009.

Panwar N.L., Kaushik S.C., Surendra K. Role of renewable energy sources in environmental protection: A review. Renewable and Sustainable Energy Reviews, 15(3):1513–1524, 2011.

Protegeropoulos C., Brinkworth B.J., Marshall R.H. Sizing and techno-economical optimization for hybrid solar photovoltaic/wind power systems with battery storage. International Journal of Energy Research, 21(6):465–479, 1997.

Seeling-Hochmuth G.C. A combined optimization concept for the design and operation strategy of hybrid–PV energy systems. Solar Energy, 61(2):77–87, 1997.

Yang H., Wei Z., Chengzhi L. Optimal design and techno-economic analysis of a hybrid solar–wind power generation system. Applied Energy, 86(2):163–169, 2009.

Yang H.X., Lu L., Burnett J. Weather data and probability analysis of hybrid photovoltaic-wind power generation systems in Hong Kong. Renewable Energy, 28(11):1813–1824, 2003.

Zhou W., Lou C., Li Z., Lu L., Yang H. Current status of research on optimum sizing of stand-alone hybrid solar–wind power generation systems. Applied Energy, 87(2):380–389, 2010.

Muselli M., Notton G., Louche A. Design of hybrid photovoltaic power generator with optimization of energy management. Solar Energy, 65(3):143-157, 1999.

Kemmoku Y, Egami T, Hiramatsu M, Miyazaki Y, Araki K, Ekins-Daukes NJ, Sakakibara T. Modelling of module temperature of a concentrator PV system. Proc. 19th European Photovoltaic Solar Energy Conference, 2568-2571, 2004.

18

Fuzzy Control of WT with DFIG for Integration into Micro-grids

Christina N. Papadimitriou and Nicholas A. Vovos
University of Patras/Electrical and Computer Engineering dpt.
Greece

1. Introduction

Few years ago Power Systems consisted mainly of large generation plants supplying distant loads through the utility grids. The last years, though, a number of factors lead this structure to change gradually. Small generators of some MW have been already dispersed (DGs) throughout the transmission grid. The distribution of smaller generation units throughout the distribution system as near as possible to the consumer loads has already begun. The DG, up to now, does not provide auxiliary services such as back-up power, voltage support and reliability of supply and its operation is kept under constant power factor equal to 1 at all times. During network disturbances, the DGs up to the present are disconnected until normal operation is reestablished. When the distributed generation penetration is high this may lead to system instability.

Therefore, the DG has to change from passive appendage of primary energy supplier to active sources remaining connected to the grid and offering ancillary services. In some countries legislation changed, so that DG remains connected during disturbances and supports the grid. Under this operating philosophy, DGs must support the grid during local disturbances, as central generation stations support high voltage systems in the transient period. This can be achieved through the control of the DGs electronic interface to the main grid and the energy storage plants. These controllable (centrally or distributed) DGs (<100kWe) connected to the distribution grid, together with the local energy storage devices and the local controlled loads comprise an active distribution grid, the so-called micro-grid. In other words, the micro-grid can be seen as a miniature of a large interconnected grid that can provide the demanded power and can also change from interconnected to islanded mode of operation (Soultanis, 2008) and vice-versa. The micro-grid concept is the effective solution for the control of grids with high level of DG penetration (Brabandere et al., 2007).

The control of the micro-grid has to be reliable, flexible and according to system specifications. The DGs of the micro-grid have to cooperate in order to cover the local load needs for active and reactive power either under local disturbances or under islanding operation mode. So, in order to achieve the full benefits from the operation of the controllable distributed generation, an hierarchical control system architecture comprising three control levels can be envisaged in a future micro-grid. The Micro source Controller (MC) is the controller of the first layer (peer-to-peer (Meiqin et al., 2008)) and uses local information to control the voltage and the frequency of the micro-grid in transient conditions. This way, any DG can be integrated into the micro-grid operating in «plug and play» mode (Nikkhajoei& Lasseter, 2009). The Micro-grid Central Controller (MGCC) optimizes the micro-grid operation and the Distribution Management Systems (DMS) optimizes multiple MGCC which are interfaced in. As the proposed controller in this chapter is based on local information it comprises the MC of the micro-grid.

The usual peer-to-peer controller of the DGs converters uses the power vs. frequency droop characteristics in order to produce the active and reactive power reference combined with classical PI controllers, as it is mentioned in the review of state-of –the- art controllers in (Meiqin et al., 2008). In (Shahabi et al., 2009), the classical active power regulation takes into account the phase-locked loop (PLL) dynamics in order to prevent major oscillations from happening when a transition between grid connected mode and autonomous operation mode takes place. In (Brabandere et al., 2007) the primary, secondary and tertiary control algorithms are designed and tested in an experimental setup. The local controller uses the classical droop equation while the secondary and tertiary controllers aim at power quality and economic optimization respectively. The converters of the DGs communicate via the Internet. In (Nikkhajoei& Lasseter, 2009) the classical control is applied giving emphasis to the storage device algorithm, so that all micro sources have unified dynamic performance. In (Nishikawa et al., 2008) the classical local controller applied to the converters of a DFIG is combined with a pitch controller for a better dynamic response of a micro-grid at the mean voltage side. In (Katirarei & Iravani, 2006), the active power regulation is achieved through the classical controller combined with a frequency restoration strategy while three different strategies for reactive power regulation are adopted. The micro-grid dynamic behavior and the parameters of the DGs are mathematically investigated.

The droop equation combined with classical control can lead, under certain operational conditions, the micro-grid to instability while the stability enhancement for these cases needs sophisticated calculations (Mohamed& El- Saadany, 2008). In addition, the non-linearity of the micro-grid makes the implementation of fuzzy logic controllers at the DGs converters an attractive proposal. Moreover, fuzzy logic facilitates the «plug and play» operation for the DGs as the needed adjustments are minor due to the flexibility of fuzzy logic and easier due to the linguistic variables used.

This chapter presents a fuzzy based local controller so that a WT with DFIG is integrated into a micro-grid which is connected to a weak distribution grid. The micro-grid under study is simulated with Matlab/Simulink software, considering a micro-grid as it is described in (Meiqin et al., 2008). The micro-grid includes a hybrid fuel cell-battery system and a WT with DFIG. Firstly the micro-grid response is recorded, when a local disturbance (step load change) happens and secondly, when a transition from interconnected mode to islanding operating mode occurs. In a few seconds the grid is reconnected and the controllers' performance is evaluated again.

In the following section the micro-grid architecture is presented. In the fourth section the WT with the DFIG system is presented and some key points about its operation are discussed. In the fifth section, the designed controller based on fuzzy logic is analyzed. In the sixth section the simulation results are presented and the last section concludes the chapter.

2. Micro-grid structure

The micro-grid concept was put forward in 2001. Since then several micro-grids with multi-energy generators (MGMEG) and storages have been built in labs in universities and institutes all over the world (Meiqin et al., 2008). The majority of the DGs are connected to the micro-grid via electronic converters e.g. voltage source inverters (VSI). Internal combustion engines, gas turbines, micro turbines, photovoltaic, fuel cells and wind turbines constitute different technology micro sources. Micro sources have different dynamics depending on the prime mover size and the technology used. In this chapter the structure of the simulated micro-grid resembles those developed in the labs and includes two different micro sources (Fuel Cell System (FCS)-Battery and WT with DFIG). The reasons that led to the selection of these micro sources and especially of the WT with the DFIG will be explained later in details. As seen in (Bathaee & Abdollahi, 2007), the micro-grid architecture that combines different DGs is more practical as the storage device equipment or controllable loads can be reduced. The FCS (with proton exchange membranes (PEM)), that is combined with a battery bank forming a hybrid system, is connected to the AC feeder via a VSI and the WT with DFIG is connected directly to the feeder. In both DGs the "peer-to-peer" controller is based on fuzzy logic and the VSI local controllers are designed in "plug and play" operation mode. The designed controllers for the hybrid system are presented in a former paper of the authors and the controllers of the WT with DFIG are presented in section 5 of this chapter. The "peer-to-peer" controllers imply that through the electronic interface the DGs can provide the local grid with the demanded active and reactive power in absence of a central controller and communication links among micro sources. The "plug and play" operation mode implies that a micro source can be added to the micro-grid without reengineering the control and protection of units that are already part of the system. It is expected that in steady state a micro-grid central controller should coordinate DG to optimize operation, minimizing active power losses and maintaining flat voltage profile. The structure of the micro-grid under study is presented in Fig. 1. The micro-grid data are given in the following.

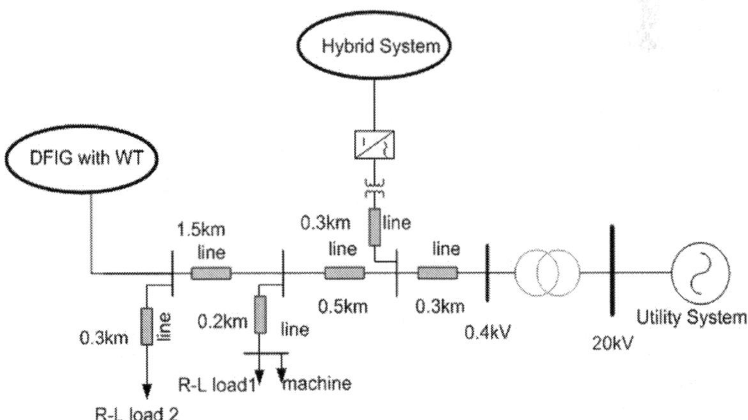

Fig. 1. Micro-grid Structure.

FCS PEM: 30 kW, 200 cells, 200A, 150V, 280cm²/cell.
DC Motor for the air supply system: 5 hp, 500 V, 1750 rpm, field: 300 V
Battery Bank: 234 HV Nickel-Metal Hybrid cells of 1.2 V, 2 Ah, 280 V.
Transformer: 50kVA, 190:380 V, 50 Hz.

WT with DFIG: The parameters of the electric models were adopted from SimPowerSystems library of Matlab/Simulink.
Doubly Fed Induction Generator: 75 kW, 400 V, 50 Hz, 1484 rpm, H=0.2, F=0.012, p=2.
WT: nominal mechanical output power: 80 kW, base wind speed: 12 m/sec, nominal rotor speed: 18rpm.
Distribution lines: AAAC type (4*185), X=0.236 Ω/km,   R=0.204 Ω/km
AC system: 380 V, 50 Hz, base p.u.:100 kW, 380 V.
R-L load1: 37kW, 13kVar, 380 V
R-L load2: 36kW, 10kVar, 380 V
Squirrel- Cage Induction Motor: 20hp, 400 V, 50 Hz, 1460 rpm

3. Wind turbine with doubly fed induction generator

3.1 Advantages of the DFIG structure

The DFIG system presents the following attractive advantages:

- The active and reactive power can be controlled independently via the current of the rotor.
- The magnetization of the generator can be achieved via the rotor circuit and not necessarily via the grid.
- The DFIG is capable of producing reactive power that it is delivered through the grid-side converter. Usually, this converter operates under constant unity power factor and it is not involved in reactive power trading with the grid. Also, the DFIG can be regulated in order to produce or consume a certain amount of reactive power. This way, the voltage control is achieved in cases of weak distribution grids.
- The converter size is not determined according to the total power of the generator but according to the decided speed range of the machine and therefore the slip range. For example, if the speed range is controlled between ±30% of the nominal speed, the nominal power of the converter is equal to the 30% of the nominal power of the generator. The selected speed range is decided according to the economical optimization and the increased performance of the system.

In the present study, the WT with the DFIG has been selected for integration into the micro-grid whose structure has already been presented. Beyond the technical advantages that the DFIG system presents, the practical reasons that led to this selection are the following:

- The micro-grid can operate under islanding mode either because of a fault at the mean voltage side or because of grid maintenance. Therefore, the microsources have to support the local grid in active and reactive power without stalling. But the microsources based on different technologies, have varying time constants. So, slow response microsources co-exist with fast response microsources in the same micro-grid structure for reliability purposes. In this study, the hybrid FCS has large time constant. The existence of the WT with the DFIG in the same micro-grid prevents the hybrid system from stalling during disturbances as the kinetic energy stored in the moving parts of the WT is exploited.
- As already mentioned, the control of DFIG is remarkably flexible through the electronic devices supporting the frequency and the voltage of the local grid. A well designed control can lead to greater power delivery (exploiting the machine inertia) than in any other type of WT.

- The DFIG technology decouples the electrical from the mechanical parts. So, in cases of frequency fluctuations of the grid, the operation of the WT with the DFIG is not affected except from the case that the control imposes a different operation. This way, the WT with DFIG can change smoothly through different operating modes when this is asked by the control system. The last is useful in the micro-grid concept.

3.2 WT with DFIG configuration

The Doubly Fed Induction Generator system can independently provide voltage and frequency regulation capabilities via the rotor current control as already mentioned. The basic configuration of the WT with DFIG is presented in Fig. 2.

The AC/DC/AC converter is divided into two components: the rotor-side converter (C_{rotor}) and the grid-side converter (C_{grid}). C_{rotor} and C_{grid} are Voltage-Sourced Converters that use forced-commutated power electronic devices (IGBTs) to synthesize an AC voltage from a DC voltage source. A capacitor connected on the DC side acts as the DC voltage source. A coupling inductor L is used to connect C_{grid} to the grid. The three-phase rotor winding is connected to C_{rotor} by slip rings and brushes and the three-phase stator winding is directly connected to the grid. The power captured by the wind turbine is converted into electrical power by the induction generator and it is transmitted to the grid by the stator and the rotor windings. The mathematical equations of the wind turbine model are given below.

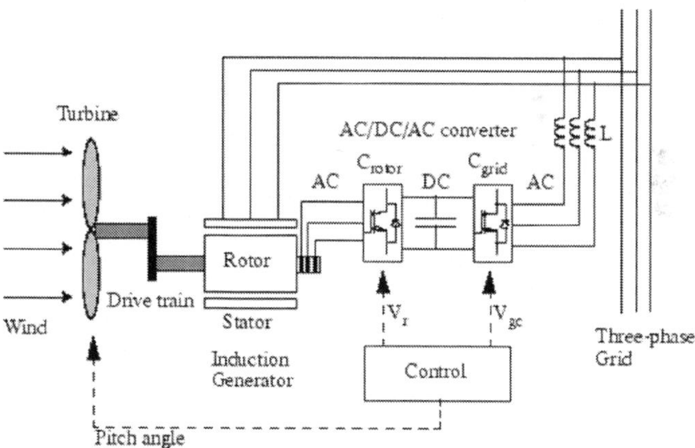

Fig. 2. The WT with the DFIG structure.

The power obtained from a given wind speed is expressed by the following equation:

$$P_w = 0.5 \rho \pi R^2 v^3 C_p(\lambda, \beta) \tag{1}$$

with ρ being the air density, R the effective area covered by the turbine blades, v the mean value of the wind speed at the height of the rotor axis, C_p the power coefficient of the wind turbine and β the pitch angle.

The tip speed ratio is defined as:

$$\lambda = \omega_w R / v \tag{2}$$

The mechanical torque is defined as:

$$T_w = P_w / \omega_w \tag{3}$$

with ω_w being the wind turbine rotor speed.
The coefficient C_P is defined as:

$$C_p = c_1 (c_2 / \lambda - c_3 \beta - c_4) e^{-c_5/\lambda} + c_6 \lambda \tag{4}$$

where the coefficient values are:
c1 = 0.5176, c2 = 116, c3 = 0.4, c4 = 5, c5 = 21 και c6 = 0.0068.
The model of the converters is the analytical model for voltage source converters given in the software package Matlab/Simulink (SimPowerSystems library). The wound induction generator model and its parameters were adopted by the same library, too.

3.3 Exploiting the kinetic energy

Among the different wind energy conversion systems, WTs with DFIG can provide the required services without the need of large costs of the power electronics hardware, if adequate control strategies are added (Shahabi et al., 2009). The control flexibility due to the DFIG electronic converter makes it possible to define various control strategies for participation in primary frequency control and voltage regulating support. A well designed control can deliver greater power exploiting the machine inertia than the delivered power from any other type of WT. So, the kinetic energy stored in the moving parts of the WT can be delivered if needed via the adequate controller.

The classical control of the wind power systems aims at the maximum power absorption for the whole wind speed range. However, if the wind turbines are part of the micro-grid, they have to provide frequency and voltage support to the local grid. There are several techniques for controlling the delivered power by the WT. In some papers, (Morren et al., 2006), during transients, the exploitation of the kinetic energy is achieved for a short time period and the optimum operation of the WT is shortly restored. In (Bousseau et al., 2006), pitch control is used so that the power delivering range increases in high wind speeds and the rotor speed increases in low wind speeds. In this study, pitch control is not applicable, as the wind speed is below the predefined limit (12m/s) and the angular position of the pitches doesn't change. Moreover, the time constant of the last action would be larger than the desirable time constant for the present study.

So, in order for a permanent control to be integrated in the WT with the DFIG, a slight de-optimization of the power production of the WT has been selected (Janssens et al., 2007).This way, a margin for excess power delivery during transients through the kinetic energy exploitation is created. In Fig.3, the means for de-optimizing the WT operation are presented graphically. The graph presents the mechanical power as a function of the rotor speed for given wind speeds and for pitch control equal to zero.

The A point corresponds to the WT operation at the optimum rotor speed. The de-optimization of the rotor speed is achieved either by imposing operation at a lower speed (point B-deceleration) or by imposing operation at a higher speed (point C-acceleration). In this study, the acceleration has been chosen as the deceleration of the rotor presents the following drawback: the delivered active power is originally degraded (the point of operation moves towards the opposite direction) while the rotor is trying to accelerate. This phenomenon is presented in Fig.4.

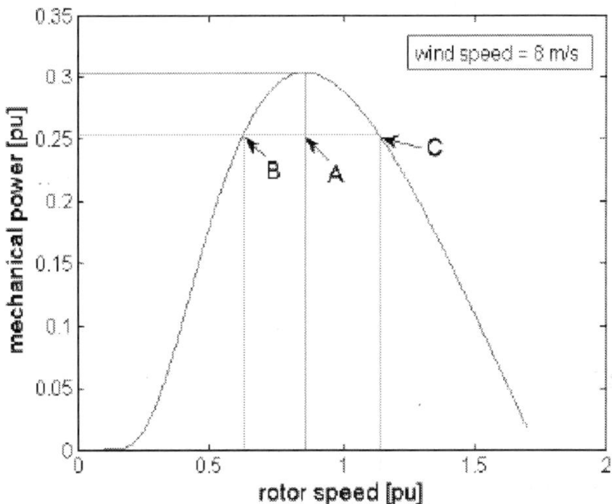

Fig. 3. Means of increase of the power margin (Janssens et al., 2008).

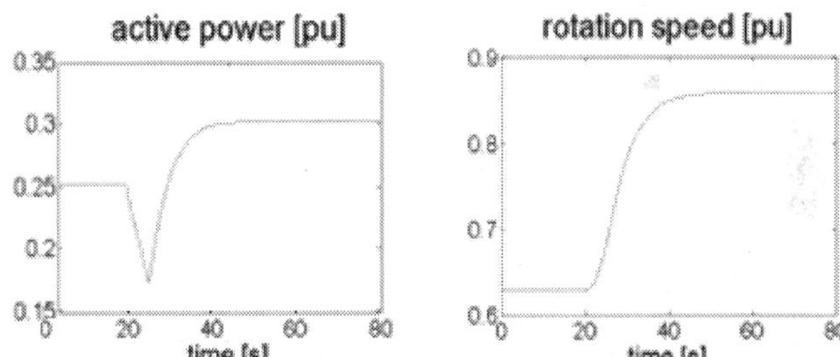

Fig. 4. Active power and rotation speed during the deceleration technique (Janssens et al., 2007).

On the contrary, during the acceleration technique, the desirable active power is delivered immediately. This advantage is of great significance as it prevents the stalling of micro sources with large time constants such as FCS during transients. This phenomenon is presented in Fig.5. The active power overshooting is noticed due to the absorption of the kinetic energy during acceleration.

4. Control design based on fuzzy logic

4.1 Why fuzzy logic?

As it is briefly mentioned in the introduction, the local controllers are based on fuzzy logic due to its flexibility and adaptiveness and due to the non-linearity of the system. The fuzzy controllers are non linear in nature and it is expected to have a robust performance under disturbances. Analytically, the four main flow subsystems of the FCS and the auxiliary

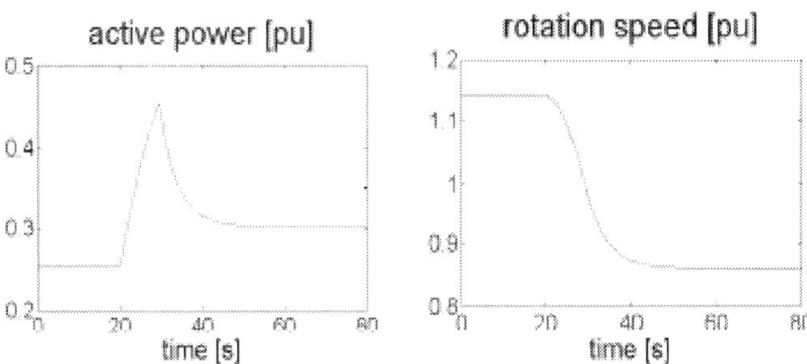

Fig. 5. Active power and rotation speed during the acceleration technique (Janssens et al., 2007).

subsystems establish a non linear FCS. In addition, the presence of non linear and cross-coupling terms of the DFIG dynamics form a micro-grid intensively non linear. Some key points of great significance for the system efficiency and performance outlined below enhance the application of a fuzzy based intelligent control. As for the hybrid FCS, it is significant for the compressor motor controller to have a good dynamic response during fast load changes so that the FCS voltage does not drop dramatically leading to oxygen starvation. Secondly, the controller for the system self-powering has to act simultaneously with the fuel flow control achieving stability and accuracy while minimizing overshooting and current rippling. The VSI controllers of the WT with DFIG and of the hybrid FCS have to meet the same requirements too. The design of the fuzzy controllers does not require a precise mathematical modeling or sophisticated computations that in many cases lack of efficiency and good performance. The last remark makes fuzzy logic suitable and practical in real systems where the engineer tunes the local controller easier via the linguistic variables (easy engineering).

The fuzzy controllers are designed from a heuristic knowledge of the system. Of course, they are thoroughly iterated by a system simulation study in order to be fine tuned. The advantage of this method is the fast convergence as it provides adaptively decreasing step size in the search of the adequate output. Besides, noisy and varying input signals do not affect the search. So, the weights of the membership functions (MFs) of every controller were chosen after qualitative knowledge of the simulation system. Triangular MFs were chosen as the most popular type providing satisfactory results. It can be seen in the following section that the MFs are asymmetrical, giving more sensitivity as the variables approach zero values. The defuzzification method that is followed is the Centre of Area (COA) defuzzifier.

4.2 Fuzzy local controllers

The current control of a machine is achieved mainly by means of vector control in a dq frame. The same technique is adopted in this chapter. The desired dimensions are firstly converted into vector components and secondly they are converted in AC components in order to be used in the PWM technique. In order to control the rotor current of a WT with DFIG by means of vector control, the reference frame has to be aligned with a flux linkage. One common way is to control the rotor currents with stator-flux orientation. If the stator resistance is considered to be small, stator-flux orientation is aligned with stator voltage. The

Fc1a: As seen in Fig.13 the input of this controller is the difference between the measured dc link voltage and the reference value ($V_{dc,ref}$-V_{dc}). The output of this controller is the deviation of the reference value of the d component of the output current (from the grid side) ΔI_{dgref}. The signal I_{dgref} is formed as already described.

The membership functions of the input and the output are shown in Figs. 14 and 15 respectively.

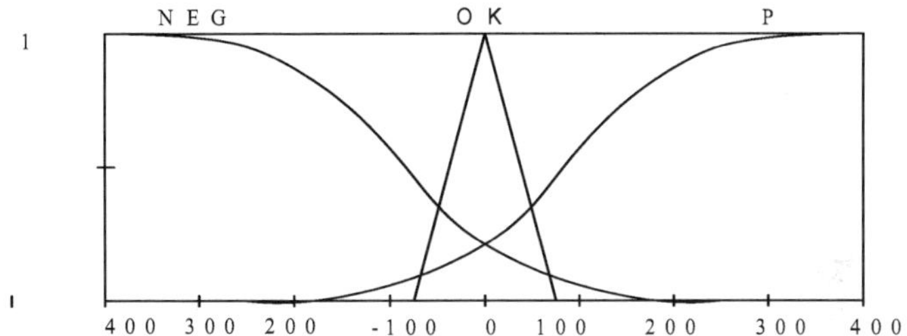

Fig. 14. Membership functions of the input signal of Fc1a.

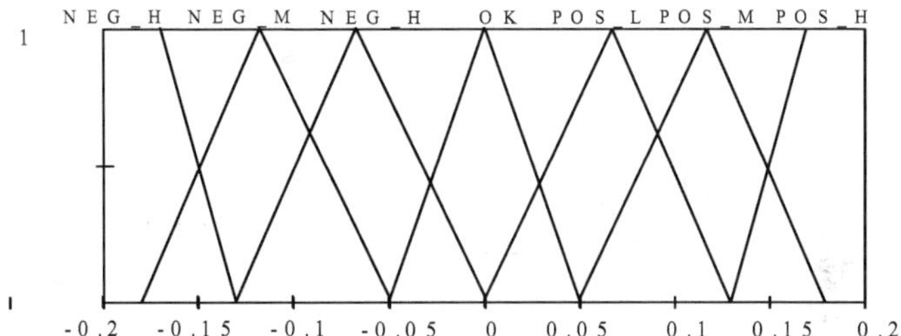

Fig. 15. Membership functions of the output signal of Fc1a.

The 7 fuzzy rules are presented in the following table:

Fc1a Input	P	P	P	OK	NEG	NEG	NEG
Fc1α Output	POS_H	POS_M	POS_L	OK	NEG_L	NEG_M	NEG_H

Table 2. Fuzzy Rules of Fc1a.

Fc2a: The input of this controller is the difference between the measured value of the q (or d) component of the output current and the reference value ((I_{qgref}-I_{qg}) or (I_{dgref}-I_{dg})). The output is the deviation of the q (or d) component of the voltage from the grid side (ΔV_{gq} or ΔV_{gd}). The control signal V_{gd} (or V_{gq}) is formed from the deviations as mention previously.

The reference value of the q component of the output current I_{qgref} is zero as the reactive power regulation through the C_{rotor} is preferred so that the electronic components rating remain small. Moreover, limiters are placed so that the currents don't exceed the electronic components specifications.

The membership functions of the input and the output are shown in Figs. 16 and 17 respectively.

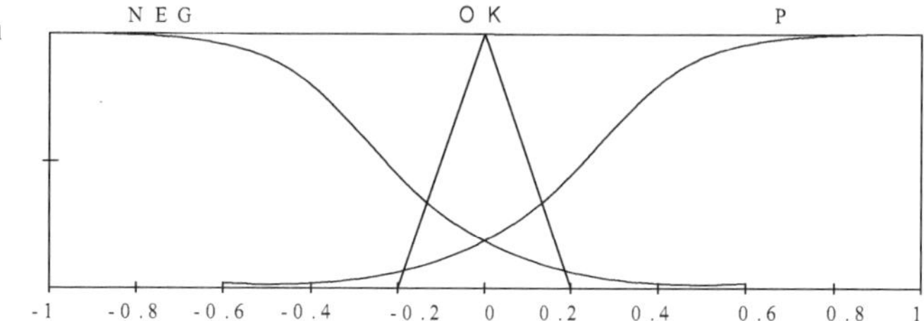

Fig. 16. Membership functions of the output signal of Fc2a.

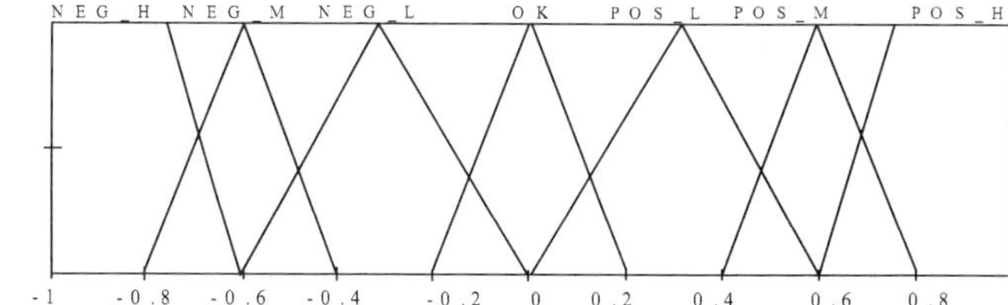

Fig. 17. Membership functions of the input signal of Fc2a.

The 7 fuzzy rules of the Fc2a are the same as those of Fc1a.

5. Simulation results

The data for the micro-grid are already given. In steady state the micro-grid is interconnected with the distribution grid and the initial steady state is the same for both cases studied. The R-L loads absorb their nominal active and reactive power and the induction motor operates at a slip of 2% and absorbs 10kW and 3kVar. 14% of the active power and almost a 100% of the reactive power of the loads are fed by the distribution grid. The DFIG feeds almost the 65% of the demanded active power and the hybrid system feeds the rest 21%. The DGs don't provide the loads reactive power during the interconnected mode of operation. The p.u. bases are: P_β=100 kW, V_β=380 V.

5.1 Local disturbances under grid-connected mode

At 0.5 sec, a step change of the mechanical load of the induction generator is imposed. The mechanical load is tripled and the DGs are offering ancillary services. The load sharing between the two DGs depends firstly on the dynamic response of each micro source and secondly on the weights of the MFs of the local controllers. In Fig.18, the measured frequency in steady state and during transient is presented. At 0.5 sec, the frequency drops due to the unbalance of active and reactive power in the system and returns to its nominal value after some oscillations within less than 0.5 sec. In Fig.19, the measured voltage at the point of common coupling (PCC) in steady state and during transient is presented. At the 0.5 sec, the voltage drops due to the unbalance of active and reactive power in the system and returns to its nominal value after some oscillations within 0.5 sec.

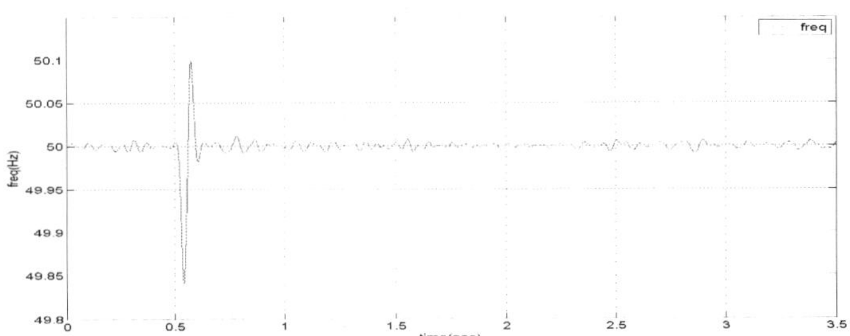

Fig. 18. The measured frequency.

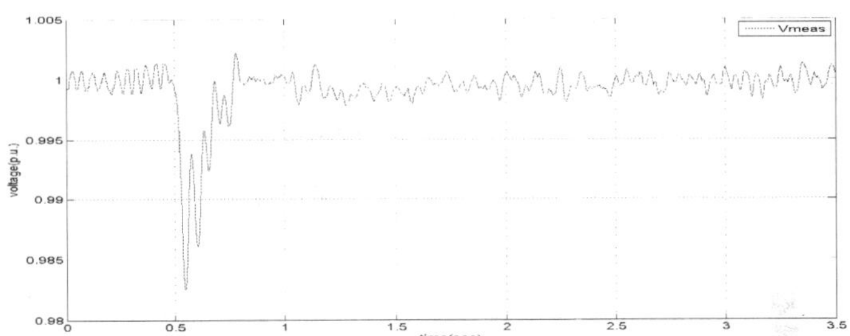

Fig. 19. The measured voltage at the PCC.

In Figs.20-22 the delivered active power by the grid, by the WT with the DFIG and by the hybrid FCS at the inverter's output are presented.

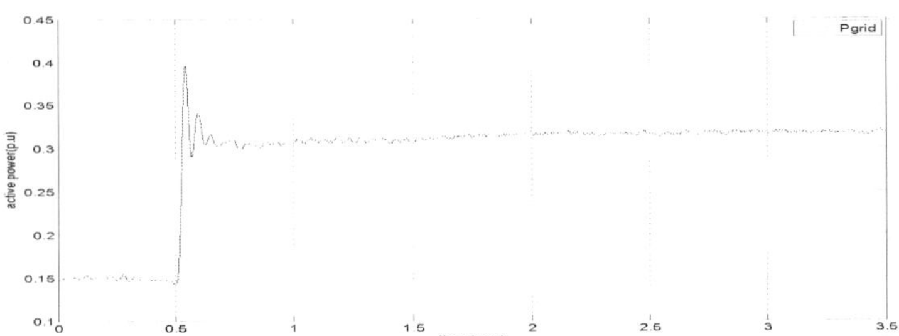

Fig. 20. The delivered active power by the weak distribution grid.

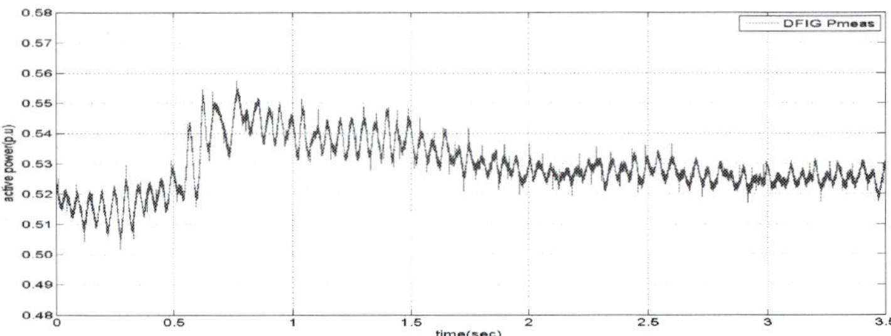

Fig. 21. The delivered active power by the WT with the DFIG.

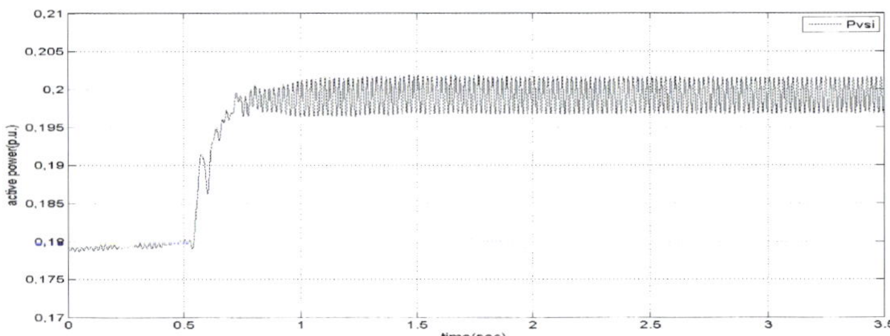

Fig. 22. The delivered active power by the hybrid FCS.

The grid (Fig.20) doubles the delivered active power and in the new steady state delivers about 30 kW. The WT with the DFIG (Fig.21) also increases the delivered power immediately to 55 kW because of the kinetic energy loss and after 1.5 sec from the disturbance it reaches a new steady state value (53 kW). Note the overshoot of the active power in the same figure. This happens due to the acceleration of the rotor technique already mentioned in a previous section. In Fig.22, the measured delivered power at the

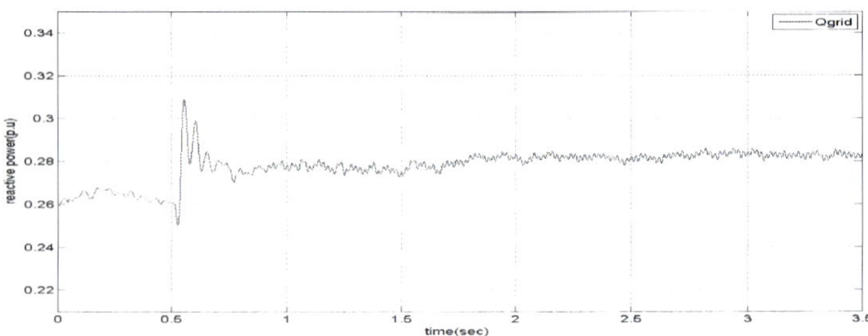

Fig. 23. The delivered reactive power by the weak distribution grid.

hybrid's FCS output is presented. Note that the fast response of the hybrid FCS is due to the existence of the battery at the dc-side. In the new steady-state the power demand has raised almost 26%. In total, the distribution grid covers the 29% of the active power demand, the WT covers the 51% and the hybrid FCS covers the remaining 20 %.
In Figs.23-25 the delivered reactive power by the grid, by the WT with the DFIG and by the hybrid FCS at the inverter's output are presented.

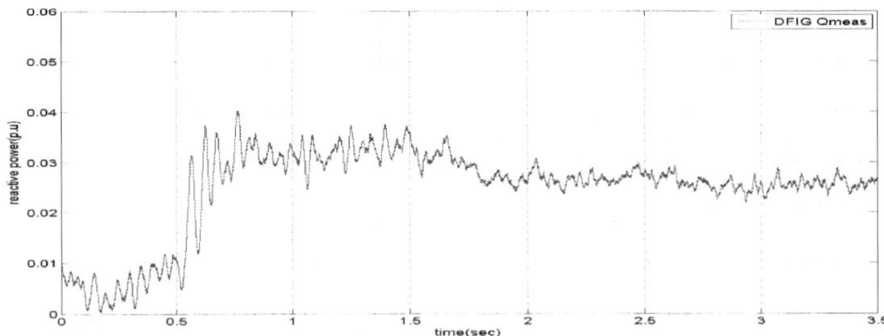

Fig. 24. The delivered reactive power by the WT with the DFIG.

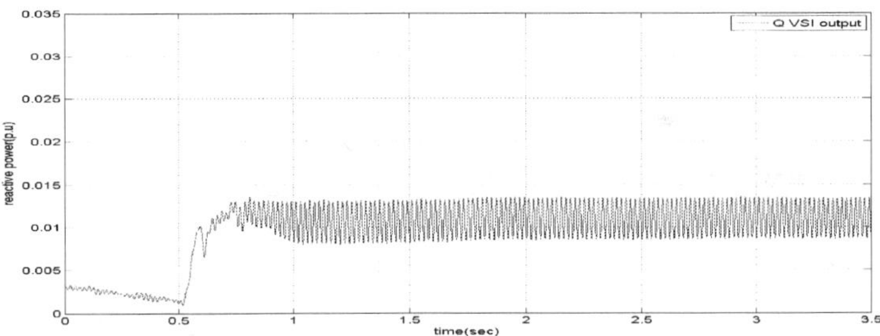

Fig. 25. The delivered reactive power by the hybrid system.

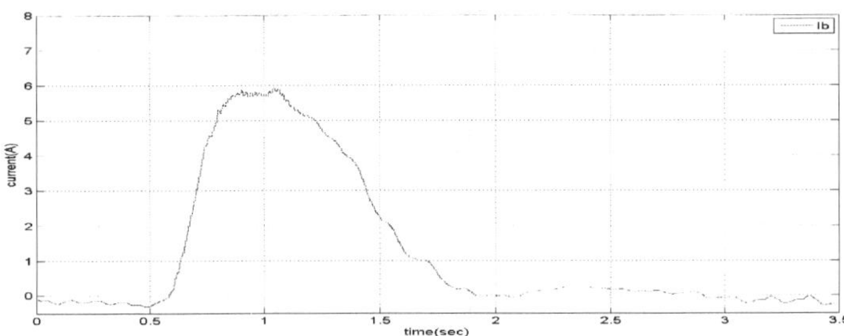

Fig. 26. The battery bank current in steady state and transient period.

In Fig.26 the battery bank current is presented. The battery bank current increases rapidly, in order to supply the battery the demanded power and returns to zero within 2 sec. In Fig.27, the FCS active power is presented. The FCS active power increases slowly in order to cover the total load demand and reaches a new steady state within 2 sec.

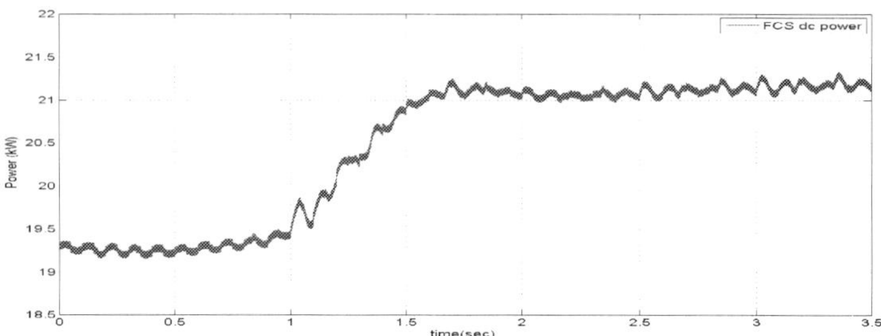

Fig. 27. The FCS active power delivered.

In Fig.28, the WT rotor speed is presented. Because of the disturbance imposed at the 0.5 sec, the rotor looses kinetic energy and reaches a new steady state.

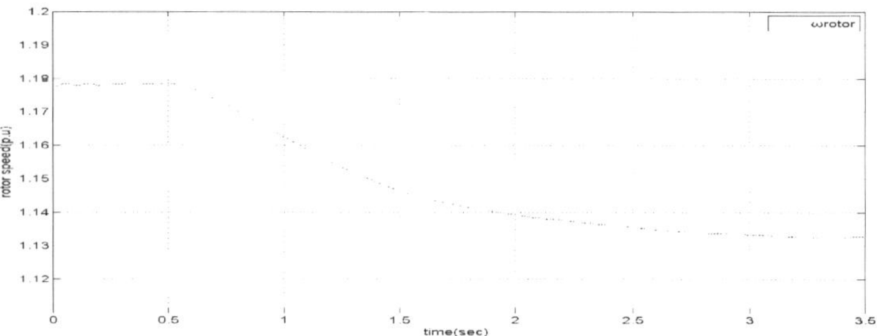

Fig. 28. The WT rotor speed in steady state and during transients.

In Fig.29, the control signals of the rotor side controller are presented in the same graph.

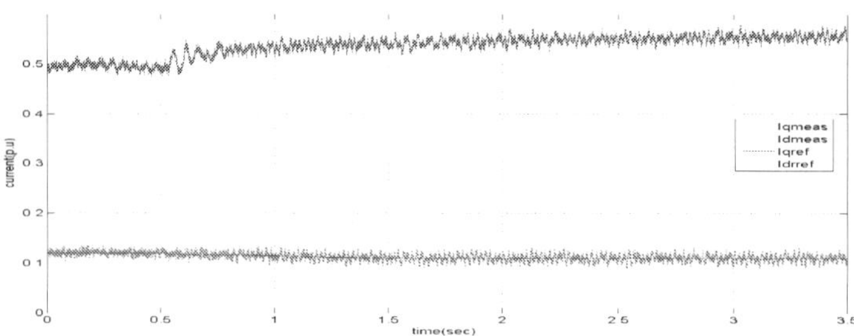

Fig. 29. The control signals of the rotor side controller.

5.2 Transition from grid-connected mode to islanding operating mode and transition from islanding operating mode to grid-connected mode

The initial steady state is the same as in the previous study case. At 0.5 sec, the grid is disconnected due to a fault at the mean voltage side or because of an intentional disconnection (e.g. maintenance work) and the micro sources cover the local demand. At 1.5 sec, while the system has reached a new steady state, the distribution grid is re-connected and finally a new steady state is reached. Note that, a micro-grid central control should lead the system to an optimal operation later.

In Fig.30, at 0.5 sec, the frequency drops due to the unbalance of active and reactive power in the system caused by the grid disconnection. The signal returns to its nominal value after some oscillations within 1sec. A small static error from the nominal value occurs but it is within the acceptable limits. At 1.5 sec. the distribution grid is re-connected with the micro-grid. An overshooting of this signal can be observed due to the magnitude and phase difference of the frequency of the two systems. Within 0.2 sec the micro-grid is synchronized with the distribution grid and the frequency reaches its nominal value of 50 Hz.

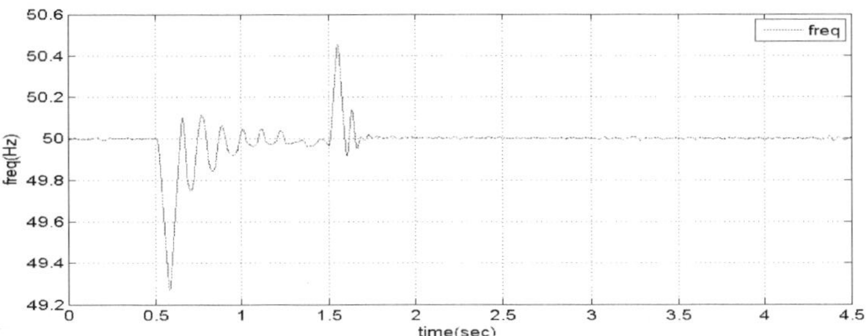

Fig. 30. The measured frequency.

In Fig.31 the voltage drops due to the unbalance of active and reactive powers in the system caused by the grid disconnection. The signal returns to its nominal value (a small static error is observed) after some oscillations within 1sec. At 1.5 sec. the distribution grid is re-connected with the micro-grid and the synchronization with the micro-grid is achieved after 3 sec.

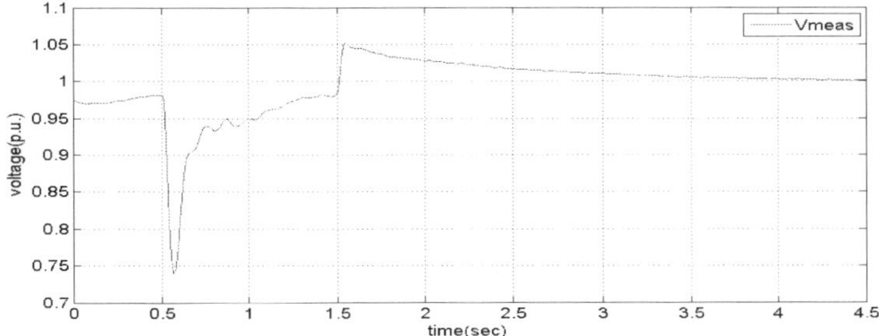

Fig. 31. The measured voltage at the PCC.

In Fig. 32-34 the delivered active power by the grid, by the WT with the DIFG and by the hybrid FCS at the inverter's output are presented. In Fig.32 the distribution grid is disconnected at 0.5 sec and is reconnected at 1.5 sec. In Fig.33 and 34, at 0.5 sec, the WT with the DFIG and the hybrid FCS increases the delivered power in order to eliminate the unbalance of power. At 1.5 sec, the grid is reconnected and the microsources are forced to regulate their delivered power so that the voltage and the frequency return to their nominal values.

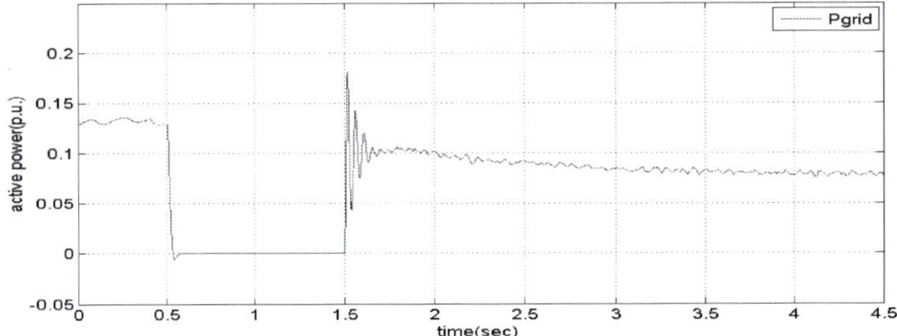

Fig. 32. The delivered active power by the weak distribution grid.

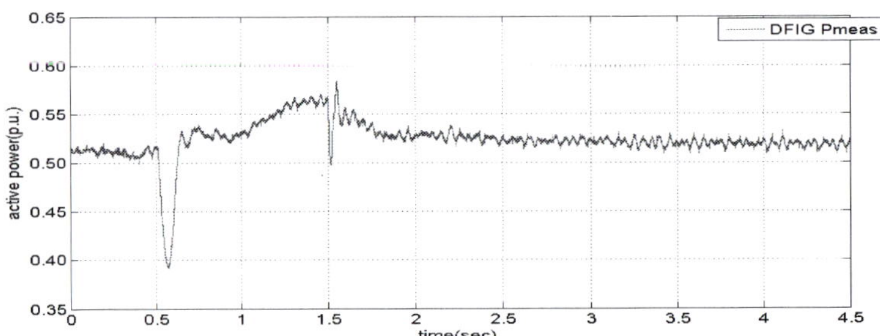

Fig. 33. The delivered active power by the WT with the DFIG.

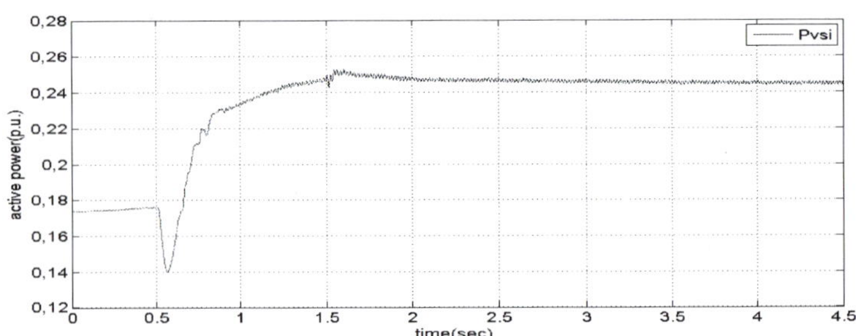

Fig. 34. The delivered active power by the hybrid FCS.

In Figs.35-37 the delivered reactive power by the grid, by the WT with the DFIG and by the hybrid FCS at the inverter's output are presented.

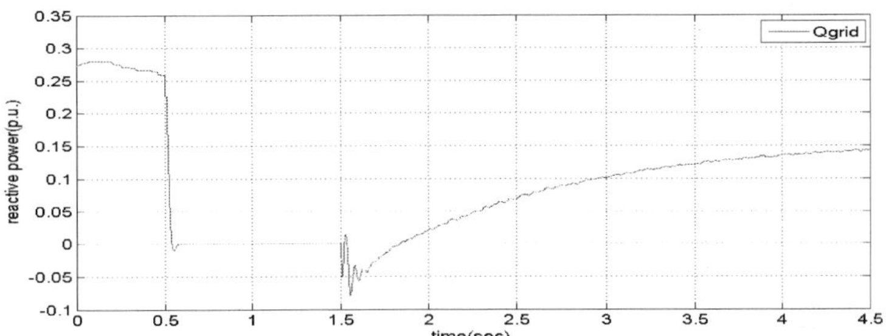

Fig. 35. The delivered reactive power by the weak distribution grid.

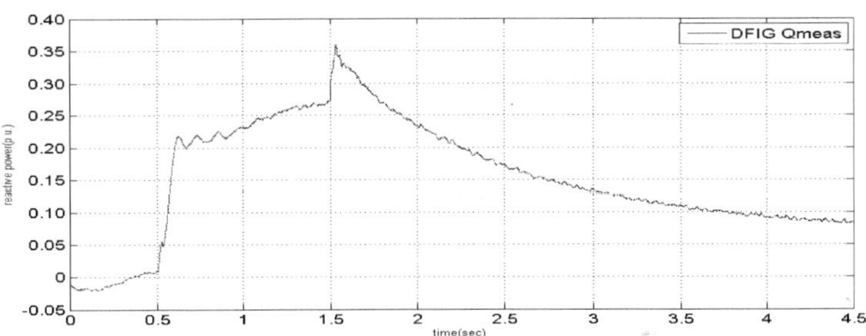

Fig. 36. The delivered reactive power by the WT with the DFIG.

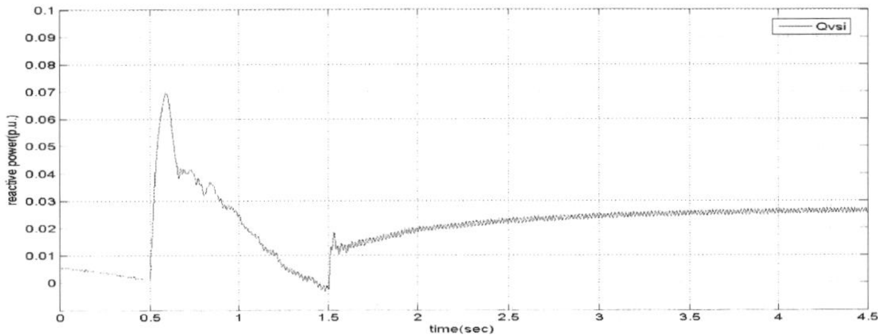

Fig. 37. The delivered reactive power by the hybrid FCS.

In Fig.38 the battery bank current is presented. The battery bank current increases rapidly, in order to supply the battery with the demanded power at 0.5 sec. At 1.5 sec, the battery bank continues to discharge and the current eventually returns to zero within 2.5 sec. In Fig.39, the FCS active power is presented. The FCS active power increases slowly in order to cover the total load demand and reaches a new steady state within 3 sec.

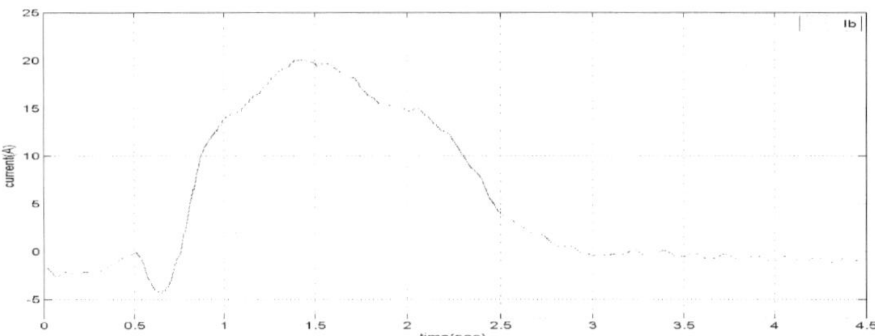

Fig. 38. The battery bank current in steady state and transient period.

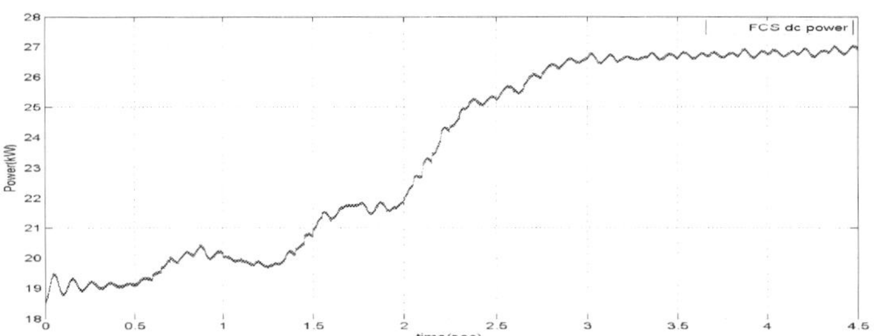

Fig. 39. The FCS active power delivered.

In Fig.40, the WT rotor speed is presented. Because of the disturbance imposed at the 0.5 sec and at 1.5 sec, the rotor looses kinetic energy and reaches a new steady state.

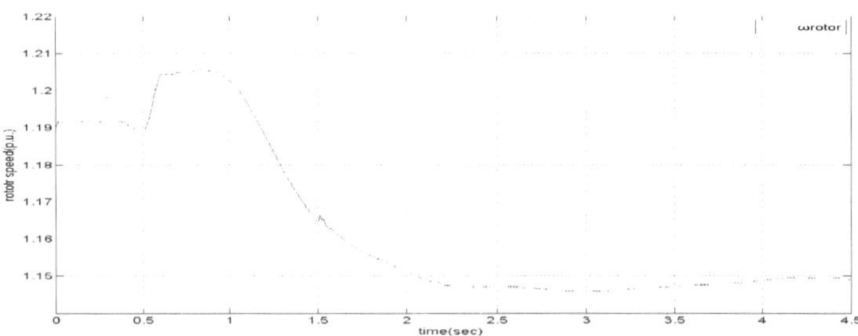

Fig. 40. The WT rotor speed in steady state and during transients.

In Fig.41, the control signals of the rotor side controller are presented in the same graph.

6. Conclusion

This chapter proposes a local controller based in fuzzy logic for the integration of a WT with DFIG into a micro-grid according to the «plug and play» operation mode. The designed

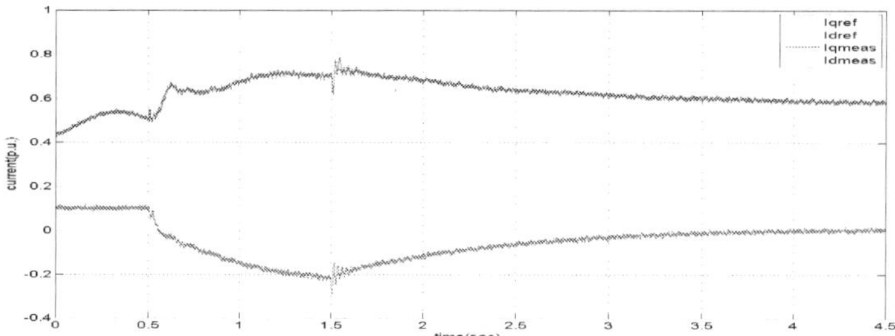

Fig. 41. the control signals of the rotor side controller.

controller is evaluated during local disturbances and during the transition from interconnected mode to islanding mode of operation either because of a fault at the mean voltage side or because of an intentional disconnection e.g. maintenance work. The simulation results prove that WT can provide voltage and frequency support at the distribution grid. The system response was analysed and revealed good performance. The proposed local controller can be coordinated with a micro-grid central controller in order to optimize the system performance at steady state.

7. Acknowledgment

The authors thank the European Social Fund (ESF), Operational Program for EPEDVM and particularly the Program Herakleitos II, for financially supporting this work.

8. References

Bathaee, S.M.T.; Abdollahi, M.H.(2007). Fuzzy-Neural Controller Design for Stability Enhancement of Microgrids, *Proceedings of UPEC 2007 42nd International Conference on Power Engineering*, pp. 562-569, ISBN 978-1-905593-36-1, Brighton, Sept. 4-6, 2007

Bousseau, P; Belhomme, R.; Monnot, E; Laverdure, N; Boëda, D; Roye, D; Bacha, S.(2006). Contribution of Wind Farms to Ancillary Services. CIGRE 2006 Plenary Session, Paris, report C6-103

Brabandere, K.De; Vanthournout, K.; Driesen, J.; Deconinck, G & Belmans, R. (2007). Control of Microgrids, *Proceedings of Power Engineering Society General Meeting IEEE*, pp. 1-7, ISBN 1-4244-1296-X, Tampa, June 24-28, 2007

Janssens, N. A.; Lambin, G; Bragard, N. (2007). Active Power Control Strategies of DFIG Wind Turbines, *Proceedings of IEEE Power Tech 2007*, pp. 516-521, ISBN 978-1-4244-2189-3, Lausanne Switzerland, July 1-5, 2007

Katirarei, F. & Iravani, M. R. (2006). Power Management Strategies for a Microgrid With Multiple Distributed Generation Units. *IEEE Transactions on Power Systems*, vol.21, No.4, (November 2006), pp.1821 – 1831, ISSN 0885-8950

Meiqin, M.; Chang, L.; Ming, D. (2008). Integration and Intelligent Control of Micro-Grids with Multi-Energy Generations: A Review, *Proceedings of ICSET 2008 on Sustainable Energy Technologies*,pp.777-780, ISBN 978-1-4244-1887-9, Singapore, Nov.24-27, 2008

Mohamed, Y. A. Ibrahim; El- Saadany, E. F.(2008). Adaptive Decentralized Droop Controller to Preserve Power Sharing Stability of Paralleled Inverters in Distributed Generation Microgrids, *IEEE Transactions on Power Electronics*, vol.23, No. 6, (November 2008) ,pp. 2806 – 2816, ISSN 0885-8993

Morren, J; de Haan, S; Kling, W; Ferreira, J.(2006) Wind turbines emulating inertia and supporting primary frequency control. *IEEE Transactions on Power Systems*, vol. 21, No. 1, (February 2006), pp 433-434, ISSN 0885-8950

Nikkhajoei, H.; Lasseter, R.H. (2009). Distributed Generation Interface to the CERTS Microgrid. *IEEE Transactions on Power Delivery*, vol. 24, No. 3, (July 2009), pp.1598 – 1608, ISSN 0885-8977

Nishikawa, K.; Baba, J.; Shimoda, E.; Kikuchi, T.; Itoh, Y.; Nitta, T.; Numata, S.; Masada, E. (2008).Design Methods and Integrated Control for Microgrid, *Proceedings of Power and Energy Society General Meeting - Conversion and Delivery of Electrical Energy in the 21st Century 2008 IEEE*, pp.1-7, ISBN 978-1-4244-1905-0, Pittsburgh, July 20-24, 2008

Shahabi, M.; Haghifam, M.R.; Mohamadian, M.; Nabavi-Niaki, S.A. (2009). Microgrid Dynamic Performance Improvement Using a Doubly Fed Induction Wind Generator. *IEEE Transactions on Energy Conversion*, vol.24, No.1, (March 2009), pp. 137-145, ISSN 0885-8969

Soultanis, N.L. (2008). *Contribution to the control and simulation of low voltage power systems with distributed generation*. PhD thesis, NTUA, Athens.